THE PICTURE OF THE TAOIST GENII PRINTED ON THE COVER of this book is part of a painted temple scroll, recent but traditional, given to Mr Brian Harland in Szechuan province (1946). Concerning these four divinities, of respectable rank in the Taoist bureaucracy, the following particulars have been handed down. The title of the first of the four signifies 'Heavenly Prince', that of the other three 'Mysterious Commander'.

At the top, on the left, is Liu *Thien Chün*, Comptroller-General of Crops and Weather. Before his deification (so it was said) he was a rain-making magician and weather forecaster named Liu Chün, born in the Chin dynasty about +340. Among his attributes may be seen the sun and moon, and a measuring-rod or carpenter's square. The two great luminaries imply the making of the calendar, so important for a primarily agricultural society, the efforts, ever renewed, to reconcile celestial periodicities. The carpenter's square is no ordinary tool, but the gnomon for measuring the lengths of the sun's solstitial shadows. The Comptroller-General also carries a bell because in ancient and medieval times there was thought to be a close connexion between calendrical calculations and the arithmetical acoustics of bells and pitch-pipes.

At the top, on the right, is Wên *Hsüan Shuai*, Intendant of the Spiritual Officials of the Sacred Mountain, Thai Shan. He was taken to be an incarnation of one of the Hour-Presidents (*Chia Shen*), i.e. tutelary deities of the twelve cyclical characters (see p. 79). During his earthly pilgrimage his name was Huan Tzu-Yü and he was a scholar and astronomer in the Later Han (b. +142). He is seen holding an armillary ring.

Below, on the left, is Kou *Hsüan Shuai*, Assistant Secretary of State in the Ministry of Thunder. He is therefore a late emanation of a very ancient god, Lei Kung. Before he became deified he was Hsin Hsing, a poor woodcutter, but no doubt an incarnation of the spirit of the constellation Kou-Chhen (the Angular Arranger), part of the group of stars which we know as Ursa Minor. He is equipped with hammer and chisel.

Below, on the right, is Pi *Hsüan Shuai*, Commander of the Lightning, with his flashing sword, a deity with distinct alchemical and cosmological interests. According to tradition, in his earthly life he was a countryman whose name was Thien Hua. Together with the colleague on his right, he controlled the Spirits of the Five Directions.

Such is the legendary folklore of common men canonised by popular acclamation. An interesting scroll, of no great artistic merit, destined to decorate a temple wall, to be looked upon by humble people, it symbolises something which this book has to say. Chinese art and literature have been so profuse, Chinese mythological imagery so fertile, that the West has often missed other aspects, perhaps more important, of Chinese civilisation. Here the graduated scale of Liu Chün, at first sight unexpected in this setting, reminds us of the ever-present theme of quantitative measurement in Chinese culture; there were rain-gauges already in the Sung (+12th century) and sliding calipers in the Han (+1st). The armillary ring of Huan Tzu-Yü bears witness that Naburiannu and Hipparchus, al-Naqqās and Tycho, had worthy counterparts in China. The tools of Hsin Hsing symbolise that great empirical tradition which informed the work of Chinese artisans and technicians all through the ages.

SCIENCE AND CIVILISATION IN CHINA

"TO MEN [he is speaking of the Jesuits in China] so qualified with *Mathematical* Knowledge, we owe the Discovery, of the before unknown Parts of the World, and from such we are to hope for the Perfection of that Knowledge, and the Discovery of the Rest. I have upon this occasion added some inquiries concerning the Literature of that Country; they are but Conjectures, grounded upon the perusal of some of their Books. A full Discovery is not pretended, however I hope they may serve as hints and incitements to others, who have better ability and other advantages to compleat it. We have hitherto not been admitted but to the Skirts, but this Discovery, when perfected, will lay open to us an *Empire* of Learning, hitherto only fabulously described; this will admit us to converse with the best and greatest of that *Empire*, that either are, or ever have been; this will Discover a new Indian Mine and Treasure, and make a new Trade to bring it hither...."

ROBERT HOOKE, F.R.S., 'Some Observations and Conjectures concerning the Character and Language of the Chinese', *Philosophical Transactions of the Royal Society* (1686), vol. 16, p. 35.

李约瑟著

中國科學技術史

冀朝鼎

SCIENCE AND CIVILISATION IN CHINA

BY

JOSEPH NEEDHAM F.R.S.

MASTER OF GONVILLE AND
CAIUS COLLEGE, CAMBRIDGE
FOREIGN MEMBER OF ACADEMIA SINICA

With the research assistance of

WANG LING PH.D.

PROFESSORIAL FELLOW IN THE INSTITUTE OF ADVANCED STUDIES
AUSTRALIAN NATIONAL UNIVERSITY, CANBERRA

VOLUME 1

INTRODUCTORY ORIENTATIONS

CAMBRIDGE
AT THE UNIVERSITY PRESS
1954

Published by the Syndics of the Cambridge University Press
Bentley House, 200 Euston Road, London NW1 2DB
American Branch: 32 East 57th Street, New York, N.Y. 10022

ISBN: 0 521 05799 X

First published 1954
Reprinted 1961 1965 1972

First printed in Great Britain by the University Printing House, Cambridge
Reprinted in Great Britain by Kingprint Ltd., Richmond, Surrey

To

LU SHIH-KUO

Merchant-Apothecary in the City of Nanking

this first volume
is respectfully and affectionately
dedicated

The Syndics of the Cambridge University Press
desire to acknowledge with gratitude certain financial aid
towards the production of this book, afforded
by the Bollingen Foundation

CONTENTS OF VOLUME 1

LIST OF ILLUSTRATIONS

LIST OF TABLES

CONTENTS OF THE SUBSEQUENT VOLUMES OF *SCIENCE AND CIVILISATION IN CHINA*

Volume 2. HISTORY OF SCIENTIFIC THOUGHT

VOLUME 3. MATHEMATICS AND THE SCIENCES OF THE HEAVENS AND THE EARTH

THE SCIENCES OF THE EARTH

[*From this point onwards, the work is still in preparation, though some sub-sections have already been written. Sub-section titles have not yet, therefore, been definitely established. It will be understood that the headings given below are only tentative; they are intended only to give a rough idea of the contents of each section.*]

VOLUME 7. THE SOCIAL BACKGROUND

[*This final volume will contain unified bibliographies and indexes of all the previous volumes* (*see vol. 1, pp. 20ff., 249ff.*).]

INTRODUCTORY ORIENTATIONS

1. PREFACE

THE HISTORY OF SCIENCE is now more and more widely recognised as an element of cardinal importance in the history of human civilisation. In the course of this development it has been natural for Western Europeans to work backward from modern science and technology, tracing the evolution of scientific thought to the experiences and achievements of Mediterranean antiquity. An abundant literature exists in which we may read of the foundations laid by Greek and Roman thinkers, mathematicians, engineers and observers of Nature. Earlier works, such as William Whewell's *History of the Inductive Sciences* of 1837, betray a bland unconsciousness of even the existence of the contributions of other peoples to the history of man's understanding of his environment. Since then, the debt of developed scientific thought to the ancestral pioneering of the Egyptians, as also to the work of the ancient inhabitants of the Fertile Crescent, Sumerians, Babylonians, the Hittites and others, has been better recognised and explored. Owing partly to circumstances which brought Europeans into close contact with Indian culture from the days of Megasthenes to those of Macaulay, some justice (still insufficient) has been done to Indian achievements, though here there are difficult problems of chronology which continue to prevent the appearance of a clear picture. And still today the contribution of the Far East, and especially of its oldest and most central civilisation, that of the Chinese, to science, scientific thought and technology, remains unrecognised and clouded in obscurity. The very term 'Far East', which I shall not use again in this book, but which springs spontaneously to the written page, exemplifies that fundamental insularity of outlook which it is so difficult for Europeans, even those who have the best intentions, to discard. The scientific contribution of Asia, and in particular of the 'Central Country', China, is the theme of this work.

What exactly did the Chinese contribute, in the various historical periods, ancient and medieval, to the development of Science, Scientific Thought and Technology? The question can still be asked for later periods, though after the coming of the Jesuits to Peking in the early 17th century, Chinese science gradually fused into the universality of modern science. Why should the science of China have remained, broadly speaking, on a level continuously empirical, and restricted to theories of primitive or medieval type? How, if this was so, did the Chinese succeed in forestalling in many important matters the scientific and technical discoveries of the *dramatis personae* of the celebrated 'Greek miracle', in keeping pace with the Arabs (who had all the treasures of the ancient western world at their disposal), and in maintaining, between the 3rd and the 13th centuries, a level of scientific knowledge unapproached in the west? How could it have been that the weakness of China in theory and geometrical systematisation did not prevent the emergence of technological discoveries and inventions often far in advance (as we shall have little difficulty in showing) of contemporary Europe, especially up to the 15th century? What were the inhibiting factors in Chinese civilisation which prevented a rise of modern science in Asia analogous to that which took place in Europe from the 16th century onwards, and which proved one of the basic factors in the moulding of modern world order? What, on the other hand, were the factors in Chinese society which were more favourable to the application of science in early times than Hellenistic or European medieval society? Lastly, how was it that Chinese backwardness in scientific theory co-existed with the growth of an organic philosophy of Nature, interpreted in many differing forms by different schools, but closely resembling that which modern science has been forced to adopt after three centuries of mechanical materialism? These are some of the questions which the present work attempts to discuss.

In the passing references of ordinary conversation conflicting tendencies may be observed. The old legendary Chinese chronology relayed by the 17th-century Jesuits still lives on, leading people to ascribe too much to East Asian origins, while at the same time there are counter-currents of marked reluctance to allow that any discovery or invention of importance could have taken place outside Europe. A few days before this paragraph was written, a producer making a documentary film on the history of ceramics and writing for advice, thought that the potter's wheel was a Chinese invention. Yet these instruments were known in Sumeria about a millennium and a half before the black-pottery makers and then the Shang people in China were using them. Similarly, during the past few years, papers on the history of certain aspects of later physics, engineering and ceramics, have been unjustly neglected because the legendary material was not recognised as such. Conversely, in an interesting work on the history of technology published in 1950, the author gave no credit to the Chinese for achievements which are indubitably theirs, such as the first knowledge of magnetic polarity, the discovery of gunpowder, and the earliest manufacture of cast iron. Chinese scientific writers themselves often ignore the contributions of their ancestors, for example, an excellent introduction to geobotanical prospecting which appeared at Peking in 1952 had no reference to the beginnings of this technique in the Liang dynasty (+6th century). The aim of the present work must therefore be to take all these matters out of the realm of shifting hearsay and 'received opinions'.

European scholars have of course long been dimly aware of that vast and complex civilisation, at least as intricate and rich as their own, away at the other extreme end of the 'heart-land' continent. In the 18th century it was frequently taken as a model for Europe, though often very incompletely understood. The vastness of the population using the ideographic as opposed to the alphabetic script, and its venerable age as the vehicle of the oldest living language with a continuous history of use, have been appreciated well enough, but the very difficulty of the characters has proved an almost insuperable barrier to understanding. Inevitably most sinologists have been of literary tastes and training, and while the progress of the world has forced Chinese scientists and technologists to be bilingual, the converse has proved so little true that the number of European or other Western natural scientists with any knowledge of the Chinese language can be counted from country to country only in twos and threes.

It is a strange revolution of history which has made the acquisition of an oriental language by a European natural scientist something highly unusual, appearing odd or even slightly reprehensible. It was not always so. Indeed, many of the great medieval translators from Arabic were natives of the British Islands. Adelard of Bath,[a] as Hitti tells us,[b] sojourned in Sicily and Syria, where in +1126 he turned into Latin the astronomical tables of al-Majrīṭī, and with his many other mathematical and astronomical translations became the first of a long line of English Arabists. The succession included Robert of Chester[c] in the same century, and Alfred Sareshel[d] and Michael Scot[e] in the following one, men who worked mostly in Spain. The difference is, of course, that in their day oriental studies (as they are called) were the only means of access to the most advanced science then existing, and this has long ceased to be the case. Now it is rather for the sake of the history of civilisation as a whole that we seek to penetrate the curtains woven on the looms of language. Fortunately, there are still a few isolated scholars, such as E. J. Holmyard and H. J. J. Winter, who maintain the tradition of Adelard of Bath. As for China, few who have attained an acquaintance with the language have ever been possessed of scientific qualifications adequate to enable them to evaluate Chinese contributions. In an earlier age, there was Father Antoine Gaubil, S.J., who spent a lifetime in China studying astronomy and its history there (1722 to 1759). In the last century there were the missionary-mathematician Alexander Wylie[f] and the navigator Leopold de Saussure;[g] more recently Henri Maspero, though by origin a humanist, studied astronomy deeply for his sinological researches. We still have among us H. Chatley the engineer, W. Hartner the astronomer, and others. So, too, Chinese botany has had its Bretschneider and Chinese medicine its Hübotter, but their work was only a beginning, and wide fields of science have never been covered, even by Chinese scholars themselves.

It is obvious from the state of the subject, upon which the labours of twenty specialists working each for a lifetime would alone begin to make some impression, that a book such as the present one can be but a reconnaissance. Its justification is that a vast and scattered literature does already exist, and that it has never before been digested into the compass of a single book, built on the framework of such questions as those set forth above. Besides this literature, in Chinese and Japanese as well as in Western languages, there is the veritable ocean of the extant original Chinese books themselves. That the appearance of a work such as the present one is now at least half a century overdue is doubtless partly owing to the fact that a certain rather unusual collocation of circumstances is required to pertain before a scholar could even

[a] Sarton (1), vol. 2, pp. 114, 167.
[b] (1), p. 588.
[c] Sarton (1), vol. 3, p. 175.
[d] Sarton (1), vol. 3, pp. 491, 561.
[e] Sarton (1), vol. 3, pp. 491, 579.
[f] His biographer, Cordier (7), tells us that he was trained as a cabinet-maker—an excellent beginning for a sinologist. He was stimulated by Prémare's *Notitiae Linguae Sinicae*, which he found on a second-hand bookstall. This was like Michael Hagerty, who did so much in the study of Chinese botany; he was a working bookbinder to begin with.
[g] Biography by R. de Saussure.

envisage undertaking it. Thus (1) he should have had a scientific training and have been engaged for a number of years upon useful and productive scientific research; (2) he should be acquainted broadly with the history of science in Europe, and should have carried out some original work in one or other aspect of it; and (3) he should have had interest in, and some acquaintance with, the social and economic background of science and technology in the successive stages of European history. But, on the other hand, (4) he should have had personal experience of Chinese life, and enjoyed the opportunity of extensive travels in that country, preferably neither as a missionary, a regular diplomat, nor a merchant;[a] (5) he should have a knowledge of the Chinese language, if not adequate for very rapid reading, at least sufficient to permit the consultation of original sources and indispensable works of reference; and (6) he should have had the good fortune of guidance from a wide range of Chinese scientists and scholars. That these strangely assorted circumstances should all have fallen to my lot is my basic defence against the criticisms, which I shall certainly deserve, from those who are, in one field or another which this book will touch, past masters. It is also the reason why I have been prepared to sacrifice much in order to accomplish the survey proposed, fearing that it might well be some time before the same collocation of circumstances recurs in another person.

For anyone who had to depend upon translations alone, a work such as the present one would be absolutely impossible, not only because so large a proportion of the necessary sources have never been translated into any occidental language, but also because many of the most reliable sinologists have been extremely careless in their use of scientific and technical terms, even when they understood them. It may be worth while to emphasise the point by giving a few concrete examples.

In the only complete translation of the *Mo Tzu* (Book of Master Mo Ti—a very important work of the Warring States period (−4th century), on social ethics, philosophy and logic) there is a reference to textile technology.[b] The translation runs: 'Women work at variegated embroidery; men work at the weaving of stuffs with inserted patterns.'[c] Now the date of the invention of the drawloom is, as we shall later appreciate,[d] a matter of much importance in the history of technology, and if this translation were taken at its face value, the mistake could easily be made of placing it in the −4th instead of the −1st century. But a glance at the text shows that nothing is said of inwoven figured patterns; the term for the men's work is *kho-lou*,[1] literally 'cut and engraved', which doubtless stands for *kho-ssu*[2] and will mean a kind of brocade made by stitching coloured threads into a fabric already woven.[e] The drawloom is thus in no way concerned.[f]

[a] 'For all story, without the knowledge of the places wherein it is performed, as it wanteth a great part of the pleasure, so it in no way enricheth the knowledge and understanding of the reader' (Sir Walter Raleigh).

[b] Ch. 6, tr. Forke (3), p. 183.

[c] 'Frauenarbeit ist die Buntstickerei; Männerarbeit das Weben eingefügter Figuren in den Stoffen....'

[d] Pp. 185, 229 below.

[e] Cf. Sect. 31 below.

[f] For a close parallel in Greek archaeology see Wace (1).

[1] 刻鏤 [2] 刻絲

Again, in the *Lun Hêng* (Discourses Weighed in the Balance), a work of much scientific importance written by Wang Chhung about +83, there is mention of wine.[a] The translation runs: 'From cooked grain wine is distilled.' But the term in the text is *niang*,[1] which means fermented, not distilled. The same applies to many parallel passages.[b] Dependence on the translation of a deservedly eminent sinologist could thus lead to an entirely unjustified belief in the knowledge and use of distillation by the Chinese in the +1st century for preparing liquors of high alcohol-content from wine. It is now more than fifty years since this error was pointed out by Dudgeon (2), who, as a medical man if an indifferent sinologist, knew the difference between fermentation and distillation. But the identical mistake is made by Feifel, translating *Pao Phu Tzu*,[c] the +4th-century alchemical work of Ko Hung, in 1944. And when it comes to wider conceptions rather than practical matters, translators have had no hesitation in reading characteristically occidental ideas into Chinese texts, as we shall abundantly see in connection with atoms,[d] laws of nature[e] and the like.[f] Conversely there is much to be said for declining to translate certain Chinese technical terms, such as *Tao*, and simply repeating them in transliterated form after doing one's best to define or explain them when they first arise.

There are many ways of writing the history of science. There is the Elementary—unsatisfying beyond a certain point—and the Anecdotal, only permissible in the hands of a master, such as E. G. Browne, in his *Arabian Medicine*. The Bibliographical style, exemplified by Aldo Mieli's *La Science Arabe*, though unquestionably valuable, can be very tantalising when one wants to know exactly what a man contributed instead of a long list of the titles of books, some probably falsely attributed to him. The Exhaustive, whether in George Sarton's unique *Introduction to the History of Science*, or in the *Origins and Development of Applied Chemistry* of J. R. Partington, is ruled out by the extensive nature of the field to be covered. The Social, adopted by Lewis Mumford in *Technics and Civilisation*, or J. G. Crowther in his *Social Relations of Science*, or in the works of Gordon Childe—all notable books—would only be serviceable for one aspect of the inquiry. Perhaps the *Histoire des Sciences* of Brunet & Mieli, the admirable monographs on herbals and optics by Charles Singer, and the *Invention of Printing in China* by T. F. Carter, come nearest to the models which I have proposed to myself.

[a] Ch. 5, tr. Forke (4), vol. 1, p. 154.
[b] E.g. Forke (4), vol. 2, p. 167.
[c] Ch. 3, tr. Feifel (1), p. 196; ch. 4, tr. Feifel (2), p. 6. For the latter passage the mistake is avoided by Wu & Davis (2), p. 237, though in general their versions are full of misunderstandings. Cf. also Balazs (1), p. 107.
[d] Cf. Sect. 26b below.
[e] Cf. Sect. 18 below.
[f] The really great sinologists have understood the necessity of practical and technical competence in the subject-matter very well. In the autobiography of Friedrich Hirth (10), one finds the following: 'Ein Kapitel aus Livius lässt sich mit Grammatik und Wörterbuch meist ohne Schwierigkeit übersetzen; nicht so ein Text des chinesischen Altertums oder Mittelalters, worin es sich oft um sehr viel mehr als die Übertragung von Wörten und Sätzen handelt. Hier muss der europaïsche Leser vor allen Dingen mit dem Milieu vertraut sein, d.h.er muss nicht nur übersetzen, sondern auch identifizieren. Erst wenn er realisiert hat, wovon eigentlich die Rede ist, haucht er seiner Übersetzung das Leben der Forscherarbeit ein. Daher müssen selbst ausgezeichnete Sprachkenner nebenbei auch Sammler sein, wenn es sich um technische Schilderungen handelt....'

[1] 釀

The present book is addressed, in fact, not to sinologists, nor to the widest circles of the general public, but to all educated people, whether themselves scientists or not, who are interested in the history of science, scientific thought and technology, in relation to the general history of civilisation, and especially the comparative development of Asia and Europe. Its intention is therefore to be comprehensive in the sense of leaving out nothing important, but not exhaustive in the sense of a paper in a learned journal, or a book such as Laufer's *Sino-Iranica*, in which the most minute points of detail are examined and turned over in footnotes occupying more space than the text itself. Unlike works of 'high popularisation' such as C. P. Fitzgerald's *China* or Sir George Sansom's *Japan*, where the origins of the evidence are generally not given, footnotes there will certainly be, but perforce (since I am to give the gist of the matter rather than a mass of detail) much in my notes remains unutilised, and such background material or confirmatory evidence will always be freely available for consultation by readers interested.

This being so, it may cause some surprise that the Chinese characters have not themselves been altogether omitted. The reason why it is, in my judgement, absolutely impracticable to do without Chinese characters, even though one may, as here, segregate them in special footnotes and indexes, is that the Chinese language is exceedingly rich in homophones. Without sight of the character intended by the transliteration or romanisation, any reader with a knowledge of Chinese finds himself in grave difficulties, and the work is rendered almost useless as a tool of research. Moreover, systems of romanisation are legion, offering endless food for sinological debate, and none which could reproduce alphabetically with unfailing sureness the manifold subtleties of the Chinese characters has yet won general acceptance. While it was stated above that this book is not addressed primarily to sinologists, but to the general educated public, that remark was not intended to express the hope that no sinologists would ever open its covers; indeed, I even dare to believe that they may find some scientific matters new and useful to them. More particularly it may be hoped that some among the younger generation of students of Chinese and other Asian languages may have, or may acquire, training and experience in the natural sciences, so that this book may stimulate them to undertake further investigations in this fascinating borderline field. It would therefore be inexcusable if they were to be subjected to the inaccuracy, and indeed the utter confusion, which inevitably arises when romanisations only, and no characters, are given. And readers, scientists or scholars, themselves Chinese, deserve no less courtesy.

For any single author, even aided, as I have been, by a few close collaborators, dependence on primary sources only in so vast a field would be quite impossible. While we have drawn upon them to the utmost, we have not hesitated to use also all available secondary sources. The fact is that the majority of those which are written in Chinese or Japanese have remained for the most part quite unknown to Western historians of science. I feel therefore that if this book constituted mainly a bibliography, it would be extremely useful. But I trust that it will be felt to be much more.

Is it not possible, indeed, to consider it a contribution to international understanding? Has not the nature of the genius of the Chinese people been greatly misunderstood by the West? The idea of this genius, so general, and so often encouraged by Chinese littérateurs trying to expound it to Western audiences, as primarily agricultural and artistic in quality, overlooks completely that long succession of technical discoveries which the West took over from China during the first thirteen centuries of the Christian era, often without the slightest realisation of where they had come from. It is to forget, for instance, that one of the most important roots of all chemistry, if not indeed the most important single one, is typically Chinese. But to multiply examples would be to anticipate what follows.

There may be some who would judge of the science and technology of the past only as it contributed in a direct genetical succession to that movement in 17th-century Europe from which modern science originated. It remains to be seen to what extent Chinese discoveries and inventions in fact did this. But even supposing it should turn out that for the most part they did not, would that be any the more reason for minimising the value of their study? For better or worse, the die is now cast, the world is one. The citizen of the world has to live with his fellow-citizens, at the ever-narrowing range of the aerofoil and the radio-wave. He can only give them the understanding and appreciation which they deserve if he knows the achievements of the sages and precursors of their culture as well as of his own. We are living in the dawn of a new universalism, which, if humanity survives the dangers attendant on control by irresponsible men of sources of power hitherto unimaginable, will unite the working peoples of all races in a community both catholic and cooperative. The mortar of this edifice is mutual comprehension, and truly the present work has been pursued in the spirit of unity of Lancelot Andrewes ('in the oriental tongues surpassing knowing'), whose 17th-century biographer said of him that 'many conceive that he might well have been (if then living) Interpreter-General, at the Confusion of Tongues'.

Today, though the 'white man' may have put down his 'burden' and even forgotten about it, are not Europeans, viewing the effects of modern science and technology in the complete transformation of the habitable globe, tempted too often to say to themselves that after all, this began in Europe with Galileo and Vesalius, and to conclude that Wisdom was born with us? A salutary correction of perspective is necessary. There *was* a Chinese contribution to man's understanding of Nature, and his control over it, and it was a great one. Perhaps the fact that Galileo and Vesalius and their like were Europeans depended not at all on any intrinsic superiority of the European peoples but upon factors of environment which did not, and could not, operate in other civilisations with a different geographical setting, and the different social evolution which this implied. At any rate, the question is worth investigating. Certain it is that no people or group of peoples has had a monopoly in contributing to the development of Science. Their achievements should be mutually recognised and freely celebrated with the joined hands of universal brotherhood.

ACKNOWLEDGEMENTS

Nearly twenty years ago there appeared in Cambridge three Chinese biochemists, candidates for the research doctorate, who worked more or less closely with the author of this book or in adjacent laboratories. Their names were Shen Shih-Chang,[1] Wang Ying-Lai[2] (now a leading member of the Institute of Biochemistry of Academia Sinica), and Lu Gwei-Djen.[3] It is to these three, and especially to the last-named, that the credit of being the hormone or evocator of the present book should be accorded. Whatever they took away with them from Cambridge, they left there a precious conviction that Chinese civilisation had played a role of hitherto unrecognised amplitude in the history of science and technology. It was during their stay in Cambridge that I was enabled, through their kindness, to acquire an understanding of the rudiments of the Chinese language, and to practise it in what must have been a series of letters both comic and tiresome. The greatest part in all this influence upon a westerner was played by Dr Lu Gwei-Djen, whose father, a distinguished pharmacist of the city of Nanking, had been learned both in traditional Chinese, and in modern, materia medica. Lu Mao-Thing[4] (Shih-Kuo[5]) had brought up his daughter to appreciate and understand modern science, but at the same time with the belief that the ancient and medieval practitioners and artisans of China had known what they were doing much better than most sinologists are usually prepared to believe. Techniques fundamental in human history had grown from that soil, certain things yet valuable might be found there if it were dug into, and at the least, the whole history of science and techniques in China would be an essential element in any comprehensive history of world achievement. This message Lu Mao-Thing's daughter was able to transmit. It is reason more than adequate for the dedication of the first volume of this book to him.

From that time until the beginning of the second world war, I continued the study of Chinese language and thought, more than fortunate in having the guidance of Gustav Haloun, the eminent sinologist who occupied the chair of Chinese at Cambridge from 1938 onwards. Not then so overwhelmed with labours as he was in later years, he was able to receive, on regular afternoons, a scientist who cycled up the rainy perspective of Sidgwick Avenue, and to give him an introduction to the difficulties of a classical Chinese text by the reading of the *Kuan Tzu* book together. This help, in different forms, continued until the time of Haloun's death at the end of 1951.[a] Only a few days before, I had telephoned him to ask a question about that strange book the *Ho Kuan Tzu*, and he had descanted impromptu for a quarter of an hour on the complexities of its composition. It is distressing indeed that I was not able to have his criticisms on a single page of the present work.

[a] A brief, but excellent, biography of him has been written by Herbert Franke (4).

[1] 沈詩章 [2] 王應來 [3] 魯桂珍 [4] 魯茂庭 [5] 仕國

As a project it had taken shape about the beginning of the war, some time before I was invited by the British Government to go to Chungking on a mission of cultural and scientific cooperation.[a] This developed, from 1942 until 1946, into the direction of a scientific liaison office attached to H.M. Embassy in that capital, which ultimately comprised a group of six British and ten Chinese scientists. The work gave unimagined opportunities for acquiring an orientation into Chinese literature of scientific and technical interest, for in every university and not a few industrial installations, there were scientists, doctors and engineers who had themselves been interested in the history of science, and who were not only able but generously willing to guide my steps in the right paths. In the following paragraphs I shall try to pay a few of the necessary debts, but the account, if complete, might be so extended that only a few of the more important names can be mentioned.

Among those who helped me in this field when I first reached Kunming at the beginning of 1943 were the historians Lei Hai-Tsung[1] and Wên I-To[2] (the latter afterwards assassinated as a leading member of the Democratic League). Dr E. R. Hughes, the Oxford sinologist, was working with Professor Wên at the time, and to him also I wish to express my gratitude for long and stimulating conversations. Among the scientists, Dr Chhien Lin-Chao[3] enthralled me with his interpretations of the physics of the *Mo Ching* (−4th century), Professor Hua Lo-Kêng[4] helped me with Chinese mathematics, and Ching Li-Pin[5] with pharmacological matters. In the capital itself, Chungking, I was able to enjoy the guidance of the outstanding archaeologist and historian Dr Kuo Mo-Jo,[6] later to occupy so eminent a position as a leader of his country in the cultural field,[b] and of Chi Chhao-Ting,[7] the learned and brilliant writer of *Key Economic Areas in Chinese History*, now one of China's most important financial experts. It is to Dr Chi that this book owes the calligraphy of its Chinese title. These friends did not fail to introduce me to many interesting authors with new interpretations of the classics such as Hou Wai-Lu;[8] while others in Chungking were also able to see that a proper understanding of Chinese social and economic history was achieved—such as Thao Hsing-Chih,[9] Têng Chhu-Min,[10] and Lin Tsu-Han.[11] In various scientific fields I received useful help from Chhen Pang-Hsien[12] and Chu Hêng-Pi[13] in medicine, from Chang Mêng-Wên[14] in biology and from Chang Tzu-Kung[15] in chemistry.

When I visited other parts of Szechuan it was the same. At Chhêngtu and Chiating there was opportunity of gaining an understanding of the difficult but vital subject of Taoism from Kuo Pên-Tao[16] and the late Huang Fang-Kang.[17] In the Abbot of

[a] Here a debt is owing to Sir George Sansom, for the suggestion that there should be such a mission was largely his.

[b] Now (1952) President of Academia Sinica, Chairman of the Cultural and Educational Commission of the Government, and Vice-Premier.

1 雷海宗　2 聞一多　3 錢臨照　4 華羅庚　5 經利彬
6 郭沫若　7 冀朝鼎　8 侯外廬　9 陶行知　10 鄧初民
11 林祖涵　12 陳邦賢　13 朱恆璧　14 張孟聞　15 張資珙
16 郭本道　17 黃方剛

Loukuantai, Tsêng Yung-Shou,[1] it was possible to meet traditional Taoism in the flesh, just as the then President of Wuhan University, the late Dr Wang Hsing-Kung,[2] manifested all the aspects of old-fashioned Confucianism. At Chhêngtu there were many scholars at whose feet one could sit, for example, Fêng Yu-Lan,[3] the philosopher, and Hou Pao-Chang,[4] the pathologist and historian of anatomy and medicine.

Some of these encounters were distinctly romantic. While at Paochi in Shensi province, I went down the Lunghai railway line one day on a platelayers' trolley to Wu-Chêng-Ssu, the final evacuation home of Honan University. It was using as one of its buildings a fine old Taoist temple on a loess bluff looking southwards across the Wei Valley (the cradle of Chinese civilisation) towards the Chhinling Mountains, about the point where the Chien tributary enters the Wei from the north. I spent the afternoon exploring the library with Professor Li Hsiang-Chieh.[5] It had been a fine collection, but successive evacuations had done some damage to the books, and the catalogue could no longer be found; there they were, many still lying tied up in bundles at the feet of the old statues of the gods, just as on the day, not so long before, when the sweating porters had dumped them down, unhooking them from their carrying-poles. Such were the circumstances in which a Cambridge biochemist was introduced by Li Hsiang-Chieh to the fact that the *Tao Tsang* (the Taoist patrology) contains a large number of alchemical works, dating from the +4th century onwards, of the greatest interest, and hardly known at all by historians of chemistry in other cultures. One does not forget such introductions.

So also, diametrically across the length and breadth of China, in the far south-east, on hot candle-lit nights on balconies over the Pingshek river, there was talk with Wang Ya-Nan[6] (now President of Amoy University) on the exact nature of medieval Chinese feudal-bureaucratic society; and other sociological themes besides, pursued among the bookshops and tea-gardens of Kukong with Wu Ta-Kun.[7] Then, when the war was over, came the climax of four years of wanderings in remoter regions, the opportunity of a short time in the legendary metropolis of Peking itself. There it so happened that the history of chemistry was again the subject of discourse, under the inspiring leadership of Chang Tzu-Kao,[8] Tsêng Chao-Lun[9] and Li Chhiao-Phing.[10] It was possible, too, to buy many precious books in that centre of Chinese literary publication, indispensable for subsequent work, such as the *Thai-Phing Yü Lan* encyclopaedia and many *tshung-shu* collections.

Readily understandable is the particular gratitude which the author feels to those Chinese scholars who aided him by gifts of books, especially in the early stages. Dr Huang Tzu-Chhing,[11] the physicist, was the first; he presented me with a copy of the *Chhi Min Yao Shu* (Important Arts for the People's Welfare), the +5th century agricultural work by Chia Ssu-Hsieh, shortly after my arrival at Kunming. The *Thien Kung Khai Wu* (Exploitation of the Works of Nature), an important book on

1 曾永壽 2 王星拱 3 馮友蘭 4 侯寶璋 5 李相傑
6 王亞南 7 吳大琨 8 張子高 9 曾昭掄 10 李喬苹
11 黃子卿

industrial technology of the early +17th century by the Diderot of China, Sung Ying-Hsing, was given to me first by the neurologist, Dr Thang Yüeh,[1] at Liangfêng, and subsequently (in the best edition) by the late Director of Academia Sinica's History Institute, Fu Ssu-Nien.[2] But our greatest benefactor was Dr Chu Kho-Chen,[3] eminent meteorologist and for a long time President of Chekiang University (now Vice-President of Academia Sinica). About the time when I was leaving China, he persuaded many friends to look out duplicates, and so not long after my return to Cambridge whole packing-crates of books began to arrive, containing among other things a set of the *Thu Shu Chi Chhêng* encyclopaedia of +1726. Dr Chu I had known first in Kweichow province, where Chekiang University had its home during evacuation, and where I became familiar with his valuable work on the history of astronomy; in his company I had met such men as Chhien Pao-Tsung[4] (with Li Nien[5] one of the two most outstanding historians of Chinese mathematics), and Wang Chin[6] (assiduous in the history of metallurgy and chemistry).

After an absence of six years I was so fortunate as to be able to spend the summer of 1952 largely in Peking. There I met many old friends again, such as Dr Chu Kho-Chen, Professor Liu Hsien-Chou[7] (specialist on the history of engineering), Liang Ssu-Chhêng,[8] the great preserver and historian of Chinese architecture, and the eminent geologists Li Ssu-Kuang[9] and Yuan Fu-Li.[10] Besides this, I was able to learn from many new ones, such as Fêng Chia-Shêng,[11] expert on the history of gunpowder and military technology, and Wang Chen-To,[12] whose brilliant archaeological researches have thrown so much light on the origin and development of the magnetic compass, and on the engineering achievements of the pre-Thang period. It was a pleasure indeed to study with him the beautiful models in which he has incorporated his conclusions, on the terrace of the old imperial Buddhist temple with its circular wall near the Pei Hai park, where, as one of the Curators of the People's Museum in the Imperial Palace, his office is located. Particular thanks are also due to Dr Li Thao,[13] who spent many hours discussing the history of medicine among the unique collection of Chinese medical books formed largely by him and preserved in the library of the China Union Medical College. Opportunity was also afforded me during this period to frequent once again the bookshops of the famous Liu-Li Chhang and the various Peking markets, and to acquire more of the texts indispensable for the study of the history of Chinese science and technology.

Throughout the period of preparation of the present book I have enjoyed the research assistance of my friend Mr Wang Ling[14] (Wang Ching-Ning[15]), whom I first met at Lichuang (Szechuan) in 1943 when visiting the Institute of History of Academia Sinica, then evacuated there. A lecture which I gave on the history of science stimulated him to work on the early history of gunpowder and firearms in China, work which he continued while Senior Lecturer in History at Futan University,

[1] 唐鉞 [2] 傅斯年 [3] 竺可楨 [4] 錢寶琮 [5] 李儼
[6] 王璡 [7] 劉仙洲 [8] 梁思成 [9] 李四光 [10] 袁復禮
[11] 馮家昇 [12] 王振鐸 [13] 李濤 [14] 王鈴 [15] 王靜寧

Peipei. In 1946 he came to England on a British Council Travelling Fellowship. From 1948 to 1950 he was supported by a special Fellowship created and financed by Mr H. N. Spalding and Mrs Spalding of Oxford, from 1950 to 1951 by the Universities' China Committee, and from 1951 to 1953 by a research grant from the Leverhulme Foundation. To all these our most sincere gratitude is offered.

Mr Wang and I both feel that it may be desirable here to explain something of the nature of the assistance which he has contributed over more than seven years. First, his training in the techniques of Chinese historical research has been of great value in our daily discussions. Then it was he who seven or eight times out of ten made the first draft of the translations given for the first time in the present book, after which there invariably followed detailed joint discussions and verifications, often necessitating many changes before their definitive form was reached. Translations by other writers were not accepted without being checked by both of us against the original Chinese texts.[a] Mr Wang spent much time, moreover, in 'scouting' and 'skimming' through texts thought beforehand to be promising, and would return from such reconnaissances with material which again required careful examination from the point of view of the history of science before its value could be assessed. Much detailed library work also fell to him, as well as indexing and listing of various kinds. Without the friendship of such a collaborator this book would have taken much longer to produce, if indeed it would ever have been possible at all, and would have contained even more mistakes than we fear it actually does.

Apart from Mr Wang, and with the exception of our ever-friendly and meticulous colleagues in the University Press, probably only one person living has read every word of the volumes now before the public. I therefore owe a great debt of gratitude to my wife, Dr Dorothy Needham, F.R.S., whose improvements to the text have been too numerous to be counted. To speak adequately of her general encouragement over the long years which this work has occupied would be to go beyond what ordinary words can convey, and it would be useless to attempt it. To her sister, Miss Muriel Moyle, lately Assistant Librarian at UNESCO, our best thanks are due for the preparation of the general indexes. If of the legion of librarians who have helped our studies so kindly, one more than any other deserves a tribute, it is Miss M. E. Fell, Assistant Librarian to the Royal Asiatic Society, whose efforts have enabled us to see many books and papers which otherwise would have been unattainable. Dr B. A. Bembridge and Mr F. C. Street have helped us by the loan of useful reference books. I am also deeply grateful to the scientific colleagues of my immediate environment for the interest and understanding which they manifest towards my task. Lastly, I believe that there could be no milieu more sympathetic and generous in which to work than that formed by the Fellows of my own College.

Among the best advantages, indeed, of working in Cambridge must be counted the great company of experts in diverse special fields who are willing to look over script which

[a] Except in a few cases where this would have been very difficult or impossible, and at the same time the authority of the translator was high.

touches upon their interests, and to suggest emendations and improvements. With these as a nucleus, we have also been fortunate enough to find some helpers further afield. The names of those who have thus given freely of their time and energy to help us are set forth in tabular form. We hope that they will each and all accept our warmest thanks.

List of Those who have kindly read through Sections in Draft

N.B. It is needless to say that no responsibility for errors of whatever sort rests with the following friends. Had it not been for their kind criticisms there would have been many more, especially as the field surveyed touches so many specialities. The following list covers the early volumes, and supplementary lists will be provided later.

Dr S. Adler (Cambridge)	All sections in the History of Thought.
Mr R. C. Anderson (Greenwich)	Nautical Technology.
Mr Courtney Archer (New Zealand)	Building Technology.
Mr Shackleton Bailey (Cambridge)	Buddhism.
Dr Etienne Balazs (Paris)	Taoism, Neo-Confucianism, Natural Law.
Dr A. Beer (Cambridge)	Astronomy and Mathematics.
Mr Eric Blackall (Cambridge)	Introduction (Language).
Prof. Derk Bodde (Philadelphia)	Natural Law.
Mr Andrew Boyd (London)	Building Technology.
Mrs Margaret Braithwaite (Cambridge)	Introduction (Language), the Mohists and Logicians, Neo-Confucianism and Mathematics.
Mr D. Bryan (Cambridge)	Introductions.
Prof. K. Bünger (Tübingen)	Natural Law.
Dr Herbert Chatley (Bath)	Astronomy, Mechanical and Civil Engineering.
Dr Chêng Tê-Khun (Cambridge)	Introductions (Prehistory, History and Contacts).
Prof. Chou Pien-Ming (Amoy)	Introduction (Language).
Rear-Admiral Day (Hydrographer of the Navy, London)	Seismology.
Mr R. D. Davies (Cambridge)	Civil Engineering (Bridges).
Mr D. W. Dewhurst (Cambridge)	Astronomy.
Prof. W. A. C. H. Dobson (Toronto)	Taoism, and Physics (Acoustics).
Prof. E. R. Dodds (Oxford)	Natural Law.
Mr D. M. Dunlop (Cambridge)	Introduction (Arabic-Chinese Culture Contacts).
Dr W. N. Edwards (London)	Geology and Palaeontology.
Mr John Ellison (London)	Physics.
Prof. R. A. Fisher (Cambridge)	Mathematics.
Mr W. Fuchs (München)	Geography and Cartography.
Dr A. R. Hall (Cambridge)	Astronomy.
Mr Brian Harland (Cambridge)	Geology and Mineralogy.
Dr K. P. Harrison (Cambridge)	Parts of the Section on Astronomy.
Dr Hsü Li-Chih (Cambridge)	Mathematics.
Mr Alec Hunter (Braintree)	Textile Technology.
Mr John Hunter (Thaxted)	Building Technology.
Mr Donald Leslie (Cambridge)	Introductions, all sections in the History of Thought, and Mathematics.
Miss Liao Hung-Ying (Mrs Bryan) (Cambridge)	Introductions.
Dr P. van der Loon (Cambridge)	Introductions.

Mr B. Lubetkin (London)	Building Technology.
Prof. Gordon Manley (London)	Meteorology, Geography and Cartography.
Mr Leslie Martin (London)	Building Technology.
Mr Scott McKenzie (Washington)	Mineralogy.
Sir B. Melvill Jones (Farnborough)	Mechanical Engineering (Aeronautics).
Dr H. Michel (Brussels)	Astronomy (sun-dials).
Mr J. S. Morrison (Cambridge)	Nautical Technology.
Mr George Naish (Greenwich)	Nautical Technology.
Dr K. P. Oakley (London)	Geology, Palaeontology and Seismology.
Dr Walter Pagel (London)	Taoism.
Prof. J. R. Partington (Cambridge)	Mineralogy, Chemistry and Industrial Chemistry.
Dr L. E. R. Picken (Cambridge)	Physics (Acoustics).
Dr Victor Purcell (Cambridge)	Magnetism in relation to Navigation.
Mr J. Ravetz (Cambridge)	Mathematics.
Dr K. Runcorn (Cambridge)	Mathematics, Physics (Magnetism).
Mr Raphael Salaman (Harpenden)	Engineering (Tools).
Dr Dorothea Singer (Par)	Taoism, Physics.
Mr Francis Skinner (London)	Building Technology.
Dr R. W. Sloley (Amersham)	Astronomy (the clepsydra).
Dr Otto van der Sprenkel (London)	Historical Introduction and all sections in the History of Thought.
Mr E. G. Sterland (Cambridge)	Mechanical and Civil Engineering.
Prof. E. S. Wade (Cambridge)	The Legalists, and Natural Law.
Dr Arthur Waley (London)	Natural Law.
Dr G. Weltfish (New York)	Introduction (Culture Contacts).
Mr E. W. White (London)	Nautical Technology.
Dr F. P. White (Cambridge)	Mathematics.
Dr H. J. J. Winter (Exeter)	Physics (Optics).
Dr W. A. Wooster (Cambridge)	Mineralogy.
Mr G. R. G. Worcester (Windlesham)	Nautical Technology.
Dr Wu Shih-Chang (Oxford)	Bone and Bronze Script (Etymologies); Introductions.
Dr Wu Ta-Kun (Shanghai)	Parts of the Historical Introduction.

In addition, we are also greatly indebted to Mr D. M. Dunlop for help with the vocalisation of Arabic and Persian names; to Mr Shackleton Bailey for assistance, equally kind, in the accenting of Sanskrit words, and to Mr J. R. McEwan for Japanese transliterations.

Over the years during which this book has been in course of preparation, the author has been honoured from time to time by invitations to deliver formal lectures, apart from those which it was natural for him to give in his own university as a member of the Faculty of Oriental Studies and the Committee for the History of Science. The first of these was one of the UNESCO Month Lectures in October 1946 in Paris; it was entitled 'The Chinese Contribution to Science and Technology', and was afterwards published in the collective volume *Reflections on Our Age*. Another UNESCO Month Lecture was given in December 1948 at Beirut in the Lebanon, 'L'Unité de la Science; l'Apport Indispensable de l'Asie'; this dealt particularly with contacts between Arabic and Chinese science. The author is indebted for many kindnesses to Mgr Maroun, at whose request the lecture was contributed. In May 1947 he delivered the Conway Memorial Lecture in London on the theme of 'Science and Society in Ancient China', and this was separately published by the Trustees.

During the Lent Term, 1949, the author held the Beard Lectureship at Ruskin College, Oxford, a position which involved eight lectures on the history of science and technology in China. His thanks are due not only to the governing body of this Trust, the members of which will, he hopes, accept the present work as a publication associated with it; but also to Professor H. H. Dubs, whose friendly sinological discussion has always been most helpful, and to Mr Lionel Elvin, then Principal of the College. The early spring of the following year was spent at Berkeley, California, where the author had been invited by President Robert G. Sproul to accept the (visiting) Hitchcock Professorship. Eight lectures were given at the University of California (seven on the Berkeley campus and one at Los Angeles) with the general title 'The Development of Scientific Thought in Chinese Civilisation'. The author wishes to thank Dean William R. Dennes for his kind welcome, and especially his colleagues in the Departments of Oriental Studies, notably Professor F. Lessing, Professor Chao Yuan-Jen, Professor Boodberg and Professor Wolfram Eberhard. This visit was followed by the tenure of the Noguchi Lectureship in the Institute of the History of Medicine at Johns Hopkins University, Baltimore, where three lectures were given with the general title 'Some aspects of the History of Chinese Science'. The author thanks Dr Richard H. Shryock and Dr Owsei Temkin for their hospitality and stimulating exchange of ideas.

In May 1950, the Hobhouse Lecture in the University of London was given on 'Human Law and the Laws of Nature in China and the West'. This was subsequently published separately by the Trustees. A year later, while Visiting Professor at the University of Lyons, France, two lectures were delivered with the common title 'La philosophie Chinoise comme philosophie Organismique'. In the summer of 1953 a lecture was delivered in Jerusalem at the Seventh International Congress of the History of Science, entitled 'Channels of Scientific and Technological Contact between East and West'. The author's warmest thanks are due to all those who invited him to place his thoughts in this way before successive audiences, and to all those who presided over the successful arrangement of the lectures.[a]

[a] Mr Wang also wishes to take this opportunity of thanking the History of Science Committee of the University of Cambridge for inviting him to deliver a course of lectures on Han Mathematics in 1953.

2. PLAN OF THE WORK

THE INTRODUCTORY PART consists of brief accounts of (*a*) the geographical background, (*b*) the history of China, (*c*) the special characteristics of the Chinese language, and (*d*) the opportunities which the centuries afforded of culture contact whereby the passage of scientific ideas or technological processes to and from East Asia may have been made possible or even facilitated. These preambles were judged necessary, since neither scientists nor historians of science could be expected to have a sufficient previous knowledge of the scene on which ancient and medieval Chinese scientists and technologists played their parts. It is hoped that such sections will save them the trouble of constant recourse to other books or encyclopaedias.

The second part takes up the question of the origin and development of scientific thought in Chinese philosophy. It is my conviction that the Chinese proved themselves able to speculate about Nature at least as well as the Greeks in their earlier period. If China produced no Aristotle, it was, I would suggest, because the inhibitory factors which prevented the rise of modern science and technology there began to operate already before the time at which an Aristotle could have been produced. But apart from the vision of the Taoists, there runs throughout Chinese history a current of rational naturalism and of enlightened scepticism, often much stronger than what was found at corresponding times in that Europe where modern science and technology in fact grew up. As we shall see, the +12th-century Neo-Confucians resembled in many ways the agnostic and organic philosophers of the European 18th and 19th centuries, though lacking completely their basis of established scientific knowledge of the world.

In the third part, which deals with the sciences, pure and applied, in due order, we attempt to answer the question, what exactly did the Chinese contribute to science, pure and applied, through the historical centuries. I planned at first to place the sciences, or pre-sciences, on the one hand, and the technologies on the other, in two quite different sections, but on maturer thought it seemed more and more impossible to separate them; they are therefore treated in close juxtaposition. The Chinese pre-sciences, both ancient and medieval, show the clearest development of experimental and observational inductive science, involving manual operations, though they were always interpreted by theories and hypotheses of primitive type. Why this was so remains to be examined.

Similarly, Chinese technology, both ancient and medieval, led to empirical discoveries and inventions many of which profoundly affected world history. It is quite clear that the Chinese could plan and carry out useful experiments for the further improvement of techniques, though again always interpreting them by theories of primitive type. It is quite clear that Chinese society, though less favourable to technological advance than post-Renaissance European society, was able to make much greater advances than the slave-owning city-state culture of the ancient Mediterranean region, or the civilisation of feudal Europe. Both these differences call for sociological comment. Whoever may trouble to read this book to the end will, I believe, be astonished at the richness and variety of the techniques which Europe adopted from China, generally with no appreciation of their origin, during the first fourteen centuries of our era. Francis Bacon wrote:

> It is well to observe the force and virtue and consequences of discoveries. These are to be seen nowhere more conspicuously than in those three which were unknown to the ancients, and of which the origin, though recent, is obscure and inglorious; namely, printing, gunpowder, and the magnet. For these three have changed the whole face and state of things throughout the world, the first in literature, the second in warfare, the third in navigation; whence have followed innumerable changes; insomuch that no empire, no sect, no star, seems to have exerted greater power and influence in human affairs than these mechanical discoveries.[a]

During the following centuries, Europeans acquired a much greater knowledge of China than was available when Bacon wrote. But those who should have known better failed to give the acknowledgement that was due. Thus J. B. Bury in our own time, in his history of the *Idea of Progress*, when describing the Renaissance controversies between the supporters of the 'Ancients' and those of the 'Moderns', shows that the latter were generally considered to have had the best of it, precisely because of the three great inventions which Bacon described.[b] Yet nowhere in his book is there even a footnote pointing out that none of the three was of European origin.

At the end of this part we shall find ourselves face to face with a truly remarkable array of scientific initiatives, outstanding technical achievements, and speculative insights. Why, therefore, did *modern* science, the tradition of Galileo, Harvey, Vesalius, Gesner, Newton, universally verifiable and commanding universal rational assent—the tradition destined to form the theoretical basis of the unified world community—develop round the shores of the Mediterranean and the Atlantic, and not in China or any other part of Asia? This is the question to which the fourth part is devoted. Its consideration involves an examination of the concrete environmental factors of geography, hydrology, and the social and economic system which was conditioned by them; though it cannot leave out of account questions of intellectual climate and social customs. At the end of this part, the problem of parallel types of civilisation will be touched upon. There is no point in outlining at this stage what conclusions, if any, can be reached.

So much for the general plan. It only remains to mention the dispositions which have had to be taken regarding technical problems inevitable in a book of this kind.

My reasons for printing Chinese characters and not only their romanised transcriptions into alphabetical form have already been given in the Preface (p. 8). To help

[a] *Novum Organum*, book 1, aphorism 129.

[b] Pp. 40, 78.

reader and printer alike, however, they have been segregated. There are therefore two kinds of footnotes, those in English, to which references are given by superscript letters, and those in Chinese, to which references are given by superscript numerals. When a Chinese book is first mentioned, but not usually at later references, the characters of its title and the name of its author are placed in a footnote of the second kind.[a] Chinese titles of books are translated wherever possible, to make them more intelligible and more easily remembered by the reader who has no Chinese.

The system of bibliographies and indexes has of necessity become complicated. The following explanations are therefore offered. There are four main lists:

A. *Chinese Books before* +1800

Each entry gives particulars in the following order:

(*a*) title, alphabetically arranged, with characters;
(*b*) alternative title, if any;
(*c*) provisional translation of title;
(*d*) cross-reference to closely-related book, if any;
(*e*) dynasty;
(*f*) date, as accurately as possible;
(*g*) name of author or editor, with characters;
(*h*) title of other book, if the text of the work now exists only incorporated therein;
(*i*) references to translations, if any, given by the name of the translator in Index C;
(*j*) notice of any index or concordance to the book if such a work exists;
(*k*) reference to the number of the book in the *Tao Tsang* catalogue of Wieger (6), if applicable;
(*l*) reference to the number of the book in the *San Tsang* (Tripiṭika) catalogue of Nanjio (1), if applicable;
(*m*) reference to *tshung-shu*[1] collections containing the book;
(*n*) indication of the edition used in the present work.

These last two pieces of information will appear only in the master-index which will form part of the last volume of the series.

B. *Chinese and Japanese Books and Journal Articles since* +1800

Each entry comprises:

(*a*) name of author or editor, alphabetically arranged, with characters;
(*b*) title, with characters;
(*c*) translation of title;
(*d*) bibliographical reference and date.

C. *Books and Journal Articles in Western Languages*

Each entry comprises:

(*a*) name of author;
(*b*) title of book or article;
(*c*) bibliographical reference and date.

[a] The Chinese characters for names of persons incidentally mentioned are usually given in the footnotes when they appear for the first time, but not if they are legendary or fictitious.

[1] 叢書

D. *Biographical Glossary of Chinese Scientists, Technicians, and Scholars down to* +1900

Each entry will give the following particulars:

(*a*) individual's name, alphabetically arranged, with characters;
(*b*) dynasty;
(*c*) dates of birth and death, or *floruit*, as accurately as possible;
(*d*) field of achievement, with brief details;
(*e*) references to biographical dictionaries or other sources from which further information can be obtained, if any.

This index will be found only in the later volumes of the series; in other volumes the general index will serve as a means of locating any individual about whom information is desired.

Except where mentioned above, each volume of the series will be furnished with all necessary indexes. In the last volume it is proposed to include a glossary of technical words and phrases.

CONVENTIONS AND ABBREVIATIONS

Ascription of Translations

When a translation made by a previous author is adopted without change (in nearly every case after checking against the original Chinese text), the citation is followed by the appropriate ascription, thus: 'tr. Hugh (7)', or 'tr. Whalley (3)'.

When a translation is so adopted, but with minor modifications which have been felt to be necessary, the ascription is: 'tr. Hugh (7), mod.', or 'tr. Fên Yu-Tan (5), mod.'.

When a translation is adopted, but had to be turned into English from the original tongue in which it appeared, we write: 'tr. Eisenhelm (5), eng. auct.', or 'tr. Piot (1), eng. auct.'. In certain cases, these also will carry the addition 'mod.'.

When the translation has been made entirely by the author and his collaborators, the ascription is simply: 'tr. auct.'.

When translations have been made by the author and his collaborators, but with the help of previous translations, even though very little of their phrasing may have been retained, the ascription runs: 'tr. auct.; adjuv. Ritti (2).'

References

Names of authors in the text or footnotes are always followed by a number in brackets corresponding to the reference in Index B or C, e.g. Smith (4), Li Tan-Yao (1) for western-language publications; and e.g. Li Tan-Yao (*4*) for Chinese or Japanese publications. The use of the ampersand (&) always indicates collaborative authorship.

It is convenient in lists of Chinese words and titles to separate Chh- from Ch- and Hs- from H-. This is therefore done in the Chinese indexes, but only for the initial words—thus Chin comes before Chha, and Hu comes before Hsin; but Hsin Chhien comes before Hsin Chien. Titles of lost books are not listed in the bibliographic index, nor are titles of maps; these can be found in the text itself by the aid of the general indexes. Double Chinese surnames are written as one word, e.g. Ssuma.

The characters for the title and author's name of Chinese books are given generally only the first time they appear, but all the Chinese bibliographies are alphabetically arranged. Where Chinese texts are divided into long chapters (*chuan*) and short chapters (*phien*), both numbered consecutively, the references given (as ch.) are always to the smaller unit. This practice is also followed in the case of certain books such as the *Lü Shih Chhun Chhiu* where the smaller units are generally numbered as parts of the larger ones. For certain classical books, e.g. *Lun Yü*, *Mêng Tzu*, etc. which became divided into very small sections in the standard editions, the system of roman and arabic numerals, as used in the translations of Legge, are followed. Some large works, such as the official histories, number chapters both in blocks according to subject-matter, and also consecutively from the beginning to the end; we invariably use the latter numbers.

Citations of Arabic and Persian books require further dispositions. Attention is drawn (p. 213) to the difficulties caused by the fact that no two Arabists seem to employ the same system of indexing for the very numerous names borne by the same person. In this book, the Arabic or Persian scholar or scientist will usually be found under that one of his names which indicates his place of origin, e.g. 'al-Andalusī', 'al-Baghdādī', etc., but this rule is not strictly followed since there are many instances where a scholar is much better known by some other name. In this case there is a cross-reference. If a given book exists only in MS form, the author's name and the title of the book (with English translation and occasionally an explanatory note) are simply repeated in Index C; further information will then have to be sought from the authority referred to in footnote or text on the page where mention of the work occurs. If the book has been printed and edited in Arabic or Persian, the usual bibliographical details are given. But if it has been translated into a European language, then a cross-reference is given to the name of the translator (always in Index C) where sufficient bibliographical details will be found, and those of the Arabic or Persian text are omitted.

We regret that, since no one can be endowed with all the facilities ideally necessary, we have been able to make relatively little use of Japanese and Russian sources. Others must supply this deficiency.

Abbreviations

B & M Brunet & Mieli, *Histoire des Sciences* (*Antiquité*).

G Giles, H. A. (1), *Chinese Biographical Dictionary*.

HCCC *Huang Chhing Ching Chieh*.

K Karlgren (1), *Grammata Serica* (dictionary giving the ancient forms and phonetic values of Chinese characters).

KCCY	Chhen Yuan-Lung, *Ko Chih Ching Yuan* (scientific and technological encyclopaedia).
PTKM	Li Shih-Chen, *Pên Tshao Kang Mu*; the Great Pharmacopoeia.
R	Read, Bernard E., Indices, translations and précis of certain chapters of *PTKM*—if the reference is to a plant see Read (1), if to a mammal, see Read (2), if to a bird see Read (3), if to a reptile see Read (4 or 5), if to a mollusc see Read (5), if to a fish see Read (6), if to an insect see Read (7).
STTH	*San Tshai Thu Hui* encyclopaedia.
TH	Wieger (1), *Textes Historiques*.
TSCC	*Thu Shu Chi Chhêng*; the Imperial Encyclopaedia of 1726 (see Giles, L. (2)).

Fuller bibliographical information on all these sources is provided under the appropriate headings in the bibliographies.

Any abbreviations not on this list will be found in the lists of abbreviations used for journals and similar publications in the bibliographies (p. 250).

At various places, special abbreviations are used only in connection with certain tabulated or other material; these are explained where they occur.

Throughout this book the cumbrous practice of writing dates as B.C. and A.D. is avoided by the use of − and + respectively. Such a convention seems also more suitable for civilisations which were never part of Christendom, and accords with the customary Chinese phrases 'before' and 'after (the beginning of the commonly accepted era)'. But it must be remembered that the astronomers, who use a similar system, insert a year 0. The B.C. dates in the present work, therefore, though bearing a minus sign, are the historical, not the astronomical, ones.

ROMANISATION OF CHINESE CHARACTERS

Anyone faced with a task such as ours is bound to adhere to some system of romanisation. Unfortunately, this is a highly controversial problem. After the earlier unsystematised efforts of the Jesuit period, in which much use was made of letters such as *x* and *q* with incorrect phonetic significance, the basic modern system came to be that of Wade (1), as modified by scholars such as H. A. Giles (2) and Soothill (1). The present position, however, is that a large number of competing systems have grown up, some stemming from Cantonese dialect romanisations such as that used at one time officially by the Chinese Post Office, others based on the studies of experts in phonetics and linguistics, such as the *Quokyu Romadzy* popularised by Chou Pien-Ming and others. The best guide to an understanding of the subject is the short paper of Karlgren (3), who points out that three adequate systems are really needed: (*a*) a strictly phonetic philological system for scientific language study; (*b*) a sinological system for dictionaries, text-books and special monographs; and

Table 1. *Romanisation of Chinese sounds*

The 24 Consonantal Initials

Wade-Giles system	Gardner system	System adopted in this book	Pronunciation
ch-	*j-* or *dz-*	*ch-*	between *chair* and *jar*
ch'-	*ch-* or *ts-*	*chh-*	as in *much harm*; strongly aspirated
f-	*f-*	*f-*	as in *farm*
h-	*h-*	*h-*	Gaelic *-ch*, as in *loch*
hs-	*hs-* or *s-*	*hs-*	a slight aspirate preceding and modifying the sibilant, the latter being the stronger. Try dropping the first *i* in *hissing*
j-	*r-*	*j-*	French *j-* as in *je* or *jaune*. A *j-* pronounced at the front of the mouth gives an impression of *r-* (cf. Polish *rz-*)
k-	*g-*	*k-*	between *k* and *g*
k'-	*k-*	*kh-*	*k* strongly aspirated, as in *kick hard*
l-	*l-*	*l-*	as in English
m-	*m-*	*m-*	as in English
n-	*n-*	*n-*	as in English
p-	*b-*	*p-*	like *b* in *lobster*, or Fr. *peu*
p'-	*p-*	*ph-*	as in Irish dialectal pronunciation of *party* or *parliament*, more strongly aspirated than anything in French, German or English
s-	*s-*	*s-*	as in English
sh-	*sh-*	*sh-*	as in English
ss-	*sz*	*ss-*	only occurs with *-ŭ*, q.v.
t-	*d-*	*t-*	nearer *d-* than *t-* in English but not quite *d-*
t'-	*t-*	*th-*	strongly aspirated *t-*, as in Irish dialectal pronunciation of *torment*
ts-	*dz-*	*ts-*	as in *jetsam*, *catsup*
ts'-	*ts-*	*tsh-*	*ts-* strongly aspirated, as in *bets hard*
tz-	*tz*	*tz-*	only occur with *-ŭ*, q.v. Sounds near to *ts'*
tz'-	*tz*	*tzh-*	
w-	*w-*	*w-*	as in English, but faint
y-	*y-*	*y-*	as in English, but faint

The 42 Vowel, Diphthong and Consonantal Finals

-a or *a*	as in *father*, the 'broad' *a*
-ai	as in *aye*, or better Italian *hai*, *amai*. English *why*
-an	somewhat like Dutch *Arnhem* pronounced by an Englishman; the *r* being unsounded. Or German *ahnung*
-ang	the *-ng* has a partly nasalising and partly gutturalising influence on the vowel. Something like German *angst*
-ao	as in Italian *Aosta*, *Aorno*. Not so fused as in English *how*

Note.

As pointed out above, the Wade-Giles system took the sounds of the Peking dialect of 'mandarin' as standard. But the national *kuo yü* pronunciation of today is not quite the same, and we have modified the transcriptions of the present book in a desire to concord with it. Thus we find the circumflex

Table 1. *Romanisation of Chinese sounds (continued)*

-ê	nearest approached by English vowel-sound in *earth*, *perch*, or *lurk*
-êi	the foregoing, says Wade, followed enclitically by *-y*. English *money* omitting the *-on-*. Generally sounded as *-ei* or *-ui* (see below)
-ei	generally indistinguishable from English *may*, *play*, *grey*, *whey*
-en	as in English *yet*, *lens*, *ten*
-ên	as in English *bun*
-êng	as in English *unctuous*, *flung*
-erh	as in English *burr*, *purr*
-i	vowel-sound as in English *ease*, *tree*
-ia	not like *yah* but with the vowels more distinct, though not so much so as in Italian *Maria*, *piazza*, and not separately accented
-iai	as in Italian *vecchiaja*
-iang	like *-ang* above, with the additional vowel
-iao	like *-ao* above, with the additional vowel
-ieh	as in French *estropié*
-ien	with vowels distinct, as in Italian *niente*
-ih	short vowel, as in *cheroot*
-in	short vowel, as in English *chin*
-ing	short vowel, as in English *thing*
-io	short vowel, as in French *pioche*
-iu	always longer than English termination *-ew*, e.g. 'chyew' instead of *chew*. *Mew* of cat, as onomatopoeically pronounced
-iung	like *-ung* below, with the additional vowel
-o	something between the vowel-sounds in English *awe*, *paw* and *roll*, *toll*
ong	as in English *dong* cut short
-ou	really *-eo*, English *Joe*
-u	as in English *too*
-ü	as in French *eût*, *tu*
-ŭ	between the *i* in English *bit* and the *u* in *shut*. Only occurs with *ss-*, *tz-* and *tzh-*, 'which it follows from the throat' says Wade 'as if the speaker were guilty of a slight eructation'.
-ua	as in *Juan* Spanish; may contract almost to *wa*
-uai	as in Italian *guai*
-uan	like *-an* above, with the additional vowel
-üan	the *ü* as above, the *-an* as in English *antic*
-uang	like *-ang* above, with the additional vowel
-üeh	as in French *tu es*
-uei	the *-u* as above, the *-ei* as above; cf. French *jouer*
-ui	as in Italian (not French) *lui*
-un	as in Italian *punto*, *lungo*
-ün	as in German *München*
-ung	as in Lancashire dialectal pronunciation of English *bung*, *sung*; not so broad as *-oong*
-uo	the *o* as in English *lone*; the whole as in Italian *fuori*

accent essential in words such as *pên* (origin) and *Chêng* (family name), but not in *jen* (a person), *chen* (true) or *Chhen* (family name). Similarly, we retain the diaeresis in *hsü* and *hsüan* but we do not write it in *yuan*. The sound represented by the inverted circumflex *ŭ* can always be recognised from the consonants which invariably accompany it; we therefore dispense with this diacritical mark.

(*c*) a popular system which could be used for creating a new colloquial literature and for setting up newspapers. For reasons which will be explained later, it seems unlikely that in the near future any serious attempt will be made in China to replace the ideographic characters altogether. It would be necessary to think very carefully indeed before taking a step which might have the effect of cutting off the younger generation from the vast treasures of Chinese literature. And we are not concerned with the first problem, which involves the use of signs absent from ordinary type-founts.

One of the chief aims of the Wade system was that it should be capable of international use, and this was why (as can be seen from Table 1) he did not employ English sounds only. Unfortunately, this international effort was not accepted by French and German sinologists, who worked out, and have continued to use, systems which are useful only for those who have these languages as their mother-tongues. Thus the French system has forms such as *tcheou* (for *chou*), *cheu* (for *shih*), *tchang* (for *chang*), etc., while the German has *liän* (for *lien*), *tschou* and so on. Wade's system was also logical in that it concentrated on a single dialect, namely the northern form of the 'mandarin' group, which now constitutes the basis of the Kuo Yü, or national, pronunciation.

Karlgren (3) proposed some improvements of the Wade-Giles system, but as they are still rather controversial I have not adopted them. My decision to adhere to Wade as closely as possible was strengthened by what is, to me, the extreme repulsiveness of the modern and more revolutionary systems. For example, in the *Quokyu Romadzy*, a phrase such as 'The World State and the Protestant Christian Students' Movement' (*Shih-chieh Ta-Thung yu Chi-tu Chiao hsüeh-sêng yün-tung*, Wade) becomes *Shihgieh Dahtung eu Gidugiaw Ciogshêng Yuindonq* (Chou Pien-Ming (2), p. 38).

The most interesting modification of Wade is that proposed by Gardner (1, 2), the main difference being that the initial consonants are changed so as to correspond more closely with the way they sound to most Western ears. Thus *p*- becomes *b*-, and aspirated *p* (i.e. *p*'-) becomes *p*-. Several contemporary sinologists have adopted similar systems (e.g. Dubs, 25), and they may be found in books such as that of McGovern on the history of Central Asia. Karlgren (3), however, gives rather convincing grounds for not adopting any of them—for example, the unaspirated *p* is not quite a *b*, but like *b* in 'lobster'. Though I have not used this simplification, I gave it up with some regret, and show the Gardner system, for comparison, in the table.

Wishing to avoid the extensive use of apostrophes characteristic of Wade and Giles, to indicate the aspirated consonantal prefixes of words, I have, at the suggestion of my friend, the late Gustav Haloun, replaced them by an additional letter *h*, as shown on the table. The system used might therefore be said to follow the Wade-Giles system as closely as possible, with the exception that *ch*'- becomes *chh*-, *p*'- becomes *ph*-, and so forth. This has the additional advantage that it permits of direct comparisons with the sounds of Indian languages, in which -*h*- to indicate the aspirate has always been used, e.g. 'Buddha', 'siddhānta'. It is therefore in some sense a return to the Müller-Legge system used long ago in the 'Sacred Books of the East' series, but only partly so, because that system did not use the symbols *ch-* and *chh-*, *ts-* and *tsh-*, *tz-* and *tzh-*. Moreover, the sound-values given to all the symbols which I use are those based on Wade's descriptions (see Table 1) and not necessarily those shown in the Müller-Legge tables. It should be noted that in that usage the aspirate *-h-* was printed in italic type, but I have not thought this worth while, as one soon gets used to remembering that the *th-* is not the English *th-*, and that the *ph-* here used has nothing to do with the Greek compound consonant sounded as an *f-*. The only other likely confusion is between words such as *shang* and *tshang*; in the former case the pronunciation is just as in English, in the latter the *-h-* is an aspirate symbol, i.e. the word is pronounced as in the example given, *bets hard*, not *bet shard*.

To avoid giving a pedantic appearance, well-known place-names are written in their most familiar forms; lesser-known ones follow Wade (cf. the note by A.R.H., 1). Names of tribes and towns are generally hyphenated with only one capital letter (e.g. 'Yüeh-chih'; 'Chiu-chhüan'), names of countries are hyphenated with two (e.g. 'Ta-Hsia').

NOTE ON THE CHINESE LANGUAGE

At the outset of this book it has been necessary to define the convention adopted for the romanised transcription of Chinese characters. Later, in the introductory account of the geography of China, there will be a brief mention of the distribution of dialect differences, and we shall have occasion to realise how powerful a factor the written language must have been in helping to assure the unity of Chinese culture in face of the great geographical barriers which dissected the scene of its development. This is therefore a suitable point, before going further, at which to place a short account of the Chinese language, aimed at sketching the linguistic medium of the scientific and technological contributions later to be discussed. The best elementary accounts of the language in English are the short books of Karlgren (4, 5), especially his *Sound and Symbol in Chinese*, to which the reader is referred.[a]

Chinese is the only language which has remained faithful to ideographic, as opposed to alphabetic, writing, for more than three thousand years. This may be (*a*) because the language was perhaps at the beginning strictly monosyllabic,[b] and (*b*) because it was 'isolating', i.e. non-agglutinative. There was no transition such as that in Egyptian from hieroglyphic through hieratic to demotic, in which a syllabic alphabet arose (Février, 1; Ferguson, 1).

The most primitive elements of Chinese were generally *pictographs*, i.e. drawings reduced to the essentials, conventionalised, and in time highly stylised.[c]

[a] The recent book of Forrest goes into the matter more deeply from the standpoint of the phonetician, giving references to the technical papers of Karlgren and others.

[b] Some scholars maintain that ancient Chinese words had more than one syllable, but that the difficulty of writing ideograms forced the language into a monosyllabic mould. One cannot be sure that every syllable of an ancient word was recorded in writing. See p. 40 below. Cf. Dubs (27).

[c] The dictionary in which as many as possible of the ancient pictographic forms of the characters are given is the great work of Karlgren (1). References to it in this book indicate the number of the relevant section thus, e.g., K451.

Naturally concrete objects such as the heavenly bodies, animals and plants, tools and implements, lent themselves most easily to such drawings.[a] A number of them are shown in the first part of Table 2, taken from a short popular article of Haloun (1). It will be seen that most of them found their way during the centuries into the list of radicals (see below), but this is by no means necessarily the case, since *hsiang* (elephant) is not a radical itself, but classed under Rad. 152 (*shih*,[1] hog), while *hu* (wine-vessel) is classed under Rad. 33 (*shih*,[2] scholar). This depended, of course, on decisions of late lexicographers.

The scope of writing was later widened to include *indirect symbols*[b] formed by various kinds of substitutions, such as making parts stand for wholes, attributes for things, effects for causes, instruments for activities, gestures for actions, and so on, in a metaphorical way. The table shows how the word *chih*, to mount, originated from the picture of two footprints pointing upwards; and how the word *fu*, meaning full, is derived from an ancient picture of a jar.

A third class of characters is composed of significant combinations of two or more pictographs, forming what might be called *associative compounds*.[c] Thus *fu*, wife, consists of the signs for woman, hand and broom; *fu*, father, consists of the ancient signs for a hand and a stick; *hao*, to love, or the adjective good, is a combination of the signs for a woman and a child. A particularly striking example is the word for male or man, *nan*, which shows the radicals for plough and for field, and implies 'that which employs its strength in the fields'. Of course the sounds of the elements in the combination are lost in the 'resulting' sound, which existed before the scribes put signs with other sounds together to represent it. We thus have a kind of equation: *li* + *thien* = *nan*. Such equations form the half-conscious background of one's mind as one becomes familiar with the language.

Closely related to these significant combinations is a small class which might be called *mutually interpretative symbols*.[d] For example, *khao*,[3] examination, was said to be derived from *lao*,[4] old, because naturally the young are examined by the old (K 1041, 1055). But in fact these two characters originally meant exactly the same thing, namely 'elder', and later there was a bifurcation or specialisation of meaning

[a] The ancient forms of the characters are therefore sometimes of considerable interest from the standpoint of the history of science and technology. Mention will be made of these from time to time in the appropriate sections.

[b] This term is Haloun's, and replaces the expression 'indicative symbols' used by Karlgren, Chiang Yi and others. The latter is unsatisfactory because it invites confusion between this classification of types of symbol, essentially genetic, and the logico-grammatical classifications which will be discussed below, Sect. 49.

[c] Karlgren, Chiang Yi and others call these 'logical compounds', but this is quite unacceptable for similar reasons.

[d] The significance of this class is much debated among Chinese scholars. One view is that it ought to contain the vast number of characters now in the sixth class (see on), and that that ought to include only instances, mostly rare and obsolete, of characters used in two different senses, though with the same orthography and the same pronunciation. This note will be better understood after the next few pages have been read.

[1] 豕 [2] 士 [3] 考 [4] 老

Table 2. *The development of Chinese script*

PICTOGRAPHS

Archaic script	Small seal	Modern script	Forms in writing	Meaning	Rad. no.
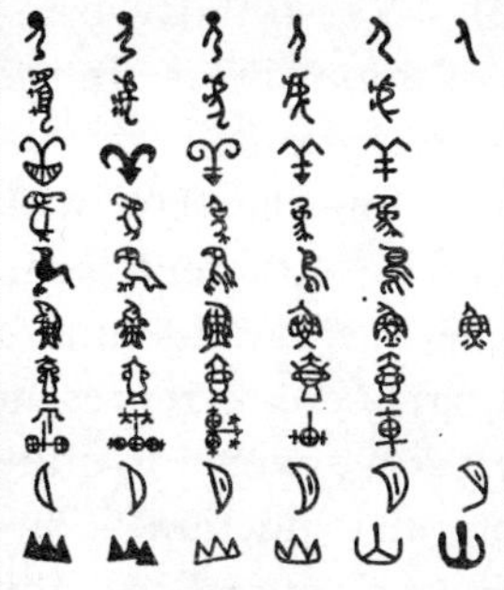	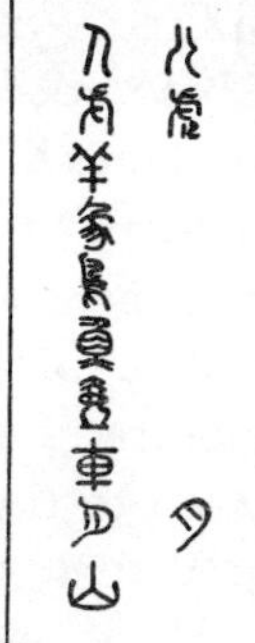	人 儿	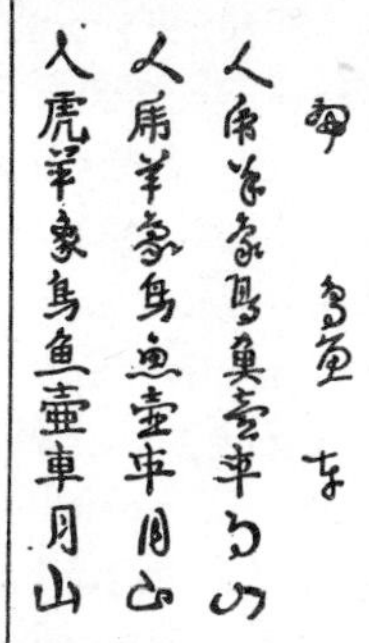	*jên,* man	9
		虍 虎		*hu,* tiger	141
		羊		*yang,* sheep	123
		象		*hsiang,* elephant	—
		鳥		*niao,* bird	196
		魚		*yü,* fish	195
		壺		*hu,* wine-vessel	—
		車		*chhê,* chariot, car	159
		月		*yüeh,* moon	74
		山		*shan,* mountain	46

INDIRECT SYMBOLS

射 *shê,* to shoot with a bow; 伐 *fa,* to attack (man being decapitated); 爲 *wei,* lead, manage, do (hand leading an elephant by the trunk); 立 *li,* to stand (a man standing); 降 *chiang,* descend (hill and two footprints pointing downwards); 陟 *chih,* to mount (footprints upwards); 至 *chih,* arrive at (arrow hitting target); 回 *hui,* revolve (meander); 曰 *yüeh,* speak (mouth and breath); 甘 *kan,* sweet (mouth and something in it); 高 *kao,* high (picture of a high building); 長 *chang,* senior, grown up, *chhang,* extended (long-haired man walking on stick); 力 *li,* strength (ard or plough); 畐 *fu,* blest (picture of a jar); 酉 *yu,* wine-must (jar and liquid inside).

ASSOCIATIVE COMPOUNDS

父 *fu,* father (hand and stick); 婦 *fu,* wife (女 woman and 帚 broom); 好 *hao,* to love, *hao,* good (woman and 子 child); 奻 *wan,* to quarrel (two women); 林 *lin,* forest (two 木 trees); 森 *sên,* umbrageous (three trees); 析 *hsi,* split (tree and 斤 axe); 牧 *mu,* tend cattle (ox and hand wielding whip); 鳴 *ming,* sing (鳥 bird and 口 mouth); 男 *nan,* male, man (employ 力 strength in the 田 fields).

DETERMINATIVE-PHONETIC CHARACTERS

耳 êrh, ear, is PHONETIC in: 珥 *êrh,* ear-pendant (determinative 玉 jade, precious stone; word cognate to 耳); 餌 鬻 *êrh,* cake (det. 食 food or 鬲 cauldron); 毦 *êrh,* plume (det. 毛 hair); 佴 *êrh,* assistant (det. 亻人 man); [illegible] *êrh,* bait (det. 虫 worm); 衈 *êrh,* a sacrifice (det. 血 blood; 恥 *chhih,* shame (det. 心 heart); 弭 *mi,* repress, ends of a bow (det. 弓 bow); DETERMINATIVE in: 聞 *wên,* to hear (phonetic 門 *mên*); 聆 *ling,* listen to, apprehend (phon. 令 *ling*); 聾 *lung,* deaf (phon. 龍 *lung*); 聰 *tshung,* acute of hearing, clever (phon. 悤 *tshung*); 聳 *sung,* alarm, excite (phon. 從 *tshung*).

立 li, to stand, is PHONETIC in: 笠 *li,* conical hat (det. 竹 bamboo); 粒 [illegible] *li,* grain of rice (det. 米 rice or 食 food); 苙 *li,* pen for animals, *chi,* hyacinth (det. 艸 herb, plant); 泣 *chhi,* to weep (det. 氵水 water); 拉 *la,* to pull, break (det. 扌手 hand); 翋 *la,* to fly (det. 羽 wings); 雴 *li, chhih,* heavy rain (det. 雨 rain); 颯 [illegible] *sa,* storm (det. 風 wind); DETERMINATIVE in: 站 *chan,* to stop (phon. 占 *chan*); 竚 *chu,* to wait for (phon. 宁 *chu*); 竣 *chün, tsun,* stop work (phon. 夋 *chün*); 靖 *ching,* quiet (phon. 青 *chhing*); 端 *tuan,* extremity, origin, end, principle (phon. 耑 *chuan*); 竭 *chieh,* exhausted (phon. 曷 *ho*).

and sound. Indeed these two words may have been originally one single word, pronounced something like *khlog*. However, perhaps all the characters in this group really belong to one or other of the remaining classes.

So far we have mentioned pictograms proper, indirect symbols, associative compounds, and the mutually interpretative symbols. About 2000 of these are still in the language today. But from very early times, since Chinese is exceptionally rich in homophones (cf. English sow, sew, so), there was a tendency to use one character with the sense which properly belonged to another of the same sound but different form. Haloun (1) believed that an inscription on a bronze of the −2nd millennium employs, as modern embroidery still may do, the picture of a bat (*fu*[1]), to represent *fu*,[2] happiness. This very strong punning tendency led to the taking over, for quite other purposes, of characters which had ceased to fulfil their original function. Thus *lai*,[3] to come, had originally the meaning of a cereal plant, as its ancient orthography clearly shows (K944). *Chhi*,[4] now the third personal pronoun, his, originally meant a

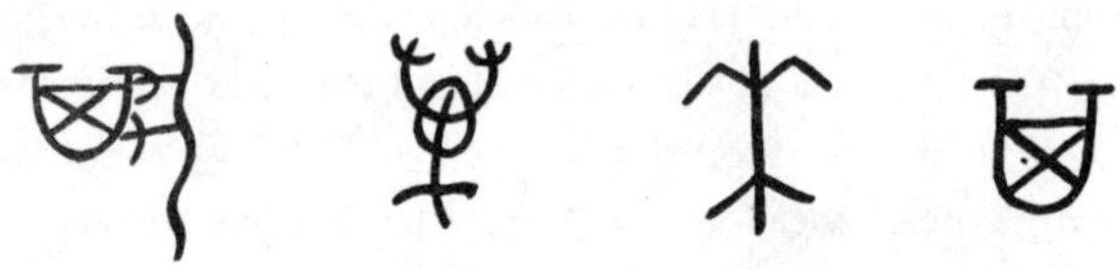

K 952 K 267 K 944 K 952

winnowing basket (K952). *Wan*,[5] ten thousand, was originally a scorpion (K267). Transference occurred because they were homophones. Such phonetic characters are called *loan characters*.

The greatest invention in the development of Chinese was that of *determinative-phonetic* characters. A determinative is an element (a radical) added to a phonetic to indicate the category within which the meaning of the word is to be sought. Thus a whole series of words with the same sound (or approximately the same sound) could be written down with no possibility of confusion. In this way: *thung*[6] (with, together) is combined with various radicals as follows:

chin[7] (metal) + *thung* = *thung*,[8] copper, bronze.
mu[9] (wood) + *thung* = *thung*,[10] the thung-oil tree (*Aleurites fordii*).
hsin[11] (heart) + *thung* = *thung*,[12] moaning, dissatisfied.
chu[13] (bamboo) + *thung* = *thung*,[14] pipe, flute.
hsing[15] (to go) + *thung* = *thung*,[16] side-street.
shui[17] (water) + *thung* = *tung*,[18] cave.

1 蝠	2 福	3 來	4 其	5 萬	6 同	7 金
8 銅	9 木	10 桐	11 心	12 恫	13 竹	14 筒
15 行	16 衕	17 水	18 洞			

This *thung* is always a phonetic, never a radical or determinative. But *shui*,[1] water, is, on the contrary, nearly always a radical or determinative, showing that the word in question has something to do with water. Take the following examples:

shui (water) + *mo*[2] (branches) = *mo*,[3] froth, foam.
shui (water) + *lan*[4] (late, end) = *lan*,[5] waves, billows.
shui (water) + *chhi*[6] (the varnish-tree, *Rhus vernicifera*) = *chhi*,[7] lacquer.
shui (water) + *chha*[8] (fork) = *chha*,[9] branching streams.
shui (water) + *mei*[10] (each, every) = *hai*,[11] the sea.

How far these combinations are the result of ingenious thought on the part of the scribes of the early Chou dynasty is a moot point; many of them certainly seem to reveal very appropriate, even sometimes rather poetical, contexts of thought, but the opportunities of fanciful interpretation on our part are too great to permit of our using these juxtapositions as more than helps in learning. We have now seen a phonetic and a radical-determinative at work; it only remains to say that some stroke-patterns can be both phonetic and radical-determinative as occasion arises. Examples of this, for *erh*,[12] ear, and *li*,[13] to stand up, will be found in Table 2.

Any one of the pictographs and symbols of the preceding classes could be used as a phonetic, that is to say, to render words with the same or very similar sounds to that which it primarily represented. But the number of determinatives was not unlimited, since the number of categories required in the primitive stages of civilisation was comparatively small. Hence the radical-determinatives came to be used as a convenient means of classifying the characters in dictionaries.

The radical system was already in full use in the −9th century and was codified about −213 by Li Ssu,[14] the premier of the emperor Chhin Shih Huang Ti (see on, p. 101). The first great dictionary was the *Shuo Wên Chieh Tzu*[15] of Hsü Shen[16] completed in +121; it recognised 541 radicals. This number remained substantially unaltered until the Ming and Chhing dynasties, when successive rationalisations reduced the number of radicals first to 360 and then to 214, at which figure they still remain today.

To the natural scientist approaching the study of Chinese, a helpful analogy is possible with chemical molecules and atoms—the characters may be considered roughly as so many molecules composed of the various permutations and combinations of a set of 214 atoms. This statement is true because nearly all the phonetics can be broken down into radicals, though this has been a late and artificial process; many had no connection originally with the radicals to which conventionalisation and stylisation have affiliated them. Hence certain characters are so difficult to find that special lists are given at the back of some dictionaries, of words 'the radical of which is not obvious'. There may be as many as seven 'atoms' in one 'molecule', and 'atoms' may repeat

1 水 2 末 3 沫 4 闌 5 瀾 6 桼 7 漆
8 叉 9 汊 10 每 11 海 12 耳 13 立 14 李斯
15 說文解字 16 許愼

(as if forming a crystal) with as many as three identical ones in one character, e.g. *sên*,[1] undergrowth, where the wood radical is repeated thrice. It will be found that some of the most complex characters cannot be dissected, e.g. *kuei*,[2] tortoise, since they are ancient pictograms stylised, and belong at the complex seventeen-stroke end of the radical list. The bearing of all this on the development of Chinese scientific terminology will be better appreciated later on (Sects. 13(*b*), 38, 39, 49).

The six classes of characters described above were distinguished first by Liu Hsin and Hsü Shen in the Han dynasty, and have been the subject of constant discussion among Chinese scholars since their time. They were called the *liu shu*,[3] (six writings),[a] hence the name of such dictionaries as the *Liu Shu Ku*,[4] compiled in the Sung by Tai Tung[5] (*c*. +1237 to +1275) and published in the Yuan dynasty in +1320 (cf. Nacken and Hopkins (36)). The names of the six classes are (in the order above described):[b]

(1) *Hsiang hsing*,[6] lit. image shapes (pictographs).
(2) *Chih shih*,[7] lit. pointing to situations (indirect symbols).
(3) *Hui i*,[8] lit. meeting of ideas (associative compounds).
(4) *Chuan chu*,[9] lit. transferable meaning (mutually interpretative symbols).
(5) *Chia chieh*,[10] lit. borrowing (phonetic loan characters).
(6) *Hsing shêng*,[11] lit. picture and sound (determinative-phonetics).

Of the 49,000 characters given in the great dictionary *Khang-Hsi Tzu Tien*[12] of +1716 not more than five per cent are pictographs and symbols; all the rest are of the sixth class.

Shang inscriptions (−14th to −11th centuries) show mostly pictographs, with some associatives and phonetic loan words. The determinative-phonetics were much developed during the Chou dynasty (−11th to −3rd centuries). Chinese thus became essentially a phonetic script employing some 1000 signs rendering the sounds of all the different syllable-words, and ready to combine with determinatives which would indicate the meaning. But unfortunately, as the language evolved further phonetically, sound combinations which started almost identical and so could be fairly represented by the same 'phonetic' became entirely unlike. For example, *kung*[13] (work) now poorly indicates the sound in *chiang*[14] (river), and the case of *chan*[15] (to divine) in *thieh*[16] (card) is even worse.[c] It is generally said that in Chinese the sound of a word as pronounced has no relation to the way in which it is written. This is true in the sense that the meaning of the written character is fixed and can be understood by

[a] This expression is mentioned in the *Chou Li*, ch. 13, p. 27 (E. Biot (1), vol. 1, p. 298), but Chang Chêng-Lang has recently shown (with epigraphic evidence) that in the Former Han it meant the practice of schoolboys in writing the sexagenary cycle characters (see on, p. 79) and not what Hsü Shen and all later scholars meant by it. This had long been suspected by scholars such as Khang Yu-Wei.

[b] A popular description of them is given by Chiang Yi. What relation they may have to logico-grammatical classes of characters is a question of great interest about which something may be said below.

[c] Cf. Karlgren (5), p. 39.

1 森　2 龜　3 六書　4 六書故　5 戴侗　6 象形
7 指事　8 會意　9 轉注　10 假借　11 形聲　12 康熙字典
13 工　14 江　15 占　16 帖

speakers of different dialects who may pronounce it in mutually incomprehensible ways. It was this 'mathematical' quality in the language which impressed the early +14th-century Persians (see p. 218) and which later attracted the attention of 18th-century European scholars such as Leibniz, perhaps stimulating, as we shall see (Sect. 49), the development of mathematical logic in Europe.[a] But in a narrower sense sound is fixed to character-form since, within the framework of a given dialect, such as the Northern Kuan-hua[1] or official pronunciation, which Wade took as the basis for his romanisation, there is a certain range of sounds within which (with certain exceptions) a character one has not previously met with will certainly fall. Thus, continuing the example given above, the character *tung*[2] (which happened to occur in the name of the author of the *Liu Shu Ku*) could only be *thung* or *tung*; it could not possibly be *ping* or *kuo*. Similarly, the word *yuan*,[3] an official, appears as the phonetic in various combinations the pronunciations of which are *yün* and *sun*; one knows that it could not participate in a word pronounced *li* or *ma*. The detailed study of these sounds and the changes which they have undergone during the centuries will be found in the books of Karlgren, especially (7), Forrest (1), etc.

One point is of interest here for the scientist. When I began the study of Chinese I adopted a method which I have not seen generally advocated, but which seemed the obvious one to use. Let each page of a notebook be devoted to one of the forty vowel-consonantal finals, thus, *-i*, *-ia*, *-iai*, *-iang*, *-iao* and so on. Then upon each page write vertically the series of twenty-three consonantal initials, thus, *ch-*, *chh-*, *f-*, *h-*, *hs-* and so forth. Four vertical columns will serve to separate the four tones.[b] Words as learnt may then be entered in their appropriate spaces as if on Cartesian coordinates, one notebook containing the characters and another their romanised equivalents, for writing and speaking respectively. This system proved extremely valuable for learning purposes.

At the time I had no idea that I was following out a method similar to that which had been used for the analysis of the sounds of characters by Chinese scholars for many centuries. They developed the system of *fan-chhieh*[4] ('cutting and joining') for 'spelling' characters.[c] Dividing all the monosyllables as just described, they could

[a] The difficulty here was that, while the semantic significance of the symbol is indeed independent of its pronunciation, the symbol itself may, and very often does, have a remarkably wide range of shades of meaning. Definitions were even less precise than those of European words. Karlgren (4), p. 85, gives the instance of *ching*.[5] Its original sense was the warp in a loom, but it came later to stand for any of the following: (*a*) the vessels in an organic body, arteries, veins and nerves; (*b*) the geographical 'warp', the meridians of longitude; (*c*) to trace the fundamental lines, to plan, to regulate, to arrange, even to rule a state; (*d*) to fix, hence the fixed stars; (*e*) a constant rule, principle, hence a canonical book, Buddhist sūtra, or prayer; (*f*) to follow a principle, to practise, to experience, to prove sufferings, to suffer; (*g*) to pass through, to pass, past, earlier, already. Everything depends on the context. Here is no mathematical logic. See Cohen (1).

[b] I say nothing about the tones here, as descriptions of them may be found in Wade and in all introductions to the Chinese language. But it is as well to be clear that they are quite inadequate for distinguishing homophones. For example, the sound *shih*, pronounced in the second tone, may mean ten, time, food, eclipse, to pick up, stone, hen roost, erect, shad-fish, or a stone shrine.

[c] Cf. Karlgren (5), pp. 67 ff.

1 官話 2 侗 3 員 4 反切 5 經

indicate the pronunciation of a given word, say *kan*, by explaining that it was composed of *k*(*uo*) + (*h*)*an*. This system appears to have originated in the Wei (Three Kingdoms period), about +270, with Sun Yen,[1] perhaps stimulated by association with Sanskrit scholars.[a] But the most important of the old dictionaries which give all the sounds of characters by this method is the *Chhieh Yün*[2] of Lu Fa-Yen,[3] published, after long preparation, in +601, i.e. in the Sui dynasty just before the Thang (see on, p. 122). The work itself as a whole is lost, but its 'spellings' were preserved in later adaptations.

Now throughout Chinese history there has been a continuous pruning and simplifying process. Not only may archaic Chinese have had some inflections[b] and other grammatical complexities afterwards lost, but it certainly had many more sounds than medieval and modern Chinese.[c] In particular, most modern Chinese dialects are greatly the poorer for having dropped many final consonants, such as *-p*, *-t*, *-k*, *-m* and *-n*. Only in Cantonese, Amoy, and other southern dialects are these retained. Thus *i*,[4] *i*[5] and *i*,[6] meaning a garment, the number one, and a bow or salutation respectively, which all sound the same in Northern Kuan-hua, are pronounced *i*, *yat* and *yap* in Cantonese. Comparative studies reveal that the character now read *fo*,[7] meaning Buddha, was in ancient Chinese pronounced *budt* (actually *b'juet* or *b'jued*). Such ancient pronunciations are preserved in foreign, e.g. Sanskrit or Tibetan, renderings of Chinese proper names or technical terms; as also in Chinese transliterations of foreign place-names or personal names. A large part of Asiatic philological research has been occupied in disentangling the old sounds (cf. Hirth (1), Laufer (1), etc.), and such studies have great importance for the history of science and technology, since they affect all conclusions about the travel of natural products, ideas, and techniques.

It is understandable, therefore, that by the +11th century Lu Fa-Yen's work could no longer be used, as the sounds had imperceptibly shifted, and therefore Ssuma Kuang[8] produced, in +1067, a systematic key to the *Chhieh Yün* in the form of a series of what are usually called 'rhyme-tables', rearranging the characters in accordance with the Sung pronunciations. This was his dictionary, *Lei Phien*.[9] I shall later suggest that the tabulation systems which thus grew up in the development of Chinese linguistics merit more attention than they have yet received from historians of mathematics, for precise tabulation was surely one of the roots of coordinate geometry (cf. Sect. 19). It was perhaps characteristic of Chinese culture that more use was made of this in historical, linguistic and philological studies than in the natural sciences, though, as we shall see, the greater part of Chinese mathematics employed a matrix or table system of notation. After Ssuma Kuang's time, many other books

[a] Nagasawa, p. 143; Forrest (1), p. 149. [b] Karlgren (6).

[c] All sinologists now accept the view, first put forward by Grube (2) in 1881, that the almost embarrassing simplicity of modern Chinese is the sign of an extremely highly developed, rather than a primitive, language. Among the Indo-European languages, English has lost many complexities, but Chinese has carried similar processes much further.

1 孫炎 2 切韻 3 陸法言 4 衣 5 一
6 揖 7 佛 8 司馬光 9 類篇

疑	羣	溪	見	泥	定	透	端	明	並	滂	幫	內轉第二
				孃	澄	徹	知	微	奉	敷	非	
		角				徵				羽		
												平
顒	蛩	銎	恭	醲	重	蹱			逢	峯	封	
				繷		統	湩	[illegible]				
												上
	[illegible]	恐	拱		重	寵	冢		奉	捧	覂	
				癑		統	[illegible]	霿				
												去
	共	恐	供	[illegible]	重		湩	艨	俸		葑	
		酷	梏	褥	毒	儥	篤	瑁	僕	蔔	襆	
												入
玉	局	曲	輂	溽	躅	棟	瘃	娼				

通志略 十二 七音一

七一 中華書局聚

Fig. 1. A sound-table from the *Thung Chih Lüeh* of Chêng Chhiao (*c.* +1150). The words are located on a coordinate system, the longitudinal axis of which (reading from right to left) is 'graduated' with initial consonants, while the vertical axis (reading from above downwards) is 'graduated' with vowels and terminal sounds. The longitudinal axis serves also for classification according to musical notes (third row from the top), while the places on the vertical axis are arranged according to the four tones of speech.

contained sound tables (cf. Fig. 1), e.g. the *Thung Chih Lüeh*[1] of Chêng Chhiao[2] in about +1150, and the *Chung Yuan Yin Yün*[3] of Chou Tê-Chhing[4] about +1250.[a] During the Ming dynasty, the study of the ancient sounds of the characters took great steps forward[b] along scientific lines, especially in the 17th century (see on, p. 145).

Perhaps this very poverty of sounds in the developed language had some bearing on the difficulties of the Chinese in forming a terminology for science. Just how few the available sounds are, is shown in Table 3 which has been constructed for the Wade-Giles system (Northern *kuan hua*),[c] and which shows by its gaps the absence of many possible sound-combinations. In fact, no less than 58·8 per cent or rather more than half the possible sound-combinations are missing. To appreciate the situation one must realise that the 49,000 characters of the *Khang-Hsi Tzu Tien* have thus only 412 sounds at their disposal. In practice this is mitigated by the four tones, which raise the number of available sounds to about 1648. There is, therefore, an average of some forty meanings to every sound. Of course, the situation is not quite so bad as this, since a very large number of the characters are obsolete, poetic or highly specialised. I remember, when beginning the study of Chinese, seeing in some dictionary a character the meaning of which was 'the songs of woodcutters returning at night', and reflecting that I should not often want to use that one, at any rate. But, *per contra*, many new characters have been invented for scientific terms in modern Chinese. Here the Chinese were at a great disadvantage. In the formation of scientific terms the West Europeans were able to draw not only upon Greek and Latin but also upon Arabic roots, adding them to languages already rather rich in Teutonic complex consonantal combinations, e.g. athwart, flowsheet, sibling, splash, clingstone. Thus many almost synonymous words were developed, which could be given slightly different shades of meaning. No such resources were available to the Chinese, although of course almost unlimited possibilities existed, and have in modern times been to some extent exploited, of forming new visual combinations in characters. We do not know whether these facts are of importance in relation to the terminology problem, which we shall examine later (in Section 49), but they seemed to call for notice here.

It is interesting that the tabulation of initials and finals found its way into European sinology at an early date, as may be seen (in degenerate form) opposite p. 176 in the 1698 edition of Louis Lecomte's *Memoirs and Observations...made in a late Journey through the Empire of China*... (cf. Pinot (1), pp. 90ff.). A full history of Western sinology has yet to be written, but brief accounts of its beginnings have been given

[a] Cf. Wylie (1), pp. 8 ff.

[b] One result of this was the formation and introduction of the Korean 'ŏnmun' (*yin wên*[5]) alphabet in +1446 by Chêng Lin-Chih and other scholars under the aegis of an enlightened prince (cf. J. S Gale, 1).

[c] Some of the regional dialects are, of course, richer in sounds, but that does not affect the argument, since Northern *kuan hua* has been the only universally spoken form among the literati for the past three centuries, i.e. from the time of the beginning of modern science and technology.

[1] 通志略 [2] 鄭樵 [3] 中原音韻 [4] 周德清 [5] 音文

Table 3. *Sound-combinations in Chinese*

Finals \ Initials	final alone	*Ch- Chh-*	*F-*	*H-*	*Hs-*	*J-*	*K- Kh-*	*L-*	*M-*	*N-*	*P- Ph-*	*S-*	*Sh-*	*Ss-*	*T- Th-*	*Ts- Tsh-*	*Tz- Tzh-*	*W-*	*Y-*
-uo	·	*	·	*	·	·	*	·	·	·	·	·	*	·	·	·	·	·	·
-ung	·	*	·	*	·	*	*	*	·	*	·	*	·	·	*	*	·	·	*
-ün	·	*	·	·	*	·	·	·	·	·	·	·	·	·	·	·	·	·	*
-un	·	*	·	*	·	*	*	*	·	·	·	*	*	·	*	*	·	·	·
-ui	·	*	·	*	·	*	·	*	·	·	·	*	·	·	*	*	·	·	·
-uei	·	·	·	·	·	·	*	·	·	·	·	·	*	·	·	·	·	·	·
-üeh	·	*	·	·	*	·	·	*	·	·	·	·	·	·	·	·	·	·	*
-uang	·	*	·	*	·	·	*	·	·	·	·	·	*	·	·	·	·	·	·
-üan	·	*	·	·	*	·	·	*	·	·	·	·	·	·	·	·	·	·	*
-uan	·	*	·	*	·	*	*	*	·	*	·	*	*	·	*	*	·	·	·
-uai	·	*	·	*	·	·	*	·	·	·	·	·	*	·	·	·	·	·	·
-ua	·	*	·	*	·	·	*	·	·	·	·	·	*	·	·	·	·	·	·
-ü	·	*	·	·	*	·	·	*	·	*	·	·	·	·	·	·	·	·	*
-u	·	*	*	*	·	*	*	*	*	*	*	*	*	*	*	*	*	*	*
-ou	*	*	*	*	·	*	*	*	*	*	*	*	*	·	*	*	·	·	·
-o	*	*	*	*	·	*	*	*	*	*	*	*	·	·	*	*	·	*	*
-iung	·	*	·	·	*	·	·	·	·	·	·	·	·	·	·	·	·	·	·
-iu	·	*	·	·	*	·	·	*	*	*	·	·	·	·	*	·	·	·	·
-io	·	*	·	·	*	·	·	·	·	*	·	·	·	·	·	·	·	·	·
-ing	·	*	·	·	*	·	·	*	*	*	*	·	·	·	*	·	·	·	*
-in	·	*	·	·	*	·	·	*	*	*	*	·	·	·	·	·	·	·	*
-ih	·	*	·	·	·	*	·	·	·	·	·	·	*	·	·	·	·	·	·
-ien	·	*	·	·	*	·	·	*	*	*	*	·	·	·	*	·	·	·	·
-ieh	·	*	·	·	*	·	·	*	*	*	*	·	·	·	*	·	·	·	·
-iao	·	*	·	·	*	·	·	*	*	*	*	·	·	·	*	·	·	·	·
-iang	·	*	·	·	*	·	·	*	·	*	·	·	·	·	·	·	·	·	·
-iai	·	*	·	·	·	·	·	·	·	·	·	·	·	·	·	·	·	·	·
-ia	·	*	·	·	*	·	·	·	·	·	·	·	·	·	·	·	·	·	·
-i	*	*	·	·	*	·	·	*	*	*	*	·	·	·	*	·	·	·	*
-erh	*	·	·	·	·	·	·	·	·	·	·	·	·	·	·	·	·	·	·
-êng	·	*	*	*	·	*	*	*	*	*	*	*	*	·	*	*	·	*	·
-ên	*	*	*	*	·	*	*	·	*	*	*	*	*	·	·	*	·	*	*
-ei	·	·	*	*	·	·	·	*	*	*	*	*	·	·	·	*	·	*	·
-eh	·	·	·	·	·	·	·	·	·	·	·	·	·	·	·	·	·	·	*
-ê	*	*	·	·	·	*	·	*	·	·	·	*	*	·	*	*	·	·	·
-ao	*	*	·	*	·	*	*	*	*	*	*	*	*	·	*	*	·	·	*
-ang	*	*	*	*	·	*	*	*	*	*	*	*	*	·	*	*	·	*	*
-an	*	*	*	*	·	*	*	*	*	*	*	*	*	·	*	*	·	*	·
-ai	*	*	·	*	·	·	*	*	*	*	*	*	*	·	*	*	·	*	*
-a	*	*	*	*	·	·	*	*	*	*	*	*	*	·	*	*	·	*	*

by Cordier.[a] After the establishment of the Jesuits in Peking early in the 17th century, it became possible for Chinese scholars to travel to Europe. The first, whose name we do not know, came with Martin Martini in 1654, and was with him on the famous occasion when he met Golius on the quayside at Leiden and their talk established the origin of the 'twelve branches' and 'twenty-four seasons' which the latter had encountered in Naṣīr al-Dīn al-Ṭūsī and Ulūgh Beg (cf. pp. 79, 217 ff. below). The second, Shen Fu-Tsung,[1] came with Couplet in 1683 and worked with Thomas Hyde at the Bodleian Library, Oxford, which still contains a copy of the *Shui Hu Chuan* dating from this time.[b]

Thus there was in the 17th century a general appreciation of the non-alphabetical character of the language, and since the existence of Egyptian hieroglyphic had long been known, it was natural that early Chinese studies were dominated by the conviction that the Chinese were of Egyptian origin. This lasted a long time, from Athanasius Kircher's[c] *China...Illustrata* of 1667 and the forgotten work of John Webb (see Chhen Shou-I), to Joseph de Guignes' boldly entitled *Memoire dans lequel on prouve, que les Chinois sont une colonie Egyptienne* of 1760, and the *De Inscriptione quadam Ægyptiaca* of one of my own forebears, John Turberville Needham in 1761.[d] During the latter part of the century, however, a disillusionment on this question set in, and de Pauw with his *Recherches Philosophiques sur les Egyptiens et les Chinois* (+1774 and +1795) devoted considerable space to denying any connection between the two civilisations, without knowing very much about either.[e]

Modern sinology did not really begin until the 19th century, and here the inaugural discourse of J. P. Abel Rémusat (2) delivered in the Collège de France in 1815 is still worth reading. I cannot refrain from quoting here what he then said about the value of the Chinese language for science and its history:

Many Westerners have been brought to believe that the Chinese have remained at the first stages of civilisation. If I may run the risk of being reproached for partiality towards

[a] (1), vol. 1, pp. 10 ff.; (9, 10); cf. Duyvendak (13).

[b] Duyvendak (15).

[c] The same Jesuit whose place in the history of bacteriology is well known (Dobell (1), p. 370; Bulloch, 1).

[d] This was the same Needham who had the controversy with Spallanzani on spontaneous generation, and whose reputation, as an epigenesist friend of Buffon's, I have tried to do something to uphold (J. Needham (2), p. 188). He discovered a vase bearing Egyptian (or Etruscan?) signs of hieroglyphic nature in the Museum at Turin and persuaded a Chinese librarian at the Vatican as well as himself that some of them bore a similarity to certain Chinese characters. At any rate J. T. Needham was a pioneer for his time and place in understanding the non-alphabetical structure of Chinese. Of course they had all been long anticipated by the 13th-century friar, William of Rubruck, who distinctly stated the case. See Yule (2), vol. 1, p. 161.

[e] De Pauw wrote in a rather aggressive way, and was on the whole more discourteous to the Chinese and their supporters than to the Egyptians and the Egyptologists. But there is much of interest in his book. Where Chinese dragons and Egyptian crocodiles are concerned, the partisans of Elliott-Smith are slain as if in a previous reincarnation. Any similarities between Egypt and China (such as the technique for the artificial incubation of hens' eggs) are put down to pure chance, Chinese alchemy and architecture are quickly disposed of, and the Great Wall is dubbed 'the greatest and most useless work ever made in the Old World'.

[1] 沈福宗

a people to whose literature I have given many years, I would like to try to bring them back to a less unfavourable opinion. Most Europeans smile at hearing of the geometry, astronomy, or natural history of the Chinese. But if it is true that the recent progress of these sciences among us has dispensed us from the necessity of having recourse to the knowledge of these far-off peoples, ought we on that account to refuse to examine the present state of their knowledge, and especially what it anciently was in a nation which has never failed to cultivate and honour it? The properties of the right-angled triangle were known in China 2200 years before the Christian era. Of this the works of Yü the Great, curbing and containing two rivers equal in swiftness and nearly so in breadth to those of America, and controlling the waters of a hundred streams so as to direct their flow in an area of over one hundred thousand square leagues, are proof more than sufficient. I would affirm, were I not reluctant to unsettle accepted ideas, that I have found, in a Chinese dictionary of date long before that of the discovery of attraction, the cause of the tides quite properly stated as the 'love of the moon for the earth'. If the astronomical theories of these people were defective, their catalogues of eclipses and comets are none the less interesting, and if one insists that the Chinese made mistakes in their calculations, at least one will admit that they had as good eyes for observing as we did. Moreover, their rural and domestic economy is sufficiently advanced to enable us to learn many useful things from it—or at least so those assure me who have made such sciences the object of their studies. As for Chinese descriptions of natural creatures, apart from the fact that these cannot be superseded until Europeans have free access to all their country, they are, as the products of so precise a people, by no means to be disdained; and I shall hope to prove, by the preparation of a botany based entirely on their writers, that they reached a level as far above that of the Latin or medieval naturalists as it was below that of a Linnaeus, a Jussieu, or a Desfontaines.

How far Rémusat's facts and judgements were correct will perhaps appear in the sequel.

I return from this digression, if such it was, to a few concluding words on the grammar of the language. Here the old work of v.d. Gabelentz has not been superseded, but apart from books already mentioned I have used also that of Wieger (5), and there is a clear and short account by H. Maspero (6). If the written character of Chinese is highly complex, the simplicity of the grammar is at the other extreme.

There are no formal word-classes, no distinctions between the parts of speech. The noun and the verb are interchangeable. Thus *shu*,[1] a tree, serves as a noun, but also as a verb in such an expression as *shu jên*,[2] planting, propagating, 'treeing' men; or, indeed, as an adjective, as in *shu lin*,[3] a (tree-) forest. There are no conjugations, declensions or inflections, the times of operations being denoted by the addition of separate words which indicate various times; personal pronouns are frequently unnecessary, and there is no gender. A most important grammatical function is fulfilled by word order; thus *shou pei*[4] means the back of the hand, but *pei shou* means putting the hands behind the back. The only ways in which Chinese is more complicated than languages familiar to us is that it has about forty 'classifiers' for collective objects, so that while one may write three men as *san ko jên*[5] (lit. three piece man), one has to write *san pên shu*,[6] three books, or *san tso lou*,[7] three buildings. It also has a number

[1] 樹 [2] 樹人 [3] 樹林 [4] 手背 [5] 三個人 [6] 三本書 [7] 三座樓

of *hsü tzu*,[1] or 'empty words' which are inserted in order to make the rhythm of the sentence run properly.[a] Old Chinese had no punctuation, a fact which may sometimes be a serious obstacle to the interpretation of texts.[b]

The statement made above that Chinese is a strictly monosyllabic language needs qualification owing to the tendency which has grown up in the spoken language to run two characters with the same meaning together in the form of doublets or couplets for avoidance of homophonic misunderstanding, for example, *khan-chien*,[2] to see (lit. look-see), or *kan-hsieh*,[3] grateful. Some dictionaries, such as that of McGillivray, are designed to list all these. The effort to avoid homophonic confusion is noticeable as soon as one participates in Chinese conversations, for one is always hearing people making such explanations as *huo-chhê ti huo*,[4] i.e. fire as in fire-carriage (i.e. locomotive). This occurs particularly with proper names. Then there are many doublets in which the second member has lost its colour; thus the word *thou*,[5] head, is tacked on to many words which indicate objects having a projecting part, e.g. *mu-thou*[6] for a piece of wood, and *erh*[7] and *tzu*,[8] both properly meaning son, are affixed to objects, presumably originally little objects, but now any objects, e.g. *cho-tzu*,[9] a table. This has little connection, however, with the *wên-li*,[10] or 'literary language' texts which the historian of science has to study. *Pai-hua*[11] usages, as they are called, began to appear in writing first in the *San Kuo Chih Yen I*[12] (Story of the Three Kingdoms) by Lo Kuan-Chung[13] of the Yuan (+13th century), and it reached full development in the *Hsi Yu Chi*[14] (Story of a Journey to the West) of Wu Chhêng-Ên[15] of the Ming (+16th century).[c] Doublet-making may be regarded as a compensation for the steadily advancing phonetic impoverishment of the language. As Forrest says,[d] the two processes must have kept pace with one another. The language did not first become unintelligible and then seek for remedies; rather the introduction of new means of discrimination between words rendered it the less needful to insist on phonetic distinctions. He sees the cause of the disintegration of the old sound system in the adoption of the language by a numerically superior population to whom its pronunciation was novel and therefore difficult.

Thus while there are many different dialectal pronunciations and colloquial idioms, the whole Chinese population possesses, in the old literary language, a kind of unified 'esperanto'. As Karlgren says:[e]

> The literary language has been an artificial thing for a thousand years and more, and for all its stylistic variations it has been essentially the same throughout the ages. Once a Chinese has succeeded in mastering it, it is the same to him, from the linguistic point of view,

[a] Or such is the traditional view, but perhaps foreign grammarians have failed to appreciate all the subtleties of the Chinese sentence structure.

[b] Cf. Karlgren (4), p. 91.

[c] Nagasawa (1), pp. 246, 295; tr. Waley (17).

[d] (1), p. 190.

[e] (4), p. 37.

1 虛字 2 看見 3 感謝 4 火車的火 5 頭
6 木頭 7 兒 8 子 9 棹子 10 文理
11 白話 12 三國志演義 13 羅貫中 14 西遊記 15 吳承恩

whether the poem he is reading was written at the time of Christ, a thousand years later, or yesterday; it is just as comprehensible and enjoyable in either case. In other countries, where the written language has followed the evolution of the spoken, a practically new literary language has been evolved in the course of a few centuries. An ordinary Englishman of today can hardly go further back than three or four hundred years in his own literature; the earliest periods he can appreciate only after special philological study. To the Chinese the literature of millennia is open; and his unrivalled love for and knowledge of the ancient culture of his country is largely due to the peculiar nature of his literary language.

And it is true that this old language, in spite of its ambiguity, has a concentrated, laconic, lapidary quality, making an impression of austere elegance, pith and virility,[a] unequalled in any other invented instrument of human communication.

[a] Karlgren (5), p. 48.

3. BIBLIOGRAPHICAL NOTES

(a) GENERAL REMARKS

THERE IS NO BOOK yet available, to our knowledge, either in Chinese or a Western language, which has covered the ground here proposed.[a] But there are a few which must be mentioned in this connection.

One of the great merits of Sarton's five-volume *Introduction to the History of Science* is that, of all books on the subject, it for the first time included detailed mention of many Chinese scientists and their work. It is also provided with an index of Chinese and Japanese names which gives their characters. But of course the encyclopaedic method there adopted does not set out to give a connected story for the development of science in a specific part of the world, still less does it attempt to present (I dare not say that anyone could answer) the problem of why modern science and technology developed in Western Europe and not in Eastern Asia. The limitations even of the generous space which Sarton fills mean that a large number of very interesting Asian contributors were perforce omitted, and, moreover, his period ends at +1400, although the succeeding two centuries (before the arrival of the Jesuits) contain some of the finest achievements of indigenous Chinese science. His great and indispensable work will always fulfil the role of a mine of suggestions for research, as well as of an encyclopaedia of information; and it is to be hoped that no one would feel (as I am sure that he himself would not) that so great an achievement would render unnecessary the elaboration of monographs such as the present one, dedicated to the study of special problems in specific parts of the world.

Next comes to mind Abel Rey's four-volume *La Science dans l'Antiquité*, in which about one-sixth of the first volume (*La Science Orientale avant les Grecs*) is devoted to Chinese science. But this treats only of the earliest astronomy and mathematics, first-hand knowledge of which was not available to Rey, and as it stands it is a very incomplete guide even to this restricted subject. Indeed, it compares so poorly with Rey's stimulating and detailed treatment of early Greek scientific philosophy that it must often have given the impression that the science of the ancient Chinese did not amount to much.

Another class of book in Western languages is represented by the older works which essayed to describe Chinese civilisation as a whole. They sometimes devoted a chapter or two to the arts and sciences, as well as to the natural products, of China. In the 17th and 18th centuries there were of course a number of these, beginning

[a] There are a few relatively short reviews, each not without its value, covering the general subject—Chatley (3), Vacca (1), and long ago Edkins (1). The paper of Chang Yin-Lin (*2*) is perhaps the most useful of these. Wu Chhêng-Lo (*1*) has constructed brief chronological tables relating discoveries, inventions and scholarly achievements in China to those of Europe.

with Trigault and Lecomte and culminating in the famous folio of du Halde (1735).[a] Then Grosier in 1819 devoted one 'book' (v) in his seven volumes to natural history, another (VI) to minerals, a third (XV) to industrial arts, and a fourth (XIII) to the sciences. So also in 1844 *The Chinese* of J. F. Davis, who had been on the staff of Lord Amherst's embassy in 1816, contained four scientific chapters in its third volume; and a parallel account was that of S. Wells Williams (*The Middle Kingdom*) of 1848. Towards the end of the century W. A. P. Martin, who was one of the first Anglo-Saxon educationalists to be invited by the Chinese to set up teaching institutions, published two books (1, 2), the *Lore of Cathay* especially giving an account of the Chinese sciences in their traditional form. Martin has the credit of having been one of the first to appreciate the high antiquity of alchemy in China. While these books may still be read with profit on the natural history of East Asia, and on the industries which the authors themselves saw at work during their travels, they are extremely unreliable on the history of science as such; sinology has in the past half-century advanced so rapidly that their interest today is mainly historical in the antiquarian sense.

It is at first sight rather surprising that there is no book in the Chinese language which deals with the history of science in East Asia as a whole, even for a particular period. The only one known to me is Chhen Wên-Thao's *Hsien Chhin Tzu-Jan Hsüeh Kai Lun*, which deals with the state of the sciences during the Chou and Chhin periods, i.e. down to the −2nd century; but it is a small unscholarly monograph subject to the common Chinese fault of failing to give exact references for quotations.[b]

This is not to say that there are not, in Chinese, a number of books describing the scientific developments of the past century, but these are outside our field.[c] Doubtless the extreme paucity of general works on the subject by Chinese themselves is due to several obvious factors: (*a*) the heavy emphasis on literary studies in traditional scholarly circles, which did not grasp that China had had a scientific history, or that it would be interesting to write one if it had;[d] (*b*) the disadvantage laboured under by

[a] Cf. Pinot (1), pp. 167 ff.

[b] While in China during the second world war, I heard that Lu Shêng-Mo[1] was attempting a similar work of broader scope, but it has not yet appeared.

[c] The Chinese language also contains two or three good books on the history of science in Europe, starting from ancient times, e.g. Chang Tzu-Kao (*1*) and Ting Hsü-Hsien (*1*).

[d] One advantage at least arises from this humanistic emphasis. Ancient or medieval Chinese texts containing information of value from the point of view of the history of science are quite unlikely to have been falsified or manipulated, because it would never have occurred to traditional Chinese scholars that any kudos was to be gained from claiming a discovery or invention at a date earlier than its true one—science, and especially technology, had no social prestige. At the same time, it is necessary to be on guard against the tendency of those who *quoted* texts to slip in touches 'modern' for their own time. Thus, for example, the *Thai-Phing Yü Lan* encyclopaedia (ch. 926, p. 2*a*) quotes, in +983, the remark made by Li Ssu to his son as they were being led to execution in −208: 'Alas, however much we might want to go hunting with our yellow dogs and our falcons on our wrists, we could not do so now.' But the source (*Shih Chi*, ch. 87, p. 23*a*) has nothing at all about falcons, though the rest is the same

1 昝勝謨

modern Chinese natural scientists themselves, who, initially set back several years by having to master at least one European language, were afterwards much too busy in keeping up with the latest advances to have any time to spare for the history of science in their own civilisation; (*c*) the fact that the Chinese were at first so dazzled by the military and political supremacy of the European powers that they attributed a false uniqueness to the scientific and technological tradition which had made that supremacy possible. The tendency, indeed, to refer to science as *hsi yang kho hsüeh*,[1] western-foreign science, something which had come inexplicably to China and had no roots in Chinese civilisation before, is still not quite dead. There were a few men who reacted against these conventional ideas, however, and I shall speak of their work in a moment.

Monographs on the history of particular sciences have, of course, appeared in Chinese in considerable numbers (see below, p. 47).

Of course modern Western sinology too has also been subject to a heavy emphasis on literary studies. After an early period in the last century when much attention was paid by men such as Joseph Edkins, G. Schlegel and E. J. Eitel to matters of scientific and technological interest, a reaction set in which has lasted until the present time. Sinologists have laboured at the translation of belles-lettres and novels, pursued artistic production and method into its minutest details, threaded the labyrinths of Buddhist theology, and sought to unravel every circumstance in the lives of poets and philosophers, monks and missionaries. The history of subjects such as law and economics has taken a very secondary place, while science and technology have been altogether forgotten. Names as great as those of Couvreur, Chavannes, Pelliot and Laufer were of course above all these tendencies. Particular praise is also due to those contemporary sinologists who have seen the value of the history of science and technology; for example in the studies of H. H. Dubs on alchemy and astronomy, E. Balazs on Taoism, W. Eberhard on calendrical science, A. C. Moule on tides, plants and animals; and those of J. J. L. Duyvendak on astronomy, geography and navigation, together with the strange history of the sailing carriage.

(*b*) SOURCES

The primary sources are naturally available in all libraries and collections of Chinese books, but libraries in Western countries are full of gaps, and it is not too much to say that the present book would have been an impossibility if care had not been taken to assemble many necessary books while the author was in China, and if good fortune had not presided over their journey to England. Those who approach the subject with knowledge of historical studies in the West will be surprised at the relatively

(cf. Bodde (1), p. 52). The passage does not therefore attest falconry in the −3rd century. It still remains curious, however, that the editors of the *Thai-Phing Yü Lan* used it in their chapter on falconry; they probably took it from some contemporary quotation, not bothering to look up the passage in the original text. For this example we are indebted to Dr A. Waley (private communication). The moral of it is that all reliance on quotation is bad, even if sometimes unavoidable.

[1] 西洋科學

small part which manuscripts play in the study of China's past. For example, while a study of the history of engineering in Europe would be very incomplete without acquaintance with many German MSS. of the +15th century,[a] printing had begun in China at least seven centuries earlier, and (broadly speaking) everything written before that time which still existed when printing started, has either been printed many times over with continual re-editing, or has been lost.

This is, of course, far from being an unmixed blessing. Karlgren (5)[b] goes so far as to say that the invention of paper in the +2nd century was a great misfortune for China. Although the Chinese were ahead of the West in printing also, this did not begin till the +8th or +9th centuries, leaving a gap of 700 years during which all manuscripts were written on perishable material instead of on the clumsy but durable parchment of the West. In Europe, on the other hand, the art of mass-production was obtained soon after the appearance of the fragile non-durable material. Fortunately, there have been finds of Chinese MSS dating from the 'interregnum' period; of these the greatest was the library of the Buddhist cave-temple monasteries at Chhien-Fo-Tung[1] (the Thousand-Buddha caves) near Tunhuang[2] in Kansu province. In March 1907, Aurel Stein obtained no less than 7000 MSS from this library, and these are now in the British Museum (description by L. Giles, 5). They are of all dates from +406 until the time when the library was walled up (shortly after +1000). In December 1907, Paul Pelliot, choosing with the knowledge of an expert sinologist, obtained a further 3000 which are now in Paris, and transferred 2000 more to the Chinese National Academy. The remainder of the original 21,000 were collected by the Chinese Government in 1910.[c] I shall refer from time to time to some of these documents which have great scientific interest, but their examination is still far from completed.

The best guide to Chinese literature in English still remains the remarkable work of Wylie (1). None the less a warning is necessary that though himself a man of no mean scientific accomplishments, his descriptions failed to do justice from time to time to Chinese books of scientific interest. Thus the +10th-century *Hua Shu*[3] (Book of Transformations), which he classed[d] as an 'ethical treatise', contains in effect a considerable amount of important material on Taoist science and philosophy of the Thang period. The numerous scientific aspects of a work such as the *Thung Chih*[4] (Historical Collections), +1150, are not recognised. There are also strange omissions, notably the *Lun Hêng*[5] (Discourses Weighed in the Balance) of +83, and the *Wu Ching Tsung Yao*[6] (Compendium of Military Technology) of +1040.[e]

[a] Sarton (1), vol. 3, p. 1550.
[b] P. 92.
[c] Salles *et al.* (1), pp. 5, 17, 25, 38, 64.
[d] P. 127.
[e] The late Ming novel *Fêng Shen Yen I*[7] (Stories of the Promotions of the Martial Genii) Wylie rightly described as dealing with 'the adventures of Wu Wang in his contest with Chou Wang[8]', but it has, in fact, some technological interest as containing imaginary anticipations of scientific war techniques, 'wish-fulfilments' as it were, of pre-industrial civilisation. See on, p. 165.

1 千佛洞　2 敦煌　3 化書　4 通志
5 論衡　6 武經總要　7 封神演義　8 紂王

For orienting oneself in the regions beyond Wylie, there is Nagasawa's history of Chinese literature (available in German), and a list of books in Wieger (3) which is sometimes helpful. Indispensable, however, is the Harvard-Yenching four-volume Index to the twenty bibliographies contained in the official dynastic histories; here at any rate an author's name and his approximate date can generally be ascertained. Recently Yang Chia-Lo has rendered a service to learning by the publication of the *Ssu Khu Chhüan Shu Hsüeh Tien*[1] (or Encyclopaedia Quatuor Bibliothecarum) which lists all the books contained in the *Chhin Ting Ssu Khu Chhüan Shu Tsung Mu Thi Yao*,[2] i.e. the catalogue of the books considered valuable by the literati in the time of the emperor Chhien Lung (+1773). About one-third of the books which were at that time sent to the capital were copied into the seven MS sets of the *Ssu Khu Chhüan Shu* (cf. Mayers, 2),[a] of which three still exist; but all which were deemed worthy, even those excluded from the imperial collection, were listed and described in the catalogue. The abridgement of it, the *Ssu Khu Chhüan Shu Chien Ming Mu Lu*[3] is not so useful, as it mentions only the books which were selected for copying. Mayers (2) estimated that 3511 books were copied, and 6462 books were returned to their owners having been examined for description in the analytical catalogue. The *Ssu Khu Chhüan Shu Hsüeh Tien*, moreover, is arranged in modern style for rapid consultation by means of an alphabetical index (odd though the romanisation is).[b]

In considering the authenticity of ancient and medieval Chinese texts, the fundamental paper of Karlgren (10) must be consulted.

Special bibliographies of Chinese books will be mentioned in the appropriate sections below; here only the bibliography of bibliographies by Nagasawa (*2*) will be mentioned.

As to secondary sources, there exist a number of monographs on particular aspects of the history of science in China. This is very fortunate, since otherwise it would be impossible for anyone to make a wide survey without falling into insuperable difficulties as soon as he left his own proper field. Here there are four obvious situations.

(*a*) There exists no monograph, either in Western language or in Chinese, dealing with the field in question.

Examples: ZOOLOGY.

MECHANICAL ENGINEERING.[c]

[a] Pp. 291 ff.; cf. Têng & Biggerstaff (1), p. 22.

[b] I owe my copy to the generosity of the editors and the National Peiping Academy.

[c] In certain of the more unsurveyed fields, such as this, there exist valuable review articles—here we have Liu Hsien-Chou's (*1*) *Chung-Kuo Chi-Hsieh Kung-Chhêng Shih Liao*, and Chatley's (2) *Development of Mechanisms in Ancient China*. But for the zoological sciences there is not even a single review.

[1] 四庫全書學典

[2] 欽定四庫全書總目提要

[3] 四庫全書簡明目錄

(*b*) There exist monographs in Western languages, but not in Chinese, dealing with the field in question.

Examples: BOTANY.

Bretschneider, *Botanicon Sinicum.*

NAUTICAL TECHNOLOGY.

Worcester, *Junks and Sampans of the Yangtze.*

(*c*) There exist monographs in Chinese, but not in Western languages, dealing with the field in question.

Examples: HYDRAULIC ENGINEERING.

Chêng Chao-Ching, *Chung-Kuo Shui Li Shih.*

GEOGRAPHY.

Wang Yung, *Chung-Kuo Ti-Li Hsüeh Shih.*

(*d*) There exist monographs both in Chinese and Western languages, dealing with the field in question.

Examples: MATHEMATICS.

Chhien Pao-Tsung, *Chung-Kuo Suan Hsüeh Shih.*

Mikami, *The Development of Mathematics in China and Japan.*

MINERALOGY.

Chang Hung-Chao, *Shih Ya.*

de Mély, *Les Lapidaires Chinois.*

It will thus become apparent as we proceed that there still remain large and important fields of research which have not been covered by any existing book, and it is to be hoped that scholars, both Chinese and Western, will be encouraged to fill such gaps. The present book has utilised the existing monographs to the utmost, and the author feels a sense of much gratitude to their writers.

(*c*) ENCYCLOPAEDIAS, DICTIONARIES AND OTHER WORKS OF REFERENCE

The largest encyclopaedia which has been at our constant disposition has been the *Thu Shu Chi Chhêng,*[1] that magnificent collection produced by imperial order in +1726, consisting of 32 sections, 6109 subsections, and 10,000 chapters (see the descriptions by Mayers (2) and O. Franke (9), and the index by L. Giles (2)[a]). I have the edition of +1888 (in some 1700 volumes), a princely gift for which I know

[a] The only serious defect of this index is that it does not give the titles and names of authors of certain books which are reproduced in the Encyclopaedia almost *en bloc*. The Encyclopaedia partakes of the nature of a *tshung-shu*, and it would be very useful to have an index which arranged the books which are embedded in it in alphabetical order. Moreover the entries selected by Giles seem rather whimsical to the historian of science, who will find among them, for example, 'tiddly-winks' but not 'tides'.

[1] 圖書集成

not how to make adequate acknowledgement. Although great reliance has been placed upon this encyclopaedia by many Western scholars, one has to bear in mind that the editors were in the habit of leaving gaps in the quotations which they made, without indicating whether, or where, they were doing so; hence it is essential to check against the original text if this is at all possible. Although the arrangement of the material is often odd, as described by Giles, there is, broadly speaking, a chronological order in the quotations given.

The same plan was adopted in the Sung *Thai-Phing Yü Lan*,[1] of +983, though the quotations are fewer and shorter (I have the edition of +1807). The Harvard-Yenching index to it is an indispensable aid. There is also the *Yü Hai*,[2] compiled by Wang Ying-Lin[3] in +1267. In these three encyclopaedias a mass of information on science and technology in old China is available.

Mention was made above of certain men who stood out against the idea, commonly accepted in China after the coming of the Jesuits, that all science was of Western origin. Reacting somewhat too far, they sought to show that, on the contrary, all the most important scientific discoveries had been made anciently in China but had been forgotten. This mood, nevertheless, led to the preparation of one useful, and several less useful, minor encyclopaedias and similar books especially devoted to the history of science and technology. The best of the class is the *Ko Chih Ching Yuan*[4] of Chhen Yuan-Lung,[5] published in +1735, in 100 chapters. It is, as Wylie (1) says, a most useful compendium, but he is also right in drawing attention to the inaccuracy of the quotations, which should always be checked against original texts. I also possess one of much less value, the smaller *Ko Chih Ku Wei*[6] of Wang Jen-Chün,[7] of +1896.[a] The titles of these books were of course taken from the famous phrase in the *Ta Hsüeh*[8] classic *chih chih tsai ko wu*[9]—the attainment of knowledge lies in the investigation of things.[b] A shortened form of it, *Ko Chih* ('Investigation Attainment') was later taken as a watchword by Chinese scientists, and used in the titles of many

[a] This book ends with a table showing which of the ancient classics contain matter of scientific interest, but it is not, as a whole, a very intelligent production. Among its favourite sources are the +18th-century *Chi Chhi Thing Chi*[10] of Chhüan Tsu-Wang[11] and the late 19th-century *Ying Hai Lun*[12] of Chang Tzu-Mu.[13] The exaggerations of the latter are to be deplored. He will seize upon an ancient statement about the motion of the earth and claim that Copernicus was anticipated by the Han people without bothering to enquire whether oscillatory or rotatory motion was meant.

[b] Legge (2), p. 222. It would take us too far at this point to discuss the history of this phrase, which has been a bone of contention among Chinese thinkers since at least the Sung. The *Ta Hsüeh* lay unremarked in the *Li Chi* (Record of Rites) which had been compiled at the end of the Former Han dynasty for more than a thousand years, after which it was taken out by the Sung Neo-Confucians and exalted into one of the Four Books which they considered as constituting the essence of the Great Tradition (Hughes (1), p. 88, (2); Fêng Yu-Lan (1), vol. 1, pp. 362 ff., 368 ff., (2), p. 191). To what extent either the original writer of the phrase, or the Sung Neo-Confucians themselves, meant it to have an interpretation along the lines of natural science will probably never be known (see on, Sect. 16).

1 太平御覽　2 玉海　3 王應麟
4 格致鏡原　5 陳元龍　6 格致古微
7 王仁俊　8 大學　9 致知在格物
10 鮚埼亭集　11 全祖望　12 瀛海論
13 張自牧

books on natural science in general (such as the *Ko Chih Tshung-Shu*[1] of Hsü Chien-Yen[2] and the *Ko Chih Chhi Mêng*[3]).

Another encyclopaedia which may sometimes be useful, though no less disagreeable to use than the others owing to its uniform lack of precise references, is the *Tzu Shih Ching Hua*[4] of +1727, which gives quotations from the classics, philosophers and histories, arranged under 280 sections.

In traditional Chinese scholarship no indexes existed, and scholars were supposed to know their texts sufficiently well to be able to refer to them for any specific inquiry. But in recent times, numerous indexes have appeared, and one may particularly mention the Harvard-Yenching series with that of the Sino-French School at Peking. For books on natural philosophy such as the *Huai Nan Tzu*[5] such indexes are quite indispensable.

Of dictionaries we have used for ancient Chinese, Couvreur (considered the best available for those whose lack of Russian prevents the use of Palladius), and for modern, Matthews. For the ancient forms of the characters Karlgren (1) is indispensable. Chinese counterparts of the *Oxford English Dictionary* are the *Tzhu Yuan*[6] and the *Tzhu Hai*.[7] The greatest Chinese dictionary of single characters is the *Khang-Hsi Tzu-Tien*.[8] For character doublets there is the *Tzhu Thung*.[9]

Biographical dictionaries are a problem. There are several standard works in Chinese, notably the *Chung-Kuo Jen Ming Ta Tzhu-Tien*,[10] which we have used much, but it is tiresome to consult as the personal names are not printed in type sufficiently outstanding from the text, and no exact dates, only dynastic periods are given. Better is the *Chung-Kuo Wên Hsüeh Chia Ta Tzhu-Tien*[11] of Than Chêng-Pi,[12] which gives wherever possible the dates of birth and death (in both Western and Chinese style) of 6851 individuals. The identification of persons when only their literary name or some other appellation is given, can be made by the use of Chhen Tê-Yün's[13] *Ku Chin Jen Wu Pieh Ming So Yin*[14] (Synonymy of Names of Distinguished Chinese, Ancient and Modern).[a] References to the places where individuals are mentioned in the dynastic histories can be found at once by Chang Hsi-Chhen's[15] *Erh-shih-wu Shih Jen Ming So Yin*.[16]

The most obvious biographical dictionary for a reader of English is that of H. A. Giles (1), *Chinese Biographical Dictionary*, but it is vexing for the historian of science, since its author was mainly interested in conventional history and literature (not to speak of superficial and apocryphal anecdotes), and omitted very many of the most important scientific names. He may be supplemented by the list in Wieger (3). For

[a] A smaller work of the same kind which we have used is Chhen Nai-Chhien's[17] *Pieh Hao So Yin*.[18]

1 格致叢書 2 徐建寅 3 格致啓蒙 4 子史精華
5 淮南子 6 辭源 7 辭海 8 康熙字典
9 辭通 10 中國人名大辭典 11 中國文學家大辭典
12 譚正璧 13 陳德芸 14 古今人物別名索引
15 章錫琛 16 二十五史人名索引 17 陳乃乾
18 別號索引

late (Chhing) personalities there is the new and splendid work of Hummel (2) and his collaborators. Occasionally Giles and Wieger may be supplemented by the older list of Mayers (1). This last book is also useful in that it gives some seventy pages of a list of the numerical categories so much used in Chinese thought and writing, the three this, the seven that, and so on. Mayers took this mainly from Kung Mêng-Jen's[1] *Tu Shu Chi Shu Lüeh*[2] of +1707, but since this is not frequently available it is worth noting that the *Thu Shu Chi Chhêng* encyclopaedia has no less than eleven chapters devoted to the same subject.[a]

Among works of a more extensive character than dictionaries there is, first, the remarkable *Chhou Jen Chuan*[3] (Biographies of the Mathematicians) by Juan Yuan,[4] a high official of the Chhing dynasty,[b] published in +1799, to which a number of supplements were later added.[c] Though including a few biographies of Western mathematicians from the Greeks onwards, it dealt in great detail with the Chinese figures, and since very often mathematics was only one of the various scientific accomplishments of individuals in a non-specialised age, the book serves as the nearest approach to a history of Chinese science ever written in China. While excellent for astronomers, however, there is little or no biological material in it. Recently, a parallel work for engineers, architects, technologists and master-craftsmen, by Chu Chhi-Chhien and his collaborators, called *Chê Chiang Lu*,[5] has been appearing in instalments. It would be highly desirable that some of the entries in these collections should be translated.

For the medieval philosophers there are the well-known works *Sung Yuan Hsüeh An*[6] and the *Ming Ju Hsüeh An*.[7] A great deal of the biographical material in these collections appears in translated form in the three great volumes of Forke (9, 12, 13).

There is only one fundamental bibliography of books and all writings in Western languages about China, namely, Cordier's *Bibliotheca Sinica*. While even in its last revision, which dates from about 1920, it does not include much modern literature, it is indispensable for all the older work.[d] Besides Cordier, there is a small bibliography down to +1876, not without value, by v. Möllendorff & v. Möllendorff (1).[e] An excellent bibliography of Chinese reference works, with annotations, has been provided by Têng & Biggerstaff (1).

We have been able to use the best atlas of China, the *Chung-Hua Min-Kuo Hsin Ti-Thu*,[8] which I fortunately found in Chhêngtu. There is often confusion about Chinese

[a] *Li fa tien*, chs. 129–40.

[b] Of course with several collaborators.

[c] On the whole work see Hummel (2), p. 402; W. Franke (1); van Hée (10), this last to be used with caution.

[d] I shall always remember the long and enjoyable days I spent during World War II, while waiting for plane transport, culling what seemed necessary for my purpose from Cordier in the library of the Royal Asiatic Society in Calcutta. The only other reader at the time was the Librarian of the King of Cambodia.

[e] Cf. the article of Hülle (1).

[1] 宮夢仁 [2] 讀書記數略 [3] 疇人傳 [4] 阮元
[5] 哲匠錄 [6] 宋元學案 [7] 明儒學案 [8] 中華民國新地圖

place-names because they were frequently changed during Chinese history; hence the necessity for using the *Chung-Kuo Ku Chin Ti Ming Ta Tzhu-Tien*[1] dictionary. Richard's geography, though rather old, can still be useful.

We possess in Cambridge an historical atlas of traditional Chinese type, the *Li Tai Yü Ti Yen Ko Hsien Yao Thu*,[2] produced by Li Chao-Lo[3] in +1838 (later edition 1879). In this the latitudes and longitudes are printed in red, as also is a base map of China showing the principal modern cities and Chhing provincial boundaries; this recurs on every sheet, the place-names and boundaries of earlier times being over-printed in black, with explanations. As one turns over the pages beginning from the Chou period the southward expansion of the empire can thus be well seen. This work was superseded by the *Li Tai Yü Ti Thu*[4] of Yang Shou-Ching[5] published in 1911 and now (1952) also available in Cambridge through the generosity of Academia Sinica. More convenient for rapid use is the *Historical and Commercial Atlas of China* by A. Herrmann. This indispensable aid, which is much more historical than commercial, is unfortunately out of print and unobtainable; its reprinting is greatly to be desired.

(*d*) CHINESE TRADITIONS OF INVENTORS

A very interesting genre of Chinese literature bearing on the history of science is formed by the books which might be called technological dictionaries, or records of inventions and discoveries. These appear to have escaped almost entirely the notice of western scholars.[a]

The oldest of the group is the *Shih Pên*[6] (Book of Origins).[b] Much of it consists of a recital of the names of legendary inventors,[c] the entries being simply of the type of *Po I tso ching*,[7] Po I invented well(-digging); *Hu Tshao tso i*,[8] Hu Tshao invented clothing; *Li Shou tso shu*,[9] Li Shou invented computations; etc. Most of the persons named are put down as ministers of Huang Ti,[10] the entirely legendary Yellow Emperor. The book is therefore valuable, not as throwing any light on the history of science as such, but for the systematisation which it brings to the body of legendary technological lore.[d]

[a] The only notice of the *Shih Pên* in Western literature is that in R. Wilhelm (1), p. 25, who does not however, indicate the intricacies of its history. Wylie (1) and Wieger (3) ignore all these books, nor does Sarton (1) consider them.

[b] A very short book which can be found in the *Han Wei Tshung-Shu*[11] collection.

[c] The rest concerns imperial genealogies and family names.

[d] There are occidental parallels, cf. Lovejoy & Boas (1), pp. 192 ff., 382 ff. The longest list of inventors in Western classical writers is that of Pliny, *Nat. Hist.* VII, lvi, 57. There is a special study of the subject by Kleingunther (1), who discusses also the Greek and Roman 'technic deities'.

[1] 中國古今地名大辭典 [2] 歷代輿地沿革險要圖 [3] 李兆洛
[4] 歷代輿地圖 [5] 楊守敬 [6] 世本
[7] 伯益作井 [8] 胡曹作衣 [9] 隸首作數
[10] 黃帝 [11] 漢魏叢書

The history of the text is obscure. A book of the same title as that which we now have is listed in the Han bibliography,[a] and was still conserved in −8. According to Pan Piao[b] it had been used by Ssuma Chhien, the great historian, as a source for the *Shih Chi*. Some later scholars such as Huangfu Mi (+3rd century) attributed it to Tsochhiu Ming, the traditional author of the *Tso Chuan* (−5th century), but this never became generally accepted. The Sui and Thang bibliographies (+6th and +8th centuries) attributed part at least of the book to Liu Hsiang, the −1st century scholar, official and alchemist, but could not distinguish his text from the enlargement made by Sung Chung,[1] a scholar of the Later Han (+2nd century). During the course of years, the book splintered into various versions and fragments, and these were collected in the Chhing (+17th and +18th centuries) by Sun Fêng-I[2] and finally prefaced by Sun Hsing-Yen.[3]

The next book of the kind was the *Shih Shih*[4] (Beginnings of all Affairs) by Liu Tshun[5] (perhaps identical with Liu Hsiao-Sun[6] to whom the Thang bibliography attributes it). Liu Hsiao-Sun was a mathematician of the Sui (fl. +605 to +616). The *Shih Shih* has about 335 entries, which include names of ancient States and official titles which had fallen into oblivion, as well as all kinds of material things and devices. It was followed by Ma Chien's[7] continuation of it, the *Hsü Shih Shih*[8] of about +960 (Ma Chien was of the independent State of Shu, Szechuan, during the Wu Tai period), which is rather longer, having 358 entries.

These books still maintain the references to legendary inventors. The *Shih Shih* frequently adduces the *Shih Pên* as its authority, and copies it (for example, in the matter of the fishing-net, the grain-mill, and the evaporation of sea-water for salt); but it also makes use of all kinds of other ancient books, such as the social evolution sections of the *I Ching*;[9] the *Shan Hai Ching*,[10] the *Po Wu Chih*,[11] and the *Hsi Ching Tsa Chi*.[12] The *Hsü Shih Shih*, on the other hand, while still mentioning legendary inventors, rarely quotes the *Shih Pên*.

It would hardly serve any purpose to give here a register of the names of the legendary inventors. Such a list could without difficulty be compiled from other ancient sources as well as the *Shih Pên*, i.e. from the *Lü Shih Chhun Chhiu*,[13] the *Huai Nan Tzu*,[14] etc. The question may be raised, however, whether the examination of such lists might not allow of the establishment of several distinct strains or currents of legend concerning inventions of high antiquity. A most interesting beginning has been made along this line by Chhi Ssu-Ho (*1*), who has tabulated the attributions made by the various ancient books, and drawn some tentative conclusions from the results. It would be desirable to carry further this line of research.

[a] *Chhien Han Shu*, ch. 30, p. 10*a*. The note describes it as containing lists of all the feudal princes and their ministers during the Chou period.

[b] *Hou Han Shu*, ch. 70, p. 3*a*; cf. Chavannes (1), vol. i, pp. cxli and ccxxxix.

1 宋衷 2 孫馮翼 3 孫星衍 4 事始 5 劉存
6 劉孝孫 7 馬鑑 8 續事始 9 易經 10 山海經
11 博物志 12 西京雜記 13 呂氏春秋 14 淮南子

The *Shih Shih*, however, by no means confines itself to the legendary inventions. Written in the +7th century, it was able to give a number of perfectly historical references to advances made by known personages, such as the invention of paper by Tshai Lun[1] in the +1st century (see Sect. 32); the development of the seed-drill plough by Huangfu Lung[2] in the +3rd (see Sect. 41); the invention of the Cardan suspension (gimbals) by Ting Huan[3] in the +2nd (see Sect. 27). Similarly, the *Hsü Shih Shih* of the +10th refers to the beginning of the use of coloured paper in the time of the first Thang emperors (+636 and +676) (cf. Sect. 32); oilskin rainproof coats dating from the Chou, the establishment of medical officers in every city in the +7th century; attribution of the perfection of dice to Tshao Tzu-Chien[4] (son of Tshao Tshao) in the +3rd, etc. These books are consequently not without value still for the investigation of the history of science and techniques in China.

This brings us to the Sung dynasty, because the *Hsü Shih Shih* was written at the end of the just preceding Wu Tai period. On the one hand the tradition now fused with that of the large encyclopaedias of which the *Thai-Phing Yü Lan* (+983),[a] already mentioned, is an outstanding example.[b] On the other hand, a number of books were produced, such as the *Shih Wu Chi Yuan*[5] of Kao Chhêng,[6] which instead of giving the maximum number of quotations concerning any given subject, concentrated on its origin, and on discoveries and inventors. This book (+1085) is thus extremely useful today. During the Ming this type of work continued to be produced.

It might seem at first sight hard to point to anything in European literature which resembled these efforts to describe the inventors and discoverers in the various arts, sciences and techniques. But the works of Polydore Vergil offer a fairly close parallel. His *De Rerum Inventoribus* ran through many editions in most European languages after its first publication in +1512, including several in English 'compendiously gathered' by Thomas Langley. This book, like the earlier Chinese lists, takes most of its material from the world of legend (in this case Greek and Latin), but mentions a few real persons, corresponding to Tshai Lun and Huangfu Lung, such as Ctesibius, 'a barber of Alexandria', who invented the organ. Frequently, even the resources of legend gave out, as may be seen from the following words, interesting since they deal with certain things the Asian origin of which Polydore Vergil never suspected:

> There be many other things, whose Authors for Antiquity cannot be known; and some, because of the negligence of men, that will not write such things. No man can tell who began

[a] The main line of descent of the *Thai-Phing Yü Lan* is, of course, from the four great Thang encyclopaedias, (*a*) the *I Wên Lei Chü*[7] of Ouyang Hsün,[8] (*b*) the *Chhu Hsüeh Chi*[9] of Hsü Chien,[10] (*c*) the *Pei Thang Shu Chhao*[11] of Yü Shih-Nan,[12] and (*d*) the *Liu Thieh*[13] of Pai Chü-I[14] (the poet), afterwards enlarged by Khung Chhuan[15] of the Sung. These are now comparatively rare books and have not been available to us.

[b] But not the only one, cf. the *Shih Wên Lei Chü*[16] of Chu Mo.[17]

1 蔡倫 2 皇甫隆 3 丁緩 4 曹子建 5 事物紀原
6 高承 7 藝文類聚 8 歐陽詢 9 初學記 10 徐堅
11 北堂書抄 12 虞世南 13 六帖 14 白居易 15 孔傳
16 事文類聚 17 祝穆

Clocks, Bells, the Ship-man's Compass, Gowns, Stirrops, Caps or Bonnets.... Water-Mills and Clavicymbals, Tallow-Candles, reclaiming of Hawks, Rings, with many others, which for the anciety, or over-sight of men, be in extream Oblivion.[a]

Later books of the same kind, such as Powell's *Humane Industry; or, a History of most Manual Arts, etc.*, as in China, diminish the amount of legend and speak more of historical characters.

The *Ko Chih Ching Yuan*, already mentioned, might be considered to constitute the ultimate development of the Chinese tradition. It was an attempt to deal with the history of discovery and invention, but this was never clearly distinguished from political affairs, social customs, and the like; and, moreover, there was little differentiation between legendary lore and historical facts, since what are now realised to be legends were then regarded naïvely as history. Still, it constitutes a rich source of quotations which indicate where the desired information may be obtained.

There is no doubt that the origin of the whole tradition was, as Maspero says,[b] the custom in Chou times and probably earlier of sacrificing to the spirits of the first inventors, who were, in fact, 'technic deities'.

[a] P. 154.

[b] (12), p. 31.

4. GEOGRAPHICAL INTRODUCTION

THE FIRST SUBJECT on which it is necessary to provide the reader of this book with some preliminary information is the geographical background—the stage on which the drama of the development of Chinese civilisation was played. Indeed, the geographical factor is more than a background; it enters vitally, as we shall see towards the end of the argument, into the pattern of the differences between the cultures of China and Europe, and all that that implies. Here, however, we are only concerned with the fundamental preparation of the stage.

Apart from the book of Richard already mentioned, the outstanding concise modern description of Chinese geography is that of Cressey. On a larger scale the best work is that of Sion (1) and Grenard (1). Other books, especially Wieger (1) and Fitzgerald (1), open with geographical descriptions which are worth reading, and Herrmann (1), Map 6/7, shows well the characteristics of the various regions—desert, loess, alluvium, ancient forest, etc. Let us briefly run over, with the assistance of Fig. 35, the principal outlines of the picture.[a]

(a) GENERAL SURVEY OF CHINESE TOPOGRAPHY

Beginning at the extreme north-east corner of the map, we see the Liao River flowing down from the north (Manchuria) into the gulf of Pei-chih-li. Facing it on the south-western side of the gulf is the mouth of the Yellow River, north of the mountainous peninsula of Shantung. In the western part of these mountains rises up the sacred mountain Thai Shan[b] [1], venerated for centuries as connected with rain-giving dragons, and the scene of many imperial sacrifices. South of the Yellow River valley at this point are further plains and slightly rolling hills of low elevation forming the basin of the Huai River, which formerly found its way out to the sea on the coast of Chiangsu province, but later silted up and became a tributary of the Yangtze River and chief supplier of the waters of the Grand Canal.

Ascending now the Yellow River,[c] we have first the North China plain upon our right hand, but about longitude 114° we enter ranges of mountains, the Shansi

[a] Figures in square brackets in what follows are references to the map in Fig. 35.

[b] An interesting, though not very scholarly, book has been devoted especially to these sacred mountains by Geil (1). There are five of them (*wu yo*[1]); Thai Shan [1] is considered the eastern one (excellent description by A. C. Moule (2) and more recent photographs in Mullikin (1). The northern one, Hêng Shan [49], is located not far from Peking [50] at the extreme north end of the Thaihang Shan [2]. The western one is Hua Shan [51] just south of Tungkuan [5] at the great bend of the Yellow River. The central one is Sung Shan [52], an outlier of the Funiu Shan [4] just south of Loyang [8]. The southern one is Hêng Shan [53] just west of the Hsiang River south of the Tungthing lake. On the cult itself, see Doré (1), part II, vol. 10, pp. 833 ff., and especially Chavannes (5).

[c] General illustrated description in Clapp (2).

[1] 五嶽

Table 4. *Chinese provinces and place-names, with corresponding characters*

PROVINCES

China Proper

Anhui	安徽
Chekiang	浙江
Chiangsi	江西
Chiangsu	江蘇
Fukien	福建
Honan	河南
Hopei[a]	河北
Hunan	湖南
Hupei	湖北
Kansu	甘肅
Kuangsi	廣西
Kuangtung	廣東
Kweichow	貴州
Ninghsia	寧夏
Pingyuan[b]	平原
Shansi	山西
Shantung	山東
Shensi	陜西
Suiyuan	綏遠
Szechuan[c]	四川
Yunnan	雲南

Manchuria

Chahar	察哈爾
Heilungchiang	黑龍江
Jehol	熱河
Kirin	吉林
Liaotung[d]	遼東
Liaosi[d]	遼西

Sinised Tibet

Chhinghai	青海
Sikang	西康

Tibet

Sitsang	西藏

Sinkiang 新疆

Inner Mongolia[e]

PLACE-NAMES

38	Alashan (Holan Shan) (mtns)	賀蘭山
48	Amoy (city)	厦門
	Brahmaputra (river)	布拉馬普德拉河
	Canton, see Kuangchow	
	Chialing chiang (river)	嘉陵江
	Chin-sha chiang (Yangtze River)	金沙江
	Ching ho (river)	涇河
	Chu chiang (river)	珠江
58	Chungking (city)	重慶
9	Chhang-an (Sian) (city)	長安
56	Chhang-sha (city)	長沙
59	Chhêngtu (city)	成都
	Chhienthang (river)	錢塘
14	Chhilien Shan (mtns)	祁連山
	Chhin (river)	沁河
3	Chhin-ling Shan (mtns)	秦嶺山
	Chhinghai (lake)	青海
	Chhung chiang (river)	沖江
23	Fangtou Shan (mtns)	方斗山
	Fên ho (river)	汾河
	Formosa, see Thaiwan	
	Fou chiang (river)	涪江
	Fu ho (river)	撫河
47	Fuchow (city)	福州
4	Funiu Shan (mtns)	伏牛山
	Gan chiang (river)	贛江
	Gobi (desert)	戈壁沙漠
	Han chiang (river, Honan)	漢江
	Han (river, Fukien/Kuangtung)	韓
18	Hanchung (city)	漢中
19	Hankow (city)	漢口

N.B. The romanisations of these and other geographical names are, it will be remembered, conventional, but orthographically incorrect.

[a] Under the last, Chhing, dynasty, this province was named Chihli, 直隸, which had the meaning of Manchu metropolitan province; the pre-Republican name must therefore be expected in writings before 1911.

[b] This new province, a triangular area between Hopei, Shantung, Honan and Shansi, was instituted by the present Government. It has subsequently been discontinued.

[c] Divided into four parts, North, South, East and West, by the present Government.

[d] These provinces were formerly united under the name of Liaoning.

[e] Now an autonomous region.

Table 4. *Chinese provinces and place-names, with corresponding characters* (*continued*)

26	Hangchow (city)	杭州
49	Hêng Shan (northern sacred mtn)	恆山
53	Hêng Shan (southern sacred mtn)	衡山
	Hsi chiang (river)	西江
	Hsiang chiang (river)	湘江
31	Hsüehfêng Shan (mtns)	雪峯山
51	Hua Shan (western sacred mtn)	華山
	Huai ho (river)	淮河
43	Huaiyang Shan (mtns)	淮陽山
44	Huaiyin Shan (hills)	淮陰山
	Huang Ho (Yellow River)	黃河
20	I-chhang (city)	宜昌
	Irrawaddy (river)	伊洛瓦底江
55	Khaifêng (city)	開封
	Khingan ling (mtns)	大興安嶺
10	Khunlun Shan (mtns)	崑崙山
28	Kuangchow (Canton, city)	廣州
	Kuei chiang (river)	桂江
25	Kunming (city)	昆明
	Kunming hu (lake)	昆明湖
61	Kweilin (city)	桂林
60	Kweiyang (city)	貴陽
7	Lanchow (city)	蘭州
	Liao ho (river)	遼河
29	Liaotung (peninsula)	遼東
	Liu chiang (river)	柳江
62	Liuchow (city)	柳州
	Liu-chhīu (islands)	琉球
37	Liuphan Shan (mtns)	六盤山
	Lo chiang (river, Honan)	洛江
	Lo ho (river, Shensi)	洛河
8	Loyang (city)	洛陽
36	Lüliang Shan (mtns)	呂梁山
	Mekong (river)	瀾滄江
	Min chiang (river, Fukien)	閩江
	Min chiang (river, Szechuan)	岷江
41	Minya Konka (mtn)	
13	Nan Shan (mtns)	南山
54	Nanking (city)	南京
35	Nan ling (mtns)	南嶺
6	Ninghsia (city)	寧夏
24	Omei Shan (sacred mtn)	峨眉山
	Ordos (desert)	鄂爾都斯
39	Paiyü Shan (mtns)	白于山
	Pei chiang (river)	北江
	Pei Chihli (gulf)	北直隸
50	Peking (city)	北京
	Phan chiang (river)	盤江
	Poyang hu (lake)	鄱陽湖
	Salween (river)	薩倫河
17	Shanghai (city)	上海
27	Shanthou (Swatow, city)	汕頭
	Sian, see Chhang-an	
52	Sung Shan (central sacred mt)	嵩山
	Swatow, see Shanthou	
32	Tahsüeh Shan (mtns)	大雪山
62	Tali (city)	大理
33	Taliang Shan (mtns)	大涼山
40	Talou Shan (mtns)	大婁山
22	Tapa Shan (mtns)	大巴山
42	Tapieh Shan (mtns)	大別山
15	Tarim (desert basin)	塔里木盆地
	Thai hu (lake)	太湖
1	Thai Shan (eastern sacred mtn)	太山
2	Thaihang Shan (mtns)	太行山
	Thaiwan (island, Formosa)	臺灣
57	Thaiyuan (city)	太原
	Tho chiang (river = Chhung chiang)	沱江
16	Tsaidam (plateau)	柴達本
45	Tunhuang (city)	敦煌
	Tung chiang (river)	東江
5	Tungkuan (city)	潼關
	Tungthing hu (lake)	洞庭湖
	Tzu shui (river)	資水
	Wei ho (river)	渭河
	Wu ho (river)	烏河
21	Wu Shan (mtns)	巫山
46	Wu-i Shan (hills)	武夷山
30	Wutang Shan (mtns)	武當山
	Yalu chiang (river)	鴨綠江
	Yalung chiang (river)	鴉礱江
	Yangtze River	揚子江
	Yellow River, see Huang Ho	
12	Yenan (city)	延安
34	Yin Shan (mtns)	陰山
	Yuan chiang (river)	沅江
11	Yulin (city)	榆林

plateau to the north, with the Thaihang Shan (mountains) [2] forming its eastern rampart; and the outliers of the great Chhinling Shan (mountains) [3] such as the Funiu Shan [4] to the south. At Tungkuan [5] the Yellow River bends suddenly at a right angle, and we must go upstream seven degrees of latitude almost due north towards the Gobi desert, after which we turn to the west again along the northern boundary of the Ordos desert. Once more the river changes its direction, and ascending it one would go south-west past the city, and through the province, of Ninghsia [6], to the great city of Lanchow [7] at the base of the Kansu corridor. Here the Yellow River is already broad and imposing,[a] though it is not far from its source, which is in the Tibetan massif near Lake Kokonor (Chhinghai).

Issuing from the Shansi hills, a small river, the Chhin, joins the Yellow River, and from the opposite side a much larger one, the Lo, famous as passing the walls of one of the former capitals of China, Loyang [8]. Then flowing into the Yellow River just where it makes its great bend is (*a*) the Wei river, which in turn receives the waters of the Ching, and (*b*) the Lo (not to be confused with the other river of the same name further east, just mentioned). All this region is saturated with history; it saw the earliest Chinese civilisation, the rise of the feudal state of Chhin to continental domination, and the successive glories of the Han and Thang capitals at Chhang-an (Sian) [9]. Then the main river of Shansi, the Fên, joins the Yellow River a short distance up its eastern south-flowing limb.

Now the whole south slope of the Wei River valley is formed by the sharp escarpment of the Chhinling Shan, a range which is itself a continuation of the mighty Khunlun Shan [10] of the Tibetan massif. The northern slopes are gentler since the ancient rock formations are covered to a depth of a hundred feet or more by yellow loess soil, i.e. compacted dust blown during the course of ages from the northern deserts (see on, p. 69). Along these northern slopes, as one can see from the train steaming up the valley, are the tumulus graves of former emperors, e.g. those of the Sui. Rather more than half of the area inside the great bend of the Yellow River is fertile land (especially if given irrigation), separated from the Ordos desert approximately by the line of the Great Wall and a series of old frontier towns, such as the romantic city of Yulin [11], where the sand blows up in drifts as high as the walls and *phailous*.[b] The eastern half of this area was the headquarters of the Communist Government during World War II, with its centre at Yenan [12]. It will be seen, therefore, that the eastern part of Kansu, the south-eastern part of Ninghsia, and the centre of Shensi, form a distinct natural province, bounded by the deserts on the north, the Shansi mountains on the east and the Chhinling Shan on the south. Hence the expression *kuan chung*[1] (within the passes) which was used with the meaning of metropolitan area by the Han people. The way into Shansi was of course up the Fên

[a] As those know, who, like the writer, have navigated upon it and crossed it by means of fragile rafts buoyed up by inflated goat or ox-skins (cf. Sect. 29).

[b] These are triumphal gateways (description in Sect. 28*d*).

[1] 關中

valley, but there were important passes to Honan on the south-east and to Szechuan on the south. Having travelled over the latter a number of times I can vouch for their impressiveness,[a] and we may pause to note one of them, the gorge called Chien-Mên-Kuan[1] (Swordgate Pass) which was the scene of several battles during the wars between the Szechuanese State of Shu and the northern State of Wei during the Three Kingdoms Period (cf. p. 113). One should note the self-contained character of this Kuanchung area, since it is a pattern repeated only too often in Chinese geography.

To the west a number of passes, such as Hua-Chia-Ling,[2] forming collectively the Ti Tao[3] (the Barbarian Ways) lead over into Kansu, giving communication between the upper Wei valley and the wild upper waters of the Yellow River. From Kansu's capital, Lanchow [7], to the north-west, runs the so-called Kansu corridor or pan-handle, by which modern provincial boundaries manifest what was one of the most ancient and famous trade-routes of antiquity, the Old Silk Road.[b] This way gave communication between China and Central Asia by means of a series of oases formed by the melting snows of the Nan Shan [13] or Chhilien Shan mountains [14] as the waters ran down to lose themselves in the sands of the Gobi. The road goes north-west, therefore, through Yümên (Jade Gate),[4] into the Tarim Basin [15] (at the extreme left-hand top corner of the map) where it divided, one way following the base of the Tibetan massif (the Tsaidam plateau [16]), the other crossing the desert to follow the base of the Thien Shan. These roads were called the Nan Shan Pei Lu[5] and the Thien Shan Nan Lu[6] respectively. We shall speak of them again in connection with east-west contacts and trade-routes (p. 170, Fig. 32).[c]

Next we may return to the Pacific coast and the estuary of the Yangtze River[d] near the modern city of Shanghai [17]. If we follow up the great stream (and it was always much more, and further, navigable than the Yellow River) passing the southern capital of Nanking [54][e] we find, between longitudes 120° and 110°, a relatively flat country intersected by ranges of small hills, and containing three rather large lakes, all south of the river. Near the sea, in Chiangsu, is Thai Hu (lake); then in Chiangsi there is the Poyang Hu, and in Hunan the Tungthing Hu. It is from the latter that the two provinces, Hupei (North of the Lake) and Hunan (South of the Lake) get their names, just as further north Hopei means North of the (Yellow) River, and Honan

[a] The most important of these passes involved carrying the road along perpendicular rock-faces by means of wooden brackets fixed into the cliff. The modern motor road is in several places fully inset into the excavated rock-face, and there are some tunnels (see the recent description of Wiens).

[b] Accounts of the modern condition of this famous road will be found in M. S. Bell (1), Cable & French (1), Langdon Warner (1) and Needham & Needham (1).

[c] The block diagram (Fig. 2), from Griffith Taylor (1), shows very well how the Jade Gate, like the Khyber Gate, was the avenue of access of anthropological types in successive migrations and invasions, into China, as into India.

[d] General illustrated descriptions in Carles (1) and Barbour (2).

[e] Another of the rambling books of Geil (2), which nevertheless contrive to give a very living picture of Chinese scenes, is devoted to the eighteen great cities, many of which have served, at one time or another, as the capital of the whole country.

1 劍門關 2 華家嶺 3 狄道 4 玉門 5 南山北路
6 天山南路

means the province south of it. Between the two western lakes the Yangtze is joined by a large river flowing from the north-west, the Han—its upper reaches are classic ground, for it takes its rise on the southern side of the Chhinling mountains near the modern town of Hanchung [18], whence passes go over both to the Wei valley and the Kuanchung domain in the north, of which we have already spoken, and to the Szechuan domain in the west.[a] The Han valley was thus throughout history the highway between the Yangtze valley (the east-central economic area, as we shall call it, cf. p. 115), and these domains.

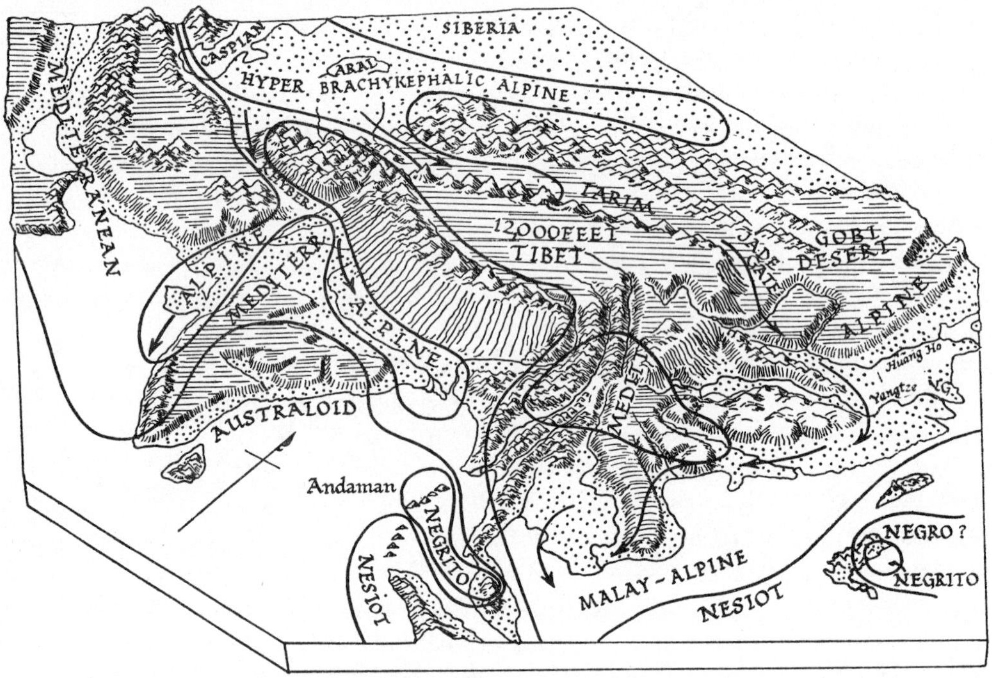

Fig. 2. Block diagram of southern Asia to show how human migrations followed topographic structure (from Griffith Taylor, 1 *a*).

Returning to the great city of Hankow [19] at the point where the Han joins the Yangtze, and making our way up the great river itself, we find, beginning at I-chhang [20] (longitude 112° approximately), that celebrated series of gorges which the Yangtze cut for itself in debouching from the inner plateau-basin of Szechuan, and which take their place beside the Grand Canyon and the African Great Rift Valley among the major geographic phenomena of the world. The mountains so cut through are the Wu Shan [21], essentially prolongations of the Tibetan massif, the Tapa Shan [22] coming down from the Chhinling Shan on the north, and the Fangtou

[a] Such as Wu-Ting-Kuan,[1] the Pass of the Five Nails.

[1] 五丁關

Shan [23] coming up from the Yunnan ranges on the south. The Szechuan plateau-basin, so enclosed, is (as its name implies) as much the Land of the Four Rivers as the similarly named Punjāb in north-western India is the Land of the Five. The four rivers are, from east to west, the Chialing (the largest of them, and the upper waters of which provide the passes to the north), the Fou, the Tho (or Chhung) and the Min. Szechuan forms another well-defined natural province, and owing to its geographical position and resources has frequently been independent in Chinese history (cf. p. 116); in World War II it provided an inexpugnable fortress for the Chinese armies, the borders of which the Japanese never succeeded in crossing.[a] Very characteristic of Szechuan is the brick-red sandstone soil, hence the name of the Red Basin first given to it by the great traveller Ferdinand von Richthofen (2). To the west the boundary is the escarpment of the definitive Tibetan high mountain block, some of the outlying peaks of which, such as Omei Shan [24], have become famous in Chinese cultural history.

The course followed by the Yangtze (often called in these regions the Chinsha chiang, the River of Golden Sand) before it enters the Szechuan plateau-basin, is a curious one, since, taking its origin far to the west and north, about longitude 92°, on the high Tibetan plateau, it descends through tremendous gorges nearly as far south as latitude 25° before receiving the waters of another equally deep-cut river, the Yalung, and then turning north-east into Szechuan. In a unique geographical formation, it is accompanied as far as its turning-point by the Mekong, the Salween and the Irrawaddy, only a few miles apart, which then continue their course towards the south.[b]

Those who have had occasion to travel by plane over the India-China route (Fig. 3) can never forget the amazing juxtaposition of these great rivers in their gorges, seen when accomplishing the stretch between northern Assam and Kunming [25], the capital of Yunnan. They are also able to appreciate the reason for the provincial name, Yunnan, South of the Clouds, for they frequently had the experience of leaving Yunnan's mountainous plateau bathed in sunlight, with blue sky overhead, and then encountering the great bank of cloud which more often than not hangs over Szechuan. Yunnan has numerous lakes, but smaller than those in the eastern provinces; and its soil is often red, but of a different type from the red soils of Szechuan. Between Yunnan and Kweichow there is no marked geographical line of separation; the tumbled mountains continue, and the latter province, traditionally the wild haunt of aboriginal tribespeople, generates eastward-flowing rivers—the Wu, which drops into the Yangtze from the south before the gorges, and the Yuan, which flows through Hunan to join the Tungthing Lake. The Tungthing Lake receives two other rivers, both native Hunanese, the Tzu and the Hsiang. We are now down again in lowland

[a] Cf. H. L. Richardson (1).

[b] Not much further west, moreover, is the Brahmaputra. All these rivers are, as Cressey puts it (1, p. 384), 'squeezed together into this canyon complex as though held by a clenched fist. In the Tibetan borderlands they are within a 400-mile zone; where they enter the sea, the outermost are 2000 miles apart in an air-line, or over 6000 miles measured round the coast.'

hilly country, and in the neighbouring province Chiangsi, the Poyang lake likewise receives two southern tributaries, the Gan and the Fu.

So much for the central regions. But besides them there is the south-eastern and southern coast of China, stretching in a vast arc from Hangchow [26] to the Indo-chinese border, and severely cut off from the inland provinces.[a] Four valleys open like amphitheatres on the sea, first the Chhienthang river which forms the Hangchow estuary in Chekiang; then, further south, the Min in Fukien, which leads down to Fuchow [47]; thirdly, the Han, which has Shanthou [27] (Swatow) at its mouth; and lastly, the river-system of Kuangchou [28] (Canton) which is much the most considerable of the four. The Canton estuary is called the Pearl River (Chu chiang), and it receives three others, the West, the North and the East (Hsi, Pei and Tung chiang), of which the only really long one is that which comes from the west; this carries the waters of the Phan chiang coming from one of the Yunnan lakes, and the Liu and Kuei (Cassia) rivers coming out of Kuangsi.[b] It has been very significant for Chinese history that the greater part of the coast-line was cut off from the interior by mountain-ranges which were poorly penetrated by passes, canals and other transportation arteries until a late time.

We may thus list the natural provinces of China as follows:[c]

(1) the Tibetan plateau and mountain massif;
(2) the Tarim basin of Sinkiang (Chinese Turkestan);
(3) the Mongolian steppe;
(4) the Manchurian plain;
(5) the south-eastern highlands of Manchuria;
(6) the mountainous Shantung peninsula;
(7) the North China Plain, including the whole of the province of Hopei, the western part of Shantung, most of Honan, and the northern part of Anhui;
(8) the mountainous plateau of Shansi;
(9) the Shensi basin;
(10) the Kansu corridor, running along the base of the Chhinghai plateau;
(11) the Lower Yangtze valley;
(12) the south-eastern maritime uplands, including the Chekiang and Fukien mountains and their seaward-looking amphitheatres;
(13) the Central Yangtze basin, covering the provinces of Hupei, Hunan, Chiangsi, and part of southern Anhui;

[a] My own travel notes remind me of several considerable passes over the Chiangsi-Fukien mountains such as Niu-Tou-Ling[1] (Bullring Pass), Sung-Mao-Ling[2] (Luxuriant Pine Pass) and Chin-Chi-Ling[3] (Goldchick Pass).

[b] The Kuangsi plateau is particularly remarkable for its karst limestone pinnacle hills which rise abruptly out of level ground. Before I saw them I often imagined that such pinnacles were the invention of painters to fit upright hanging scrolls, but in fact they are a real and remarkable element in Chinese scenery (see Fig. 4, from Koester).

[c] Roxby (2) has a simpler formulation.

[1] 牛頭嶺 [2] 松茂嶺 [3] 金雞嶺

Fig. 3. The mountainous country between India and China; flying above the gorges of the Salween and the Mekong rivers.

Fig. 4. Karst limestone pinnacles; a typical Kuangsi landscape (from Koester, 1).

(14) the red plateau-basin of Szechuan;
(15) the Kweichow plateau;
(16) the southern maritime amphitheatres, especially Kuangtung, backed by
(17) the Kuangsi platform;
(18) the south-western highlands of Yunnan, with their deeply incised river-gorges, stemming from
(19) the Sino-Tibetan Alps of Sikang.

(*b*) THE GEOTECTONICS OF CHINA

Up to the present point, the description has kept to a relatively superficial level. But the only way to make any sense of the vast jumble of mountains which constitute China is to hear what the geologists have to say, and fortunately one of the most eminent scientists in the field, Li Ssu-Kuang (J. S. Lee), has given us the first comprehensive book (1) on the structure of the Chinese subcontinent.

There are three main features to be considered: (*a*) the north-south (or more correctly north-east/south-west) fold mountains; (*b*) the east-west mountain ranges which interfere with this tectonic grain; and (*c*) other folds which greatly complicate the pattern.

First, the north-south folds. Proceeding from the Pacific Ocean westwards, Li Ssu-Kuang points out that after the deeps (such as the famous Tuscarora Deep) the backbone of Japan, the Liuchu islands, and Formosa (Thaiwan) form an obvious ridge, though small in bulk. Next to this comes the trough represented by the straits of Formosa and of Japan, and then the great folds represented by the coastal mountains[a] of Kuangtung, Fukien and Chekiang, which are clearly continuous across the Yellow Sea with the mountains of Korea, and further north with the Shikota-Alin mountains of the Siberian coast. A branch of this structure extends southwards from Manchuria and Korea down the Liaotung peninsula [29] and appears again in the mountains of Shantung. Then west of these ranges comes the very large belt which Li calls the Neo-Cathaysian Geosyncline (the name is derived from elaborate considerations on the 'Palaeo-Cathaysian' and 'Meso-Cathaysian' palaeo-geography of China into which we need not here enter). This great belt comprises the whole of the natural provinces nos. 4, 7, 11 and 13 in the foregoing list, and therefore most of the 'northern' and 'east-central' key economic areas, which we shall discuss later (p. 115). It is important to note that it has access to the sea only at two points, the coast of Hopei and the coast of Chiangsu—in all other directions it is cut off from direct maritime contact.

As we proceed from east to west we are mounting a series of gigantic steps which lead up to the high Tibetan massif as their ultimate platform. The Neo-Cathaysian Geosyncline is bounded on the west by still another geanticline running from the

[a] See Li Ssu-Kuang's Fig. 8, showing drowned valleys near Hongkong where the range peters out.

north of Manchuria to the south of Yunnan, and forming the next step or escarpment. It begins in the far north with the Great Khingan range (Fig. 5) and continues with the Thaihang Shan [2], which, as we have seen, form the eastern rampart of the Shansi plateau. Immediately south of the Yellow River comes the bastion of the Funiu Shan [4], then south of the Han River, the Wutang Shan [30]; then, south of the Yangtze, the Hsüehfêng Shan [31] on the borders of Kweichow and Hunan, whence derive the Yuan and Tzu rivers draining into the Tungthing Lake. The southern end of this geanticline is confused with the Nanling ranges running at right angles to it.

The last step of the series may be considered as constituted by the borders of the Tibetan massif itself, namely, the Nan Shan in the north [13], the Tahsüeh Shan [32] west of the Szechuan basin, and the Taliang Shan [33] along the course of the upper Yangtze (Chinsha chiang). Just as the Great-Khingan Shansi Kweichow fold forms the raised front of the Mongolian, Shensi and Szechuan-Yunnan blocks respectively, so the mountains at the edge of the Tibetan massif prelude the highest platform of the Central Asian alps. All these fronts have bolder escarpments on their eastern than on their western sides.

We can now take up the second of Li Ssu-Kuang's points. The Cathaysian ranges are interrupted by no less than four important mountain chains running east and west. This is the reason why the western half of the country became divided up into the four domains represented by nos. 9, 14, 15 and 16 in the list of natural provinces; latitudinal segments separated by fold-zones.

The most northerly of these latitudinal fold-zones is the Tannu-Kentai mountain range (see Fig. 5) well to the north of the Gobi desert and Mongolia. Then comes the Yin Shan [34], a range about 41° N. which divides the Mongolian block from the Kuanchung block; it can be seen running along the north side of the northernmost course of the Yellow River in its great bend, through Suiyuan province. Greatest of all the latitudinal fold-zones is the Chhinling mountain chain [3], which may be said to begin in the Khunlun mountains [10] of Tibet, and to die away in the east at the edge of the central north-south geanticline. The Chhinling Shan comprise a great mass between the Shensi (Kuanchung) block to the north and the Szechuan block to the south. Finally, the Nanling mountains [35], though much broken up and disturbed by subsidiary folding, divide off the southward-looking Kuangtung amphitheatre (no. 16) from the main Neo-Cathaysian Geosyncline. These east-west mountain ranges may not, in fact, cease to exist in the area of the Neo-Cathaysian Geosyncline, but possibly reappear on the other side of it, in the anticlines of the coast and of Japan.

Lastly comes the controversial part of the analysis, namely, the shear-folds which Li has recognised as interfering with the already cross-grained structure of the country. The most important of the shear-fold types is what he calls the ϵ-type from the pattern in plan, which resembles this letter. He recognises five of these ϵ-type folds in China.

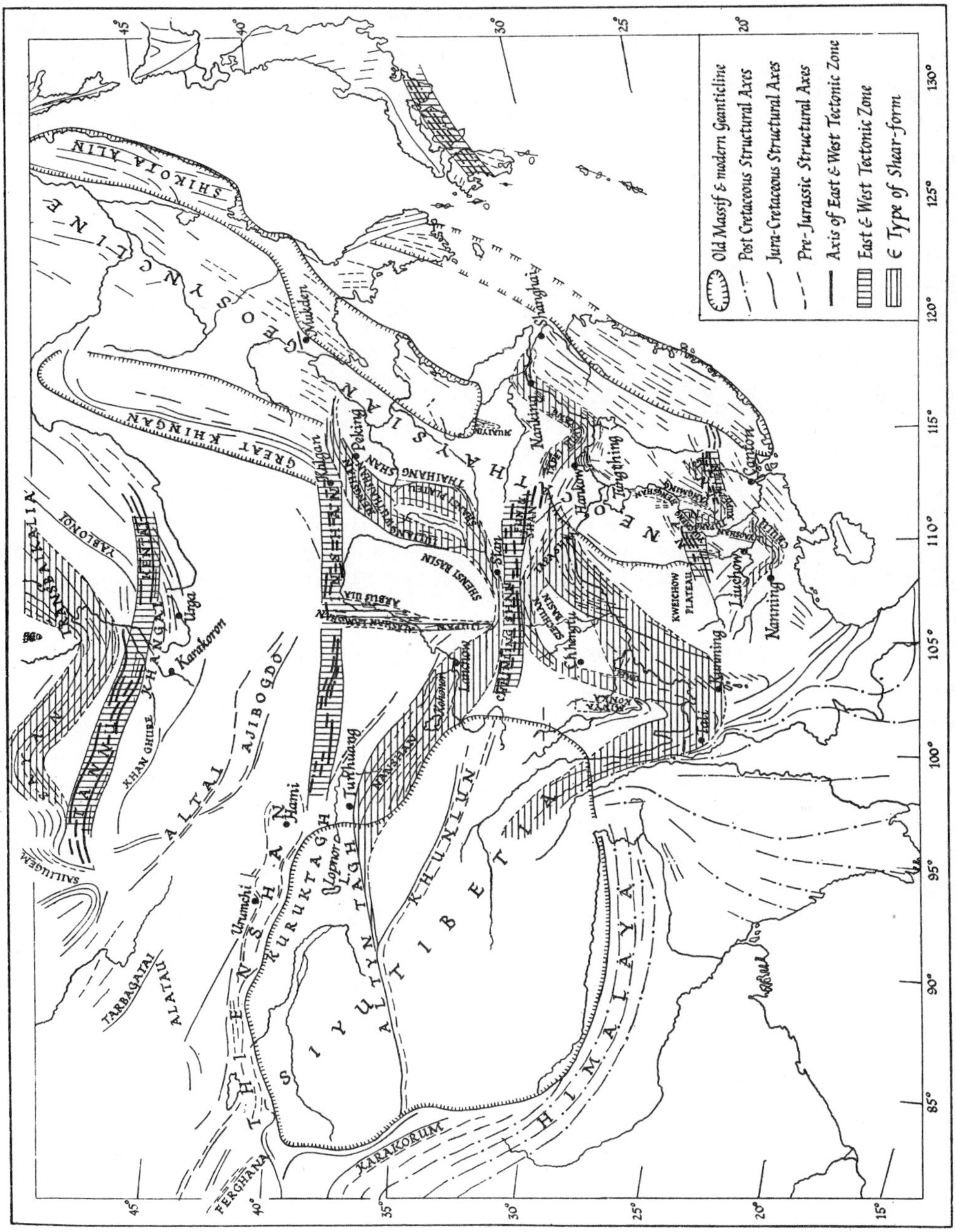

Fig. 5. A simplified tectonic map of eastern Asia, showing structural lines produced by compressive and shearing stresses. From Li Ssu-Kuang (1) (J. S. Lee).

The broadest of these systems is that in the north-west of the country in which the west half of the bow is formed by the Nan Shan [13], the centre by the Chhinling Shan [37], and the east half by the Lüliang [36] and Thaihang mountains [2] of Shansi. The backbone or 'arrow', indicating the direction of shear, is the Liuphan Shan [37], with its tails, the impressive Alashan (Holan Shan) [38] and Paiyü Shan [39] mountains, all trending north and south.

The highest and most bulky of these systems is the famous 'Yunnan arc', the bend of which the upper Yangtze follows.[a] Here the west half of the bow would be formed by the Sikang ranges (in the gorges between which run the several great rivers already mentioned), and the centre by the hills just north of the Kunming lakes. The eastern half divides to sweep round the Szechuan basin to the north and the south; the northern and western prong forming first the Taliang Shan [33] and then fusing with the Chhinling Shan [3] to form the Tapa Shan [22]; the southern prong forming first the Talou Shan [40], then the Fangtou Shan [23], and finally dying out in Hupei. The backbone of the system is represented by several north-south ranges in the neighbourhood of longitude 102° E.; of which the largest includes the peak known as Minya Konka [41] (altitude about 24,900 ft.).[b]

The other ϵ-shear-folds, though of the same general type, and with the backbone always pointing a little west of south, are all considerably smaller. One is represented by the Tapieh Shan [42] (west half) and the Huaiyang Shan [43] (east half) to the north of the Yangtze in the Neo-Cathaysian Geosyncline itself. Li Ssu-Kuang claims to have detected its backbone in the Huaiyin hill-ranges [44] adjacent to the Shantung massif. Two further ϵ-shear-folds are said to occur in southern Hunan and in Kuangsi respectively, and far to the north (outside China proper) there is a sixth on the Amur River.

Other writers have generalised the geological structure of China in different terms, for instance, as a miscellaneous collection of irregular stable masses (plateaux, basins, plains or seas) surrounded and separated by marginal fold mountains. Whatever tectonic theory be used to explain the pattern, it is clear that the structure of China and Central Asia is distinguished from that of most other regions by having a complex network of high mountain ranges separating a number of flatter areas. The significance of these complexities for the development of Chinese civilisation is very great. Without anticipating here what may later have to be said in comparison of China and Europe,[c] it is certainly clear that Europe was much less cut up by mountain ranges and much more accessible by inland seas—the Mediterranean, the Adriatic, the Aegean, the Black Sea, the Baltic, the North Sea, etc. Europe's main mountain massif, that of the Alps, was central, radiating in different directions (Carpathians, Apennines, etc.), and apart from the Pyrenees and the Scandinavian backbone, which were excentric to the foyers of civilisation, communications were not interfered with by

[a] Other geologists consider that this system is not an ϵ-fold at all, since the central north-south ranges appear to continue on the south, like the great folds to the east and west, and parallel with them.

[b] Cf. Burdsell (1).

[c] Sect. 47.

high mountain barriers. From the mouth of the Garonne on the Atlantic at Bordeaux to the opening of the Volga on the Caspian Sea at Astrakhan, or indeed to the Tarim Basin in Chinese Turkestan, there was no serious mountain barrier. In Cambridge we often have occasion to say that East Anglia has no shelter from the cold winter winds of the Urals. Thus only an appreciation of the geographical difficulties which Chinese civilisation had to surmount can make one realise the immensity of the task of unification which Chinese culture and Chinese language successfully accomplished.[a]

(c) HUMAN GEOGRAPHY OF THE NATURAL PROVINCES

Having now examined the map of China, and having touched upon the geological analysis of it, we must now say a few words about the human aspect. Recently a particularly graphic description of the human geography of China has been given by Winfield (1),[b] but we shall follow Cressey in outlining briefly the human characteristics of the natural provinces which were enumerated above.

The Tibetan plateau and mountain massif (no. 1) is mostly an inhospitable waste of frozen high desert with few inhabitants. The life is largely nomadic, the shepherds moving their herds of yak and goats from one area of sparse alpine grasses to another.[c] It has been suggested that the requirement of Lamaist Buddhism that every family should contribute one son to the great monasteries, and also the system of polyandry, have had the contraceptive effect necessary where the resources were so few. The nomadic shepherds are tent-dwellers, but in the south where settled agriculture is possible, farmhouses have been built. Even today the high valleys have very little contact with the outside world, and their life centres round the monasteries, which also play a part in what caravan trade there is.

The Tarim basin with the Mongolian steppe (nos. 2 and 3) is a vast area of arid grassland and true desert, where life has always been mainly nomadic. In Mongolia and Sinkiang there are a million square miles where the rainfall averages less than 15 in. The Huns, the Mongols, and the host of lesser peoples who have inhabited this area, have always depended upon their flocks and herds, and built up cultures which have never coalesced with the Chinese civilisation of intensive agriculture (cf. Lattimore, 1). The integration of the pastoral and the agricultural still today presents the greatest problem of China's north-west. Life in these steppes is entirely dependent upon grass, which in turn rests upon the rainfall. Sheep, horses, camels, and some cattle, are the beasts which constitute the wealth of the Mongols, and provide them with food, clothing (skins and furs) and shelter (felt-covered *yurts*). Even the fuel for the fires consists of the dung of animals (*argol*). Wells and oases exist; those of the Old Silk Road beneath the Nan Shan [13] have already been mentioned. If the water supply is sufficient, a Chinese outpost hsien-city will probably have developed, or in

[a] Cf. Febvre (1), pp. 359 ff., and Sorre (1), vol. 2, p. 107.

[b] See especially his first two chapters. Dudley Buxton's book is also useful, but less systematic than Cressey.

[c] Description in Kingdon Ward (1).

Sinkiang itself, a township of Turki farmers. Many places which formerly must have supported a large population, such as the monasteries in the cave-temple oasis of Tunhuang [45], or the long-buried city of Lou-Lan excavated by Stein (Fig. 32), have been overtaken by the desert and now contain few or no inhabitants. But the beauty and romance of this arid frontier area are very great, and captivate all who, like the present writer, have had the good fortune to know it.

The Manchurian Plain (no. 4) is quite a different type of area. Formerly purely pastoral, this grassy steppe has in modern times been greatly developed agriculturally, in spite of the rather hard climate, with only five months of the year frost-free. Kaoliang, soya-beans, millet, wheat and corn grow luxuriantly. But the development of Manchuria is a new thing, and the region has therefore minor importance from the historical point of view. Similarly, the mountains of eastern Manchuria (no. 5), though today one of the great timber-producing areas in the world, were throughout the historical periods the haunt of primitive east-Siberian tribes, or the hunting-grounds of the Manchu nomads.[a]

With the Shantung mountains (no. 6) we enter historic ground—part of the ancient feudal state of Chhi. The area is rather barren, and with severe winters; little irrigation is possible, but staple crops which can do without extra water flourish well (wheat, kaoliang, millet, beans, barley and sweet potatoes). A special kind of wild silk, based on the oak, not the mulberry (cf. Sect. 31), is important. Deforestation has been severe.

Next we come to the North China Plain (no. 7); the northern part of the great Neo-Cathaysian Geosyncline. It is largely composed of silt carried down through geological ages by the Yellow River, which has built up its bed above the level of the surrounding country. During historical times the Yellow River has frequently changed its course, its outlet to the sea oscillating between points north and south of the Shantung peninsula (see Fig. 35). But in spite of the destruction caused by its periodical floods, the vast plain teems with human activity. It is what we shall later describe (p. 115) as the northern economic area. The alluvial loess soil is itself fertile, given intensive manuring (including, as in all China, the use of human excreta), but the weather is uncertain, precipitation averaging only 21 in. and dangerous droughts in the key month of June being not infrequent. The basic crop is winter wheat, but a wide variety of other plants are grown (millet, kaoliang, barley, beans and the textile crops hemp and cotton). Too much rain is as bad as too little, for even a moderate excess has led to serious floods. This vast and complex agricultural community has had in the past too small a factor of safety, too little resilience against recurrent climatic abnormalities, but great public works are now in progress to give it security.

Shansi, Shensi and Kansu provinces (nos. 8, 9 and 10) show a totally different picture. Cressey groups them together as the loess highlands, and his map, reproduced here as Fig. 6, shows how the loess covers the whole of what we have already

[a] The best recent book on Manchuria is that of Fochler-Hauke (1).

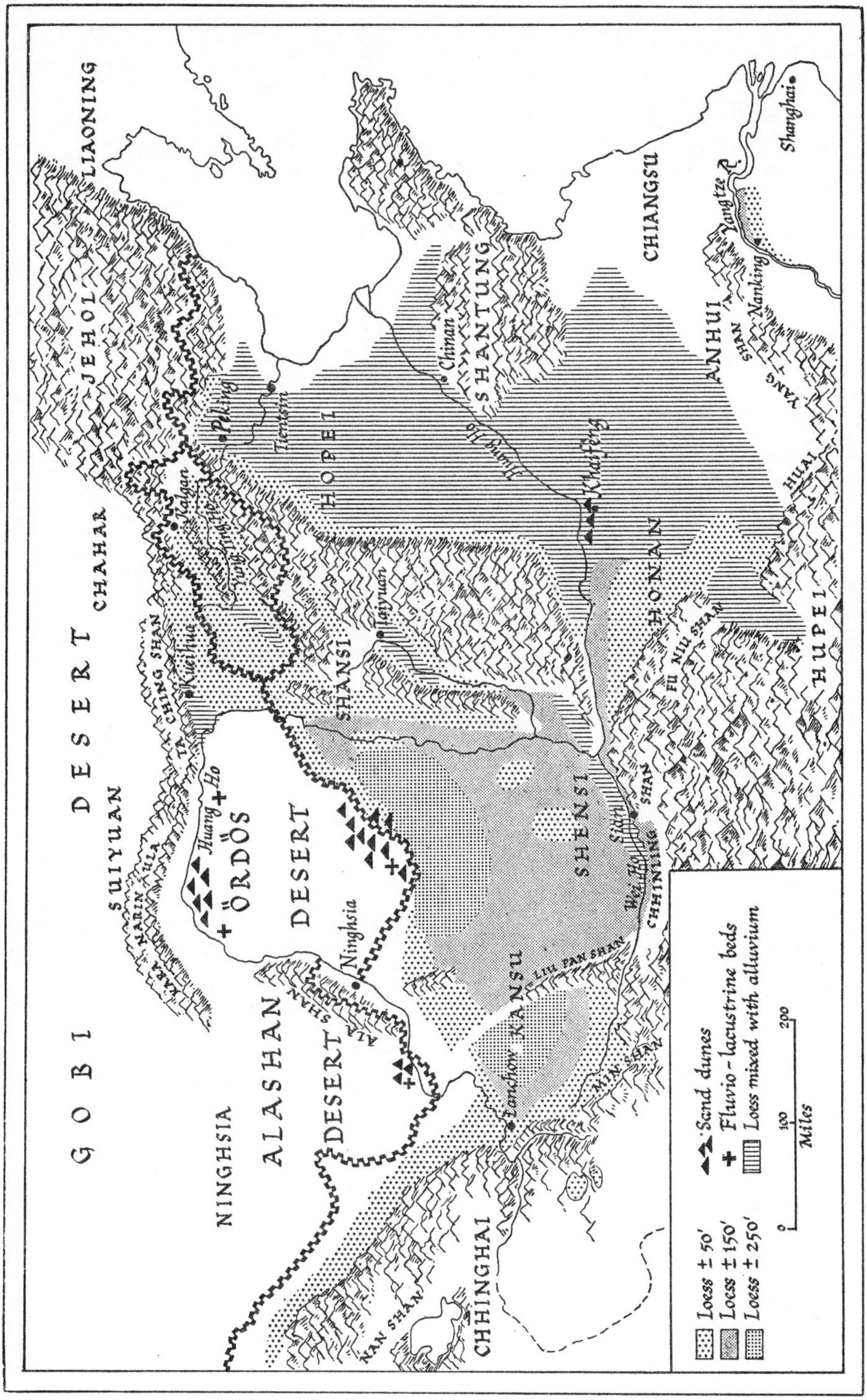

Fig. 6. Distribution of the loess in North China (after Cressey, 1).

described as the Kuanchung domain, together with the plateau of Shansi itself.[a] Cressey says:

Sprinkled over the countryside as though by a gigantic flour-sifter, a veneer of fine wind-blown silt blankets over a hundred thousand square miles of the north-western provinces. Although this formation is described by the German word *löss*, derived from deposits along the Rhine, it would not be inappropriate for these far more extensive accumulations to be known by their Chinese name of *huang thu*[1] or yellow earth. The material consists of very fine silt, yellowish brown in colour, so fine that when rubbed between the fingers it disappears into the pores of the skin without noticeable gritty material. The porosity exceeds 45 per cent, enabling loess to hold considerable moisture, while the fineness of the openings facilitates capillary action.

The thickest deposits over bedrock are probably about 350 ft.[b] One of its most striking characteristics is its capacity for forming vertical-walled canyons dozens of feet high; the vertical cleavage is fixed by the natural development of a surface cement. Roads in the loess country have sunk themselves into such rutted canyons, and the easy excavatability of the loess has invited the construction of innumerable cave-dwellings by the people (Fig. 7). Such homes (as I may say from personal experience) are very convenient, cool in summer and warm in winter;[c] but cause much loss of life by collapsing in earthquakes. Loess provides an unleached soil of exceptional fertility, on which crops may be raised year after year without fertilisation. Its capacity for holding moisture makes good harvests possible in spite of limited rainfall. One can see, therefore, why the loess lands were the oldest foci of ancient Chinese agriculture. It has been judged that if they had a more favourable climate and were less dissected, they might be some of the finest agricultural lands in the world.

In this area wheat, millet, buckwheat and kaoliang are the staple cereals. It is particularly suitable for fruit, producing excellent apricots, peaches and melons; and also some fruits not known elsewhere, such as the Kansu *thieh-tzu*, a fruit with the shape of an apple, the skin and texture of a good pear, and a strong quince taste.[d] Tobacco is also much grown, especially around Lanchow.

We return now to the east, to the Lower Yangtze Valley (no. 11), which, with its innumerable canals[e] and rivers, is again of a different character from the regions which we have been examining.[f] The greater number of China's estimated 200,000 miles of

[a] See Koester (1).

[b] Detailed accounts of this unique formation and the resulting landscape, which was first described by v. Richthofen (1), will be found in Barbour (1, 3), Kanter (1), Obruchev (1) and Schmitthenner (1).

[c] Further description in Fuller & Clapp (1).

[d] I have eaten it often, to my great pleasure, but as in the case of other local Kansu words, could never ascertain the characters by which its name should be written.

[e] Good illustrations in G. F. Anderson (1) and F. H. King (1).

[f] See the geographical sketch of Chiangsu by Hu Huan-Yung (1).

[1] 黃土

Fig. 7. Cave-dwellings in the loess country (Shansi province).

Fig. 8. Terraced rice-fields in Yunnan, seen from above, looking down towards a small steep-sided valley (photo. Hurlimann).

canals (though not all the most important of them) are in the Yangtze plain, i.e. the southern part of the Neo-Cathaysian Geosyncline. These canals provide transportation and irrigation, and the mud dredged periodically from their bottoms is used as fertiliser. The region is less liable to floods than the North China Plain, since the great lakes already mentioned act as reservoirs for the waters of the Yangtze, but extensive flooding has none the less occurred from time to time, especially in the Huai valley. Farming is everywhere very intensive, no land being unused, and even the grave mounds giving pasture for the water buffaloes. This is the heart of the rice-growing country, though there is subsidiary cultivation of mulberry for silk, and of cotton. In the winter months, when the paddy-fields are dry, they are used for a second crop of wheat, beans, rape-seed or barley.

We are still in the rice area, which, after all, comprises all the parts south of the southern borders of Shantung, Honan and Shensi provinces,[a] when we come to the south-eastern maritime uplands of Chekiang and Fukien (no. 12). But here the picture is diversified by the existence from ancient times of important fishing and lumber industries. China's sailors have traditionally been Fukienese and Cantonese. The climate is warm and wet, the annual precipitation rising above 60 in. Besides rice, sugar-cane, sweet potatoes, beans, rape-seed, bamboo-shoots, peanuts and tobacco are grown. But the region is particularly renowned for its tea, grown for instance in the Wu-i hills [46] (hence the old term Bohea). Another important aspect of human geography is the fact that in this south-eastern area a multitude of mutually incomprehensible dialects are spoken. The two principal cities of Fukien, Fuchow [47] and Amoy [48], for example, each have their own. This is in contrast with the wide sway of the 'mandarin', *kuan hua*,[1] or *kuo yü*[2] type of dialect in the Great Geosyncline and westwards, ranging over the whole of China except the coastal region.[b] Thus Chekiang has a number of so-called Wu dialects, Fukien has many more, and then further round the coast comes Cantonese—all entirely dissimilar to *kuo yü*. This persistence of local (and there is some reason to think, more ancient) pronunciations is a direct consequence of the topographical structure already seen, whereby the greater part of China's coast is cut off by mountains from the interior.[c] Obviously connected with these facts is the circumstance that most of the emigration from China has come from these seaward-looking amphitheatres, so that it is the Cantonese and Fukienese who have established themselves in the important colonial groups in Malaya, Siam, Burma, Indonesia, Australia and America.[d]

There is less to say about the Central Yangtze basin (no. 13), which follows mainly the type of the Lower Yangtze and terminates the Great Geosyncline by abutting on

[a] See Cressey's Figs. 47, 48, 49.

[b] Exception made, of course, for the aboriginal languages in Kweichow, Yunnan and Szechuan; and for the totally different languages of Tibetan and Mongol in the borderlands.

[c] Further details in Chao Yuan-Jên (1); and see also Forrest (1), p. 199.

[d] Of course a large literature exists on this subject; I can only indicate a few pointers by referring to the reviews of Duyvendak (2) and Wu Ching-Chhao (1), and a very recent one (with bibliography) by Freedman (1). The excellent book of Purcell (1) is now the standard account.

[1] 官話 [2] 國語

the Nanling range to the south. The climate is temperate rather than subtropical, and the agricultural pattern is that characteristic of all south China. There is a large tea production in Hunan, and Chiangsi province acquired great fame as the centre of the porcelain industry.

The red basin of Szechuan (no. 14) is one of the most attractive and fruitful geographic regions in China. The area is very hilly, and there is little flat land except at the tops of the dissected hills or as flood plains in valley bottoms, but generations of intensive effort have made up for this by terracing the hills to an extraordinary extent. Slopes less than 30° are usually terraced from top to bottom, and only if they exceed 45° are they left unutilised. The terraces are generally narrow and seldom continuous for more than two or three hundred feet on the same level; they are connected by intricate irrigation channels. The ribbon-like mirrors of the flooded terraces present a picture which cannot be outdone for beauty anywhere in the world (Fig. 8). Rice is the main crop, but maize, sugar-cane, tobacco, beans, sweet and Irish potatoes, many vegetables, millet and kaoliang are also grown. In the winter wheat and rapeseed, beans and peas come up; and many farmers are able to grow three crops a year. There is some cotton, and much silk, tea and tung oil. The province maintains an extremely dense population, and its economy is not subject to the vicissitudes and dangers of production in the plains and basins to the east.

South of Szechuan lie Kweichow and Yunnan[a] (nos. 15 and 18), highland plateaux and mountain blocks derived from the Tibetan massif at the basis of the great fold which carries the Yangtze away to the north-east. There is not much level land, probably between five and ten per cent. Climatically the region has the moderate and agreeable character of all places of high altitude in tropical or subtropical latitudes. So wherever level land is available there is an intense cultivation, following the general lines of the south.[b] These provinces have a large population of primitive aboriginal tribesfolk, who have retained their own dress and customs, and practise subsistence agriculture in secluded valleys and uplands. The general character of the region is shared by Sikang (no. 19), which, however, is much more mountainous.

Lastly we come to the seaward-looking river-valleys of Kuangtung, backed by the plateau of Kuangsi (nos. 16 and 17). The climate of the 'viceroyalty of Liang-Kuang' (both the Kuangs), as it used to be called, is definitely tropical. There is a long hot summer of high humidity, then a rather dry cool winter, followed by two transitional months of mist and fog. Farming is predominantly rice cultivation, and dry slopes are not much utilised, nor is there much terracing; but it was not necessary, for in this part of China famines have always been unusual, and three crops a year are common. Sugar-cane, tobacco, oranges (as also in Szechuan) are important products, and much silk is made.

[a] Davies (1), Fitzgerald (2, 3, 4, 5).

[b] On the failure of traditional Chinese culture to make full use of uplands see Gourou (1).

5. HISTORICAL INTRODUCTION. THE PRE-IMPERIAL PHASE

BEFORE EMBARKING UPON the main subject of the book, it has seemed desirable to add to the short geographical description a brief outline of Chinese history, since many readers interested primarily in science and the history of science could not be expected to be already familiar with the nature of Asian kingdoms, the rise and fall of dynasties, and the broad movements of peoples in the lands beyond the Central Asian steppes.

There are, in the English language, three books which a critical friend, giving them his highest praise, has described as the 'least bad' of all books on this difficult and intricate story: those of Fitzgerald (1), Richard Wilhelm (1) and Carrington Goodrich (1). The last-named, which is of recent date, is the first book on Chinese history which has attempted to give its due share of space to the development of technology. To these has recently been added a new account (in German) brilliantly compressed into a very small compass, by the son of one of the former, Hellmut Wilhelm (1). Another recent book, that of Eberhard (9), is especially valuable for its accounts of the numerous northern half-sinified States and Dynasties. The two volumes of Grousset (1), though emphasising religion and art even more than those of Fitzgerald and Richard Wilhelm, will nevertheless be found valuable, more especially as they give the Chinese characters for many proper names.

The four-volume work of Cordier, though classical, is now rather out of date and therefore less useful, except for certain special matters; partly because it narrowly concentrates attention on dynastic and military events, and partly because it devotes half its space to comparatively modern times, which are the least interesting from the point of view of the present book. I suppose that the best history of China in any Western language is the work of O. Franke (1) in five volumes, in German. Its encyclopaedic quality ensures that it pays some attention to matters of scientific interest, but unfortunately it is difficult to read—a single paragraph may often cover a couple of pages—and it is a hard labour to excavate the author's general conclusions. Hirth's book (3) on ancient Chinese history (down to the end of the Chou dynasty) contains a great deal of detail, and was the work of a great sinologist, but so much has been discovered since the time when it was first written (1908) that it is now an unsafe guide except for the reader who is already well informed. Among the more interesting books of recent date on Chinese history, published in China, which I have come across, are the general history of Chou Ku-Chhêng (*1*); and the social histories of Têng Chhu-Min (*1*) and Sun Yü-Thang (*1*). Economic interpretations are given in those of Fan Wên-Lan (*1*), Lü Chen-Yü (*1*), and Chien Po-Tsan (*1*).

(a) A SKETCH OF CHINESE HISTORIOGRAPHY

Perhaps it need hardly be said that China is better provided with original sources than any other Eastern, and indeed most Western, countries. Unlike, for example, India, where the chronology is still very uncertain, the Chinese have one of the greatest historiographical traditions of the world.[a] It is frequently possible to be sure, not only of the year, but also of the month, and even the day, when a certain event occurred—for example, when Li Shao-Chün[1] proposed to the Emperor Han Wu Ti[2] a subsidy for his alchemical researches (see Sect. 33). Through every dynasty, with varying titles (a learned account of which may be found in Chêng Chhiao's[3] +12th-century *Thung Chih Lüeh*,[4] for example), there were official historians who recorded recent and contemporary events, and ultimately produced complete dynastic histories. The degree of objectivity and lack of bias which they showed has recently been defended and demonstrated by Dubs (1) and Hughes (4).

The dynastic histories (*chêng shih*[5])[b] are arranged on a fairly uniform pattern, beginning with the *ti chi*[6] or imperial records, which deal with the successive emperors in turn. The *chih*[7] or memoirs, which follow, consist of chapters on *li*,[8] the calendar; *li*,[9] rites and ceremonies; *yo*,[10] music; *hsing*,[11] jurisprudence; *shih huo*,[12] economics and commerce; *chiao ssu*,[13] official state sacrifices; *thien wên*,[14] astronomy; *wu hsing*,[15] lit. the five elements, but really concerning unusual occurrences or prodigies and the inferences drawn from them; *ti li*,[16] geographical notes; and lastly *i wên*,[17] literature. More important information is often given in the third section, that of the *lieh chuan*,[18] or biographies of the most eminent persons of the period. A tabulation of the Twenty-Four official histories is given in Wylie.[c] They begin with the *Shih Chi*[19] of Ssuma Chhien[20] (−145 to −79), the father of Chinese historiography, who started with the legendary period, and brought his work down to his own time.

Besides the official histories, there are many other classes of works classified as history by the Chinese. The *pien nien*,[21] or annals, are supposed to be modelled on the *Chhun Chhiu*[22] (Spring and Autumn Annals), enrolled among the classics and associated with the name of Khung Fu-Tzu[23] (Confucius) himself. The other most famous books of the class are the *Chu Shu Chi Nien*[24] (Bamboo Books)[d] dealing with the −4th and

[a] Cf. O. Franke (2, 10); R. Wilhelm (1), pp. 21 ff.; Chu Hsi-Tsu (*1*); Gardner (3); Chang Hsin-Hai (1). The best work in Chinese on the history of Chinese historiography is probably that of Chin Yü-Fu (*1*).

[b] The most ancient forms of the character *shih* (K 975) seem to show a hand underneath the symbol for 'middle' or 'central'. Chu Hsi-Tsu (*1*) suggests that this implies the conception of historical impartiality; but it may also have referred to the central location of the official recorders at the capital.

[c] (1), p. 13. The reader is referred to the whole relevant chapter of Wylie's unique guide to Chinese literature.

[d] Tr. E. Biot (3).

1 李少君	2 漢武帝	3 鄭樵	4 通志略	5 正史
6 帝紀	7 志	8 歷	9 禮	10 樂
11 刑	12 食貨	13 郊祀	14 天文	15 五行
16 地理	17 藝文	18 列傳	19 史記	20 司馬遷
21 編年	22 春秋	23 孔夫子	24 竹書紀年	

−5th centuries, but of the genuineness of which there is doubt; and the late +11th century *Tzu Chih Thung Chien*[1] of Ssuma Kuang.[2] A third class is the *chi shih pên mo*[3] or 'complete records' which derive from another of the classics, the *Shu Ching*[4] (Classic of History), dealing with events of the Shang and early Chou periods (the late second and early first millennium before our era) and now generally considered to have been compiled in the late Chou period.[a] The authors of these books, of which there were many in later ages, do not follow a set pattern, nor adopt an annual chronological order, but treat a particular subject, such as the troubled times after the end of the Thang dynasty, or the pacification of Thaiwan (Formosa).

Other types of books are the *pieh shih*[5] or 'separate histories', among which is included the *Thung Chih* already mentioned; and the *tsa shih*[6] or miscellaneous histories. Besides these there are of course many collections of official documents, and many specialised biographical collections, together with anthologies of extracts and contemporary records.

It is one of the greatest deficiencies of world scholarship that practically none of the Chinese histories has been translated into European languages. There are, it is true, the translations by Couvreur (1) and Legge (11) of the *Tso Chuan*,[7] an ancient commentary dealing with the times before the Warring States (*c.* −722 to −468), now thought to have been written between −400 and −250,[b] though certainly retouched by Chhin and Han scholars. About a third of the *Shih Chi* has been translated in the famous version of Chavannes (1). The only comparable work is the initial part of the *Chhien Han Shu*[8] (History of the Former Han Dynasty) by Dubs (2), of which two volumes have appeared and two more are ready. Wieger's *Textes Historiques*, if only on account of its bulk which gives it the air of a mine of information, is at first attractive to the Western student; but for those without any knowledge of Chinese it can be really dangerous;[c] it is in fact a collection of passages taken mostly (though by no means exclusively) from the *Thung Chien Kang Mu*,[9] the reconstruction and condensation of the *Tzu Chih Thung Chien* already mentioned, by Chu Hsi[10] the great Sung philosopher, and his school.[d]

[a] A good many chapters, however, are certainly forgeries of the +3rd and +4th centuries.

[b] Cf. H. Maspero (1; 2, p. 592), Karlgren (8).

[c] Wieger's translations are very tricky, since (*a*) the references to the original source are almost never given; (*b*) his renderings are often careless; (*c*) he is liable to insert asides, sarcastic remarks, and all kinds of verbal capers in the French without indicating that he is departing from his text; and (*d*) he writes with a bias which might be described as a conviction of total depravity in Chinese culture before the arrival of the theology of his own faith upon the scene. It is necessary to draw attention to these shortcomings since in the present introductory volume, but rarely thereafter, we make numerous references to Wieger's *Textes Historiques* (abbreviated as *TH*) for the convenience of the reader who has only European languages. In the later volumes, devoted to the diverse technical subjects in detail, precise references to Chinese sources, by chapter and page, will always be found.

[d] There had been an 18th-century translation of the *Thung Chien Kang Mu*, that of de Mailla, but it is not usable today for the reasons given by Hirth (1), p. 29.

1 資治通鑑　2 司馬光　3 紀事本末　4 書經
5 別史　6 雜史　7 左傳　8 前漢書
9 通鑑綱目　10 朱熹

This is the place to mention August Pfizmaier's translations from the Chinese histories, little known because buried in the publications of the Vienna Academy of Sciences, and little used because the precise textual sources were not clear until recently identified by R. L. Walker. We thus have twelve chapters of the *Tso Chuan* (Pfizmaier, 1–12), and thirty-four chapters of the *Shih Chi* (Pfizmaier, 13–36), only sixteen of which duplicate Chavannes (1). The rest may be tabulated. The German

	Chapters	Reference
Chhien Han Shu	19	Pfizmaier (32–34, 37–51)
Hou Han Shu	3	Pfizmaier (52–3)
Chin Shu	10	Pfizmaier (54–7)
(Liu) *Sung Shu*	4	Pfizmaier (58)
Chhen Shu	7	Pfizmaier (59)
Pei Chhi Shu	11	Pfizmaier (60)
Sui Shu	20	Pfizmaier (61–5)
Hsin Thang Shu	21	Pfizmaier (66–74)

translations of Pfizmaier, however, are as dangerous for those who cannot check them against the Chinese text as are the French ones of Wieger; though less biased, he was earlier in date and worked in a very isolated environment. So we find *sha-mên*[1] translated as 'shaman' instead of Buddhist monk ('śramaṇa') in Pfizmaier (56); and *ming chia*[2] in Pfizmaier (71) as 'the most famous schools' instead of the 'School of Logicians'. Nevertheless, this large body of material can still be helpful in various ways, such as the quick location of passages.

It is in many ways unfortunate that the only exhaustive analysis of a single Chinese dynasty, including a mass of translated material, in a Western language, concerns a dynasty which was not really Chinese at all—the Liao; we refer to the monumental book of Wittfogel, Fêng Chia-Shêng *et al.* (1). This dynasty is one of the least interesting from the point of view of science and technology; it was based upon the conquest of a large part of the population of North China by a nomadic people, the Chhi-tan.[a]

Apart from these translations, and of course an appreciable number of fragmentary renderings made by different scholars at various times (e.g. Hirth (1), Carter (1), Hambis (1), de Saussure (1), etc.), some of which we may use in due course, there is no access to Chinese historical literature except by the hard way. Historians of science can claim no exemption. But while recognising that there is undoubtedly an immense amount of material to be obtained from the official histories and their related works, especially in economic and social matters (and as regards astronomical and calendrical science), it seems nevertheless true that the historian of science may

[a] See below, p. 133.

[1] 沙門 [2] 名家

find less than he might expect. To such an extent was Chinese literary culture uninterested in science that—apart from astronomy and meteorology, the 'orthodox' because prognosticatory, sciences—it is probably not in the histories that the most valuable information on the development of science in China will be found. It is rather in the class of literature which the Confucian scholars termed 'Miscellaneous', namely, in the surprisingly large number of technological and scientific books which have survived through the centuries; either singly or buried in those favourite 'Everyman' collections of the Chinese,[a] the *tshung-shu*,[1] of which the earliest (and one which we still possess) is the *Pai Chhuan Hsüeh Hai*[2] (The Hundred Rivers Sea of Learning) compiled by Tso Kuei,[3] a Sung scholar of whom practically nothing else is known.[b] Apart from the books in these *tshung-shu* which are mainly devoted to scientific subjects, there are also others of the type known as *pi chi*[4] or *pi than*,[5] 'pen jottings' or 'pen conversations'; often devoted to personal experiences of the writer, or to memoirs of the preceding century. A wealth of information is contained in these memoirs, very few of which have ever been translated; we shall draw on them considerably in the present book. Indeed, they often contain statements crucial for the history of science, and one may feel confident that their texts have never been intentionally interfered with, partly because the Confucian scholars considered them too unimportant, and partly because until modern times it would never have occurred to any Chinese scholar that the slightest interest attached to placing a piece of scientific knowledge or a technical process earlier than its proper date.

It is now time to refer, before going further, to the Table of Chinese Dynasties, which will be found useful whatever section is being read. Unfortunately for the scientist, who does not wish to become involved in history of a too complex kind, the Chinese dynasties do not follow a perfectly consecutive course. The immediately following pages will explain why this happened. But the table will orientate the reader in that long time-scale of close on four thousand years.[c]

At this point it may be mentioned that the Chinese counted, as in medieval Europe, from the accession of their kings or emperors. But in addition to the *miao hao*[6] or dynastic title of the emperor, his reign was generally split up (from about −165 onwards) into a series of shorter periods, or *nien hao*,[7] 'reign periods', lasting sometimes as long as thirty years, but commonly much shorter. These reign periods serve to date events quite accurately; tables are available showing their correspondence with the modern system.[d] Dates are, however, often also given in terms of the sexagenary

[a] The best index to them and their contents is that of Shih Thing-Yung (*1*).

[b] He is said to have obtained the idea from Lu Kuei-Mêng,[8] the Thang Taoist poet, who was the first to collect his own writings together. Tso Kuei collected those of others.

[c] On 'periodisation' see Lei Hai-Tsung (2).

[d] We use the tables of Wang & Lyman (1); there are also the older ones of Mayers (1), and the useful lists of Ezerman & van Wettum (1) (Chinese emperors) and Schlegel (1) (Japanese mikados and shoguns). The most elaborate standard tables are those of M. Chang (1) and Arendt (1). One may also need to use those in *TSCC*, *Huang chi tien*, chs. 211–12. We hyphenate reign-period names.

1 叢書 2 百川學海 3 左圭 4 筆記 5 筆談
6 廟號 7 年號 8 陸龜蒙

Table 5. *Chinese Dynasties*

夏	HSIA kingdom (legendary?)		*c.* −2000 to *c.* −1520
商	SHANG (YIN) kingdom		*c.* −1520 to *c.* −1030
周	CHOU dynasty (Feudal Age)	Early Chou period	*c.* −1030 to −722
		Chhun Chhiu period 春秋	−722 to −480
		Warring States (Chan Kuo) period 戰國	−480 to −221
First Unification	秦 CHHIN dynasty		−221 to −207
	漢 HAN dynasty	Chhien Han (Earlier or Western)	−202 to +9
		Hsin interregnum	+9 to +23
		Hou Han (Later or Eastern)	+25 to +220
	三國 SAN KUO (Three Kingdoms period)		+221 to +265
First Partition	蜀 SHU (HAN)	+221 to +264	
	魏 WEI	+220 to +265	
	吳 WU	+222 to +280	
Second Unification	晉 CHIN dynasty: Western		+265 to +317
	Eastern		+317 to +420
	劉宋 (Liu) SUNG dynasty		+420 to +479
Second Partition	Northern and Southern Dynasties (Nan Pei chhao)		
	齊 CHHI dynasty		+479 to +502
	梁 LIANG dynasty		+502 to +557
	陳 CHHEN dynasty		+557 to +587
	魏	Northern (Thopa) WEI dynasty	+386 to +535
		Western (Thopa) WEI dynasty	+535 to +554
		Eastern (Thopa) WEI dynasty	+534 to +543
	北齊 Northern CHHI dynasty		+550 to +577
	北周 Northern CHOU (Hsienpi) dynasty		+557 to +581
Third Unification	隋 SUI dynasty		+581 to +618
	唐 THANG dynasty		+618 to +906
Third Partition	五代 WU TAI (Five Dynasty period) (Later Liang, Later Thang (Turkic), Later Chin (Turkic), Later Han (Turkic) and Later Chou)		+907 to +960
	遼 LIAO (Chhitan Tartar) dynasty		+907 to +1125
	West LIAO dynasty (Qarā-Khiṭāi)		+1144 to +1211
	西夏 Hsi Hsia (Tangut Tibetan) state		+990 to +1227
Fourth Unification	宋 Northern SUNG dynasty		+960 to +1126
	宋 Southern SUNG dynasty		+1127 to +1279
	金 CHIN (Jurchen Tartar) dynasty		+1115 to +1234
	元 YUAN (Mongol) dynasty		+1260 to +1368
	明 MING dynasty		+1368 to +1644
	清 CHHING (Manchu) dynasty		+1644 to +1911
	民國 Republic		+1912

N.B. When no modifying term in brackets is given, the dynasty was purely Chinese. Where the overlapping of dynasties and independent states becomes particularly confused, the tables of Wieger (1) will be found useful. For such periods, especially the Second and Third Partitions, the best guide is Eberhard (9). During the Eastern Chin period there were no less than eighteen independent States (Hunnish, Tibetan, Hsienpi, Turkic, etc.) in the north. The term 'Liu chhao' (Six Dynasties) is often used by historians of literature. It refers to the south and covers the period from the beginning of the +3rd to the end of the +6th centuries, including (San Kuo) Wu, Chin, (Liu) Sung, Chhi, Liang and Chhen.

year cycle, each year being denoted by two characters; the first one of the ten *kan*[1] (the so-called 'celestial stems') and the second one of the twelve *chih*[2] (the so-called 'horary characters' or 'branches'). More will have to be said about this system later.[a] Here it is only necessary to refer to the availability of tables from which dates can be calculated with precision, unless there is an uncertainty as to a probable date greater than sixty years.[b]

All authorities on the chronology of ancient China are, as Creel[c] says, in general agreement concerning both the relative and absolute dating of events after −841. Before that time there is considerable uncertainty. Matthias Chang, following the traditional chronology, places the beginning of the Chou period at −1122; others, following critical reconstructions of the *Chu Shu Chi Nien* mentioned above (Bishop, 1; Wang Kuo-Wei, *1*), make it begin at −1050 and −1027 respectively. The differences increase for the earlier Shang and the semi-legendary Hsia dynasties. Creel (1) concludes that it is not possible at present to be certain of dating earlier than the −9th century; fortunately this does not matter very much for the history of science.

(*b*) CHINESE PREHISTORY AND THE SHANG DYNASTY

During the past thirty years knowledge of China's most ancient history has undergone revolutionary changes. A vast literature has grown up, to which, however, since it is primarily of background interest to the present theme, we shall only give certain key-references. J. G. Andersson (1–5), at one time a leading member of the Chinese Geological Survey, and colleague of the then Director, Ting Wên-Chiang, later Secretary-General of Academia Sinica, was a pioneer in the work. He describes graphically[d] how it was that his interest was first aroused. Most of the assured conclusions of today will be found, with interesting discussions, in the books of Creel (1, 2) and in concise but brilliant summaries by Chêng Tê-Khun (4), Bishop (2), Ting Wên-Chiang (1) and Li Chi (1, 2).[e]

Parallel with these investigations in material archaeology there was another current of scholarship which occupied itself with the interpretation of the ancient legends of China. This is especially associated with the name of Granet (1–3), whose often daring conclusions, though criticised severely by many (e.g. Karlgren, 2), remain highly stimulating and suggestive.

The first inhabitants of Chinese soil whose remains are known to us were those of the race to which 'Peking Man' belonged. The discoveries of these bones made at Chou-Khou-Tien by Phei Wên-Chung[3] in 1927 and exhaustively studied by Weiden-

[a] Cf. below, Sect. 20*h*. For conversion from sexagenary dates to western dates consult Hoang (1, 4); Nagel (1); Havret (1); Kliene (1), etc.

[b] We use the tables in the *Manuel du Sinologue* (see Anon. 1), and Wieger (1, p. 7) also has a table, but less convenient.

[c] (1), p. xvii.

[d] (5), p. 12.

[e] The geographical and pedological aspects have been treated by Roxby (1) and Bishop (3).

[1] 干(甲乙丙丁戊己庚辛壬癸) [2] 支(子丑寅卯辰巳午未申酉戌亥) [3] 裴文中

reich, showed that *Sinanthropus pekinensis* lived at the beginning or middle of the Pleistocene.[a] The position of this form stands at an intermediate level, being rather later and more human than *Pithecanthropus erectus*, but earlier than Neanderthal Man. It is quite certain therefore that there was a Palaeolithic population in China able to make rough stone implements.[b] Creel[c] points out that these facts have rendered the attempts, formerly made, to ascribe to late foreign migrations the origins of Chinese civilisation, rather unplausible. Weidenreich indeed has claimed (1) that mongoloid and even specifically Chinese anatomical traits are to be found in the *Sinanthropus* bones. Late Pleistocene[d] sites of hunting cultures have been found in the Ordos Desert (Shensi) by Licent and Teilhard de Chardin,[e] and by Phei Wên-Chung (1) at the *Sinanthropus* site near Peking, on higher levels.

After the Palaeolithic, however, there is a remarkable break of continuity. Andersson[f] has written:

> The deeper we penetrate into the study of those remote times, the more we are impressed by the inflexible riddles barring our way. Foremost of these is the 'Neolithic hiatus'. During the loess period (Palaeolithic) the climate of Northern China was so arid that the region, apart from residual lake areas, may have been largely depopulated. After the loess period followed the Phan-Chiao stage of vertical river erosion, during which the loess cover was largely dissected, and small canyons cut locally into the solid rock. This period, which may correspond approximately to Mesolithic and Early Neolithic, was a time of abundant rainfall, which in that part of the world must mean a genial climate. In other words, the region certainly abounded in game and must have formed a pleasant habitat for primitive man. However, in 1943 the situation is, as far as I know, the same as it was in 1925—no indisputable Mesolithic or early Neolithic site has so far been found in Northern China.

This is still true today, save that in Manchuria all stages have been found (de Chardin & Phei Wên-Chung, 1). Then suddenly, at a time which Andersson dates as about −2500, the apparently empty land begins to support a large and busy population. There is evidence of hundreds, even thousands, of villages, inhabited by a people of agricultural as well as pastoral economy, acquainted with carpentry, textiles and ceramics.[g] Only further archaeological work in the field will throw light on this curious gap.

The first important culture, which has been revealed by excavations along a belt from west to east comprising Kansu, Shensi, Shansi, Honan and Shantung, is that

[a] I.e. of the order of −400,000.

[b] By pebble-shaping and flaking. Bone was also worked. In China, just as in Europe, Palaeolithic and Neolithic stone axes were confused in medieval times with meteorites, and so called 'thunder-axes' (de Mély (5); see below, Sects. 21, 23).

[c] (2), p. 40.

[d] *c.* −25,000.

[e] Black & de Chardin (1); Black, Licent & de Chardin (1); de Chardin & Yang Chung-Chien (1).

[f] (5), p. 296.

[g] In following the particularities of the Early Neolithic cultures which occupy the period from about −2500 to −1600, and which will here be briefly mentioned, it is useful to have available, for comparative purposes, a general handbook of prehistoric archaeology, such as that of Furon (1).

designated as the Yangshao[1] culture. The chief cereal of the Yangshao people was almost certainly millet (*Panicum miliaceum*), and towards the end of the period, rice; since neither of these plants is thought to be indigenous to China, there must have been a diffusion from south-east Asia. Bones of the dog and the pig, and later of sheep and cattle, have been identified from these sites. Bones of the horse have also been found, but one cannot be sure that it was domesticated, since a true wild horse, *Equus przevalskii*, still exists, or did so until recently, in Mongolia, and may then have been found wild in northern China. The most outstanding characteristic of the Yangshao people, however, was their painted pottery, of which abundant illustrations will be found in the works already quoted. This was probably made from about −2300 by the 'coiling' process, and not with a potter's wheel; it bears magnificent designs in a variety of colours and belongs to the finest type of Neolithic ceramics.[a] It continued to be made in Kansu until nearly the end of the first millennium before our era.

It is at this stage that certain traits are found which point to a wide community of culture throughout the northern latitudes below the Arctic Circle,[b] i.e. Northern Asia and Northern America. This culture area could almost be called the Shamanism area (see on, Sect. 10*h*). A typical implement common to all parts of this vast area is the rectangular or semilunar stone knife, quite unlike anything known in Europe or the Middle East, but found among Eskimos and Amerindians as among Chinese and in Siberia. Creel points out[c] that such knives were common in the Shang dynasty, and continued to be made (of iron) down to recent times in China.[d] Another characteristic of this northern culture area is the use of pit-dwellings or earth-lodges,[e] the beehive shape of which may have descended to the peasants' houses of the Thang period which may be seen painted on the frescoes of Tunhuang. The sinew-backed or composite bow seems to have been an invention of this area. If America was peopled by migrations across the Behring Straits at the beginning of the Neolithic, we might have an explanation of some of those strange similarities which exist between Amerindian and East Asian civilisations; but this is a very difficult problem, to which we may later return.

K 252 K 252 K 855

There are two types of pottery vessels of the Chinese Neolithic which merit a moment's attention,[f] because of their wide divergence from anything known elsewhere, and because of the ancient association of cooking processes with the earliest chemistry. One of these, known as the *li*,[2] is a cauldron or pot with a tripod base, the

[a] Cf. Bishop (4); Chêng Tê-Khun (4). The latter, however, believes that the potter's wheel was used.
[b] Bishop (2), p. 303; Creel (2), p. 45. [c] (1), p. 174.
[d] Andersson (3) has demonstrated their persistence into the peculiar metal rattle still used by knife-sharpeners in the streets of Chinese cities of our own time to announce their presence.
[e] Creel (2), p. 56.
[f] On the prehistoric pottery in general see Wu Chin-Ting (1).

1 仰韶 2 鬲

legs of which are hollow (see Fig. 9).[a] Some have thought that this must have been derived from the fusion of three long and narrow but bulging vessels with the addition of a chamber above them. The other is an arrangement for cooking by steam, the

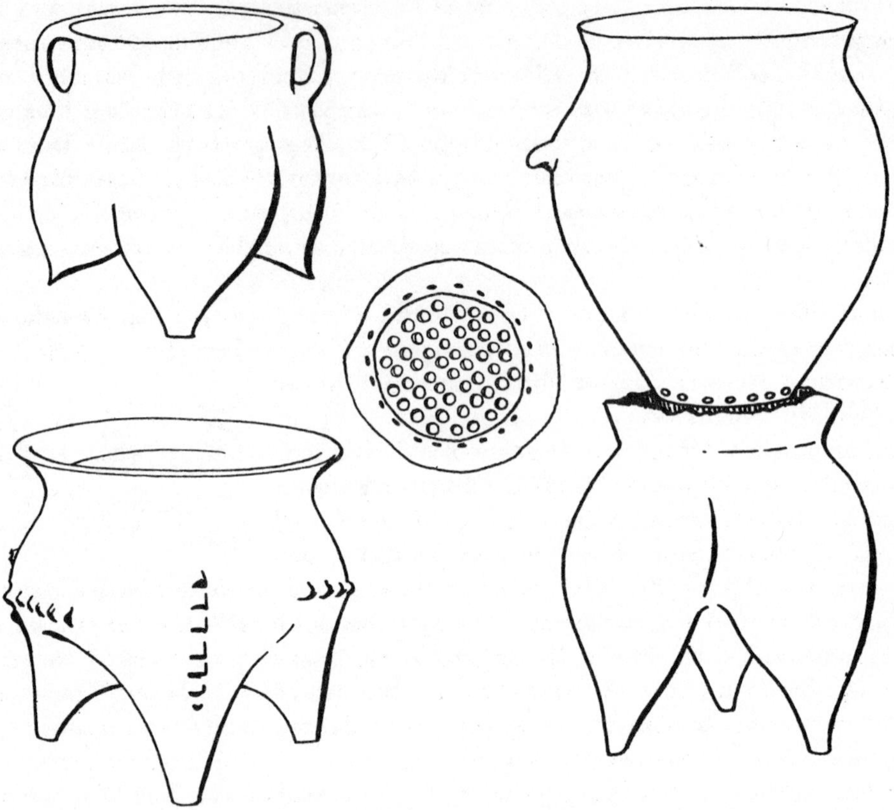

Fig. 9. Ancient Chinese pottery vessels. On the left two types of *li* (from de Tizac, 1); on the right a *tsêng* above a *li* (from Andersson, 6). The combination of the two vessels, the upper one of which has a perforated bottom (shown in the middle above), was called a *hsien*. All these examples are from the −3rd millennium. Their form gave to such vessels a large effective heating surface and permitted separate cooking of several foodstuffs at one time, but it must have required great skill on the part of the potter. Later on, certain bronze vessels, such as the *ting*, perpetuated the shape. In Shang times the pictograph of *hsien* came to denote, by extension of meaning, a religious sacrifice.

hsien[1] (Fig. 9),[b] in which a lower vessel, which may or may not be in the shape just described (the *li*), is surmounted by another (the *tsêng*[2]) having a perforated floor, the

[a] The character is derived from an ancient pictogram, K855; as also is the following, *hsien*, K252.
[b] Alternatively *yen*, perhaps best pronounced *chuan*.

1 甗 2 甑

whole being topped by a cover. These vessels are found in the earliest Neolithic sites known, and continued to be made through the Shang dynasty.[a]

In north-eastern China (Honan and Shansi) the Yangshao phase was followed by another Neolithic culture known as that of Chhêng-Tzu-Yai[1] or of Lung-Shan[2] (site-names). This, still lacking all metal, was characterised by a smooth black earthenware of fine texture and high finish. The Lung-Shan men were certainly more advanced than those of Yangshao; the four animal species previously mentioned were fully domesticated, and probably also now the horse. At this point, too, appear other techniques, long known in the Middle East, but new to China, particularly the potter's wheel, and the building of towns and houses made of tamped earth (*terre pisé*). It is possible that the Lungshan people knew the use of wheeled vehicles, but the absence of metal parts makes this uncertain.

This takes us down to about −1600. Within a hundred years of that date either way appeared, with apparent suddenness, that mature bronze-age culture which came to be embodied in Chinese history as the Shang dynasty. Most of what we know about it derives from the excavations at its metropolis, Anyang[3] in Honan, carried out under the auspices of the staff of the Historical Institute of Academia Sinica, 1929–33. But from the beginning of the present century information had been filtering through, and the story of this, one of the most romantic in all modern archaeology, has been well told by Creel (1) and Yetts (1). For some decades during the end of the 19th century farmers tilling their fields near Anyang often found curious pieces of bone coming to the surface. These were bought by a man of the village, who sold them to drugstores as 'dragon-bones' (cf. Sect. 23*c*) for medicine. By 1899 Chinese scholars realised that they were inscribed with very ancient writing, and in 1902 Wang Kuo-Wei (*1*) and Lo Chen-Yü (*1*) recognising fully their significance, declared them to be such materials as no scholars since the beginning of the Han had possessed. They pushed back, in fact, all the philology, linguistics and history of China for nearly as much as a thousand years. About the same time some of the bones were purchased by Western scholars such as Couling and Chalfant, and in 1904 Sun I-Jang completed (*1*) the first attempt at deciphering the oracle-bone inscriptions, as they were now recognised to be. Since that time an enormous amount of work, largely by Chinese scholars such as Tung Tso-Pin[4] (*1*), Wang Kuo-Wei[5] (*2*) and Kuo Mo-Jo[6] (*3*), but also by Westerners such as Menzies (1) and Hopkins (cf. the dictionary, including bone forms of characters, by Karlgren, 1), has been done on this most ancient writing. We know now that some of the characters have remained almost unchanged through the 3500 years which separate us from the time of the king Phan Kêng[7] (*c.* −1300), who was the founder of the city of Anyang. In later sections we shall scrutinise from time to time some of the bone forms of characters which are interesting for technological reasons (cf. Fig. 10).

[a] The *li* appeared first in Honan and Shantung, spreading later to Shansi and Kansu.

[1] 城子崖 [2] 龍山 [3] 安陽 [4] 董作賓 [5] 王國維
[6] 郭沫若 [7] 盤庚

The oracle-bones were employed for a method of divination, 'scapulimancy', which appears to have been peculiar to this culture-area, and may have originated a little before the Shang. It consisted in heating the shoulder-blades of mammals or the carapaces of turtles with a live coal or a red-hot bronze poker, the reply of the gods being indicated by the shape or direction of the cracks produced.[a] It was a fortunate thing for posterity that the Shang diviners (*chen jen*[1]) kept such ample records of the results, probably as secret dossiers. There were two kinds of replies, 'yes or no' or 'lucky or unlucky'; in general, the answers were not inscribed on the bone beside the questions, as they would be indicated by the shape of the cracks, but sometimes the subsequent verification of the prophecy was recorded. Classifications have been made of the questions asked; among the most important were: (*a*) to what spirits should certain sacrifices be made; (*b*) travel directions, where to stop and how long; (*c*) hunting and fishing; (*d*) the harvest; (*e*) weather; (*f*) illness and recovery, etc.

The other outstanding characteristic of the Shang period was the extensive use of bronze for all kinds of uses, ritual, war and luxury, less perhaps for tools and implements. It has been repeatedly remarked that the magnificent workmanship of the sacrificial vessels, which were often cast in commemoration of some honour bestowed on a prominent man by the king, and bear inscriptions to that effect, surpasses that of all later work. An important use of bronze was for the metal parts of wheeled vehicles, which were becoming common. Its spread is seen in Fig. 11.

There now appears the beginning of wheat cultivation, which must have spread from the Middle East, where the plant is native (Vavilov, 1). Bishop (2) has pointed out that the area of wheat-culture in antiquity is almost identical with that of the use of bronze. Shen Tsung-Han[2], a leading agriculturist, has stated that the wheat varieties grown in earlier times in China were the same as those found along the lines of communication in Central Asia and in the Middle East.[b]

Most archaeologists are of the opinion that the Shang people certainly, and probably also the Yangshao people, were predominantly agricultural, and Creel, fairly representatively, denies[c] that the Chinese people ever passed through a pastoral period. Milk and milk products have been absent from their diet since time immemorial, and pastoral metaphors are exceedingly rare, if present at all, in the most ancient Chinese writings which we possess. Nevertheless, I always find it difficult to account otherwise for the fact that so many words in the Chinese language which have the meaning 'good' or related meanings, are based on radical no. 123, *yang*, sheep.[d]

[a] It is extraordinary, but by no means inconsistent with the duration of culture traits in East Asia, that in our own time the lamas of Mongolia still make use of divination by means of scapulimancy (Larson (1), p. 102). The Franciscan friar William of Rübruck saw it still going on in the +13th century (*TH*, vol. 2, p. 1677; Yule (2)). It is also known from other parts of the world (Andrée, 1).

[b] In a letter to C. W. Bishop.

[c] (2), pp. 78 ff.

[d] *Yang*,[3] sheep (K732); *hsiang*,[4] felicitous (K732); *mei*,[5] beautiful (K568); *ta*,[6] grain-sprouts breaking through (K271); *tsang*,[7] dense foliage (K727); *hsiu*,[8] food (K1076); *i*,[9] righteousness (K2). Laufer (3), p. 48, is among those who have drawn attention to this.

[1] 貞人 [2] 沈宗瀚 [3] 羊 [4] 祥 [5] 美 [6] 達 [7] 牂
[8] 羞 [9] 義

PLATE III

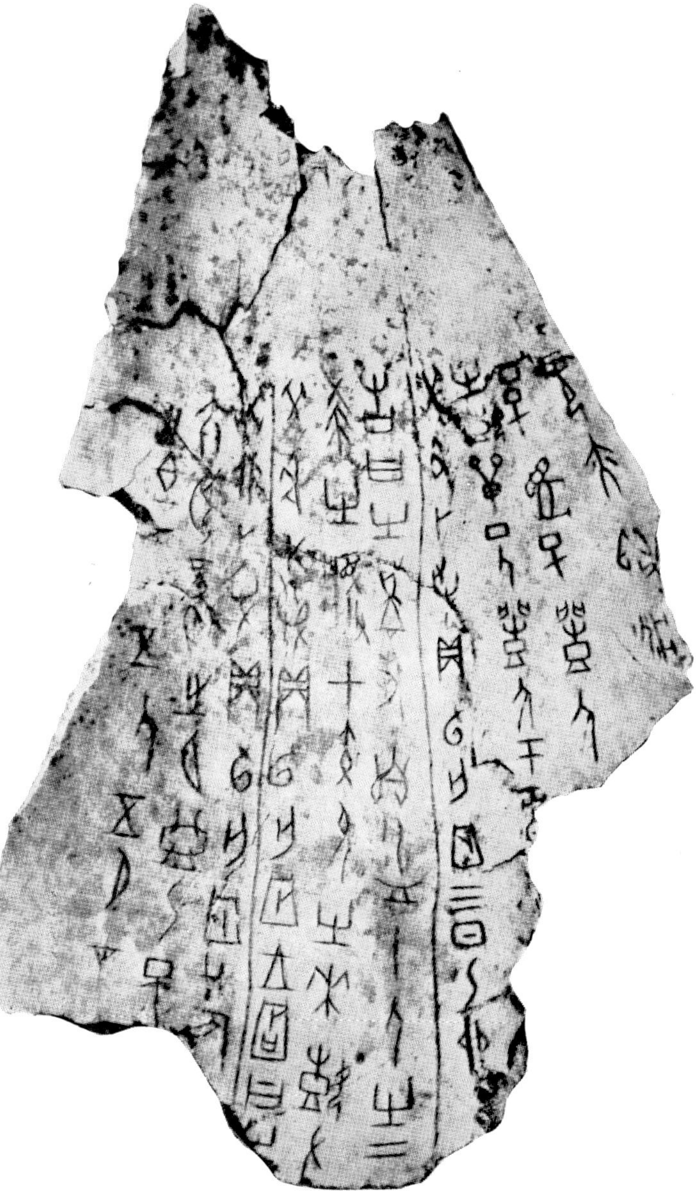

Fig. 10. Oracle Bone (Shang period). Specimen described by Lo Chen-Yü (*Yin Hsü Shu Chhi Chhing Hua*, Pl. 5) and Kuo Mo-Jo (cf. Hopkins, 30). The inscription in the left of the three panels has been thus interpreted: 'On the day *chia-chhen* a great and violent wind, and the moon was eclipsed...on the day *i-ssu*...five men, (in the) fifth month...' An alternative interpretation (Tung Tso-Pin) is that a special rite, the *yen*-sacrifice, was performed on the evening in question.

So far as can be made out,[a] the social organisation of the Shang time was a kind of bronze-age proto-feudalism, such as we shall have occasion to discuss in a little more detail later on. Quite probably the area under the jurisdiction of the Shang king was small, perhaps not more than one or two hundred miles at most in any direction from the capital.[b] Traces of the family-system and of ancestor-worship are discernible, human sacrifice was certainly known, and slaves at least were immolated at the burial

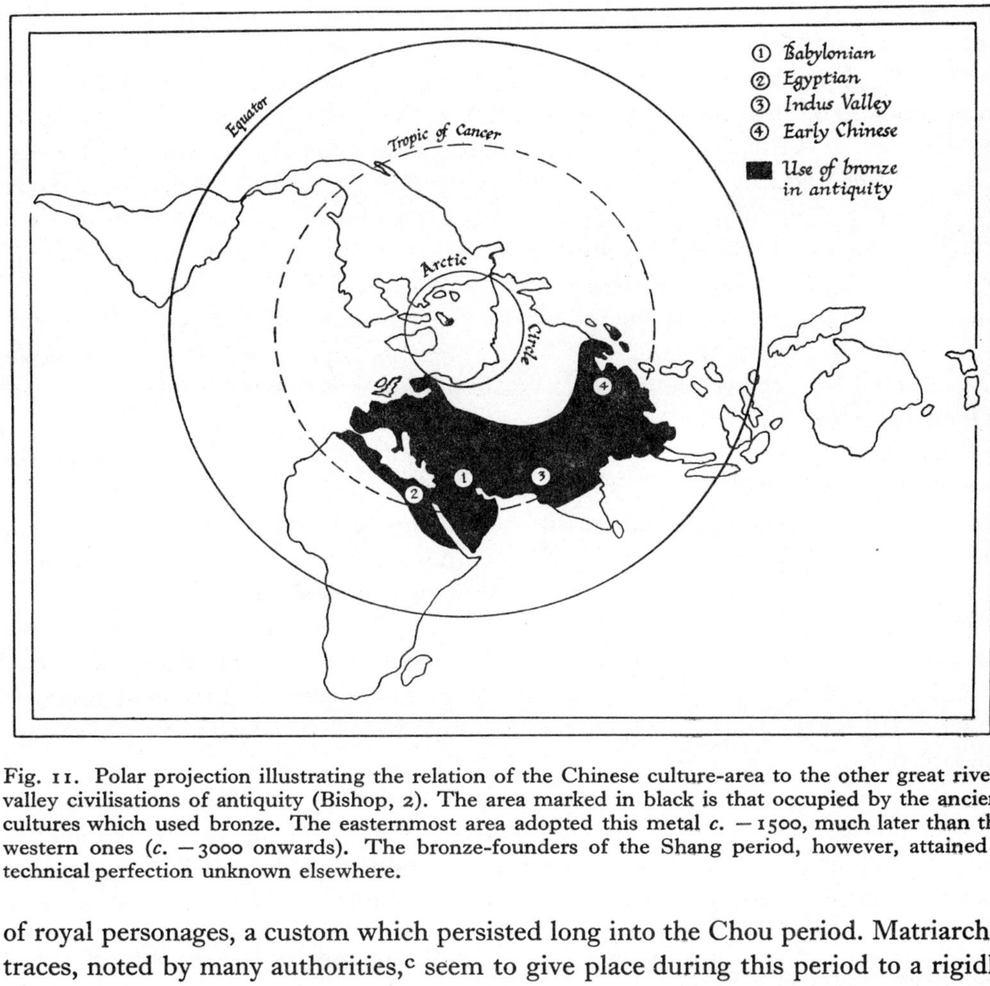

Fig. 11. Polar projection illustrating the relation of the Chinese culture-area to the other great river-valley civilisations of antiquity (Bishop, 2). The area marked in black is that occupied by the ancient cultures which used bronze. The easternmost area adopted this metal *c.* −1500, much later than the western ones (*c.* −3000 onwards). The bronze-founders of the Shang period, however, attained a technical perfection unknown elsewhere.

of royal personages, a custom which persisted long into the Chou period. Matriarchal traces, noted by many authorities,[c] seem to give place during this period to a rigidly patriarchal society.

Another feature of the Shang age was the use of cowrie-shells as a medium of exchange. There is no question that this is the origin of the fact that so many words with the general significance of 'value' are based on radical no. 154, *pei*, which

[a] Creel (2), p. 129.
[b] Herrmann (1), Map 9.
[c] E.g. Creel (2), p. 173.

originally meant a cowrie-shell.[a] Where the cowries (*Cyprea moneta*) came from is still uncertain, but perhaps from the Pacific coast south of the Yangtze estuary. Their long journey to the centre of Shang civilisation is remarkable, and parallels that of another commodity, namely, the carapace and sternum of the land-tortoise for divination. The animal used seems to be of a now extinct species *Pseudocadia anyangensis* (Sowerby, 1).

East Asia has often been called a 'bamboo' civilisation; there is evidence that the appreciation of the manifold uses of these plants was already present in the Shang time.[b] Among their applications was the making of books of tablets, for which wood was also employed. These were probably similar to the books of the early Han which are still preserved; the lines of characters were written on the slips of wood or bamboo, and the slips held together with two lines of cords (cf. Stein, 6). Hence the character *tshê*,[1] which, as we know from its bone forms (K845), originally depicted a written book (Yetts & Hopkins). So also *tien*,[2] which now means a dictionary, then showed a book lying on a table, i.e. on a place of honour. The writing-brush has also now been shown to be definitely Shang.[c]

K 845

There exist documents which purport to be of Shang time; notably certain parts of the *Shih Ching*[3] (Book of Odes[d]), and five books of the *Shu Ching*.[e] But critical examination of them leads to the conclusion that the former cannot be earlier than about −800, and that the latter cannot be pre-Chou, i.e. earlier than the beginning of the first millennium.[f] It has been suggested that those of the *Shu Ching* were indeed Chou propaganda, and that the Chou people systematically destroyed what remained of Shang writings. In any case, while the discovery of some undoubtedly genuine Shang book by excavation is not impossible, it is not very likely to happen, and we have to be content with the conclusion—indicated by the complete absence of any writing in Yangshao and other Neolithic remains—that it must have been at the beginning of the Shang time (*c.* −1600) that Chinese writing developed from its first pictograms. While its antiquity is thus much less than Mesopotamian cuneiform and Egyptian hieroglyphic writing, it lives still today while they have long been dead.

One of the most interesting and important of the effects produced by the Anyang discoveries was the realisation that the 'sceptical renaissance' of modern China had been pushed rather too far. They have given us a sounder picture of China's past than that of either the credulous medieval scholars, or the excessively sceptical moderns.

[a] *Pei*,[4] cowrie-shell (K320); *pao*,[5] precious (K1059); *ku*,[6] merchant (K38); *mai*,[7] to buy (K1240); *mai*,[8] to sell (K1240); *chih*,[9] hostage, matter or substance, to pawn (K493).

[b] Creel (1), p. 24.

[c] See Creel (1), pp. 42ff.

[d] Cf. Nagasawa (1), p. 74.

[e] Cf. Nagasawa (1), p. 120.

[f] Creel (2), pp. 54, 63, 69, 76, 81, 93.

1 册 2 典 3 詩經 4 貝 5 寶 6 賈 7 買
8 賣 9 質

Throughout our era it was firmly believed that China had a recorded history of five thousand years. The prehistoric legends[a] began with Phan Ku,[1] the creator of heaven and earth. He was followed successively by the Emperor of Heaven, the Emperor of Earth and the Emperor of Humanity. Later there appeared Fu Hsi,[2] the patron of hunting and animal husbandry, and Shen Nung,[3] the patron of agriculture and medicine. Huang Ti,[4] the Yellow Emperor, was supposed to have reigned in the −27th century; and in the −24th and −23rd came the 'three model kings', Yao,[5] Shun,[6] and Yü.[7] Yao abdicated and gave over his responsibilities, not to his son, but to Shun, an able and virtuous fisherman and farmer; Shun likewise abdicated, and gave the throne to a commoner, Yü. This was Yü the Great Engineer, who succeeded in 'mastering the waters' by hydraulic conservancy works, and so rendering the country safe from floods and well irrigated for agriculture. He it was who founded the dynasty of the Hsia. Like all subsequent dynasties it continued until by misgovernment and tyranny it 'lost its mandate', whereupon risings led by virtuous commoners or small feudal lords took place, leading to the foundation of a new dynasty, as in the case of the Shang and the Chou successively. One can see now that a great deal of this legendary material was the backward reflection of events and practices of historical times, and one can read in Nagasawa's book the backgrounds and ramifications of the key phrases:[b] abdication or non-hereditary selection, *shan-jang*;[8] revolt against tyrants, *fang-fa*;[9] loss of the mandate of a dynasty, *ko-ming*.[10]

The modern viewpoint[c] began its victorious spread about 1918 with the work of Hu Shih and Ku Chieh-Kang, whose interpretation, notwithstanding the opposition of traditional scholars, became in due course generally accepted.[d] It ended practically in a denial of the historicity of anything prior to the coming of the Chou about the −11th century. Ku Chieh-Kang (5) urged that the so-called ancient history of China had been composed of successive layers or accretions which had been gradually built up, the order of accretions and the chronological order being exactly the opposite. Thus, for example, in the early Chou period Yü the Great was considered to have been a god who made the earth rise above the surface of the water; only in the late Chou did he become a human king, and not the founder of the Hsia dynasty until the Warring States period. Then Yao and Shun, who took their place earlier than

[a] The best account of these will be found in the chapter which the Thang scholar Ssuma Chen prefixed to the *Shih Chi*, and which Chavannes translated with much commentary (1), vol. 1, pp. 3 ff.

[b] (1), pp. 23 ff.

[c] Though this phrase is here used, we shall see later on (Sect. 14*i*) that in fact a sceptical attitude to the authenticity of ancient texts was already widespread among scholars as early as the Sung dynasty. Thus, for example, both Wang An-Shih and Ma Tuan-Lin considered that the *Tso Chuan* (p. 75) had been written in the Han and not in the Chou (see Karlgren, 8). But recent research has shown that whether or not Tsochhiu Ming was the chief author, the work cannot possibly have been put together later than the −3rd century. The grammar of the *Chhun Chhiu* and *Tso Chuan*, however, is so different from other classical writings, that they cannot have had any connection with Confucius or the scholars of the State of Lu, as was traditionally supposed.

[d] See the convenient account by Lin Mou-Shêng (1).

1 盤古 2 伏羲 3 神農 4 黃帝 5 堯 6 舜
7 禹 8 禪讓 9 放伐 10 革命

Yü, first appear in history much later, i.e. not till the time of Confucius (−6th century). Huang Ti, the Yellow Emperor, though earlier in legend, comes even later in fact, and was probably invented by the Taoists.[a] Indeed, each successive ruling house or dominant philosophy was inclined to place an essentially invented character at the head of the list of ancient culture-heroes, in order to heighten its own prestige. Moreover, each school of thought secreted legends in order to justify its own position; the Mohists (cf. Sect. 11) emphasised the abdication-legends in the interests of peace and good government (Ku Chieh-Kang, 1); Mencius emphasised the people's choice of Yü instead of the son of Shun on account of his democratic theories; the Taoists invented Huang Ti who governed through inactivity, and so on.

Moreover, a pioneering contribution was made by a European sinologist, Haloun (6), who concluded, independently, that much of the legendary material must be of late date, since it could be shown to incorporate cosmological speculations (e.g. the five-element theory) of the Warring States and Han periods.

In their attack on the historicity of the legends, the modern scholars have been abundantly justified,[b] but the Shang dynasty suffered somewhat in the process. The *Shih Chi*, written about −100, gives (chapter 3), with all verisimilitude, an account of thirty Shang kings, beginning with Chhêng Thang[1] (whose date would be about −1500) and ending with the 'wicked tyrant' Chou Hsin[2] (of about −1050). It was commonly maintained that Ssuma Chhien could not have had adequate historical materials for his account of what had happened more than a thousand years earlier. One may judge of the astonishment of many, therefore, when it appeared that no less than twenty-three out of the thirty rulers' names were to be clearly found on the indisputably genuine Anyang bones (cf. the lists of Wang Kuo-Wei, *2*; and Hopkins, 1). It must be, therefore, that Ssuma Chhien did have fairly reliable materials at his disposal—a fact which underlines once more the deep historical-mindedness of the Chinese—and that the Shang dynasty is perfectly acceptable. One curious fact about the Shang kings is that the second character in their names is nearly always one of the cyclical 'stem' or 'branch' series already referred to.

What, then, is the position of the Hsia? Andersson[c] hints that it may have to be considered simply a semi-legendary memory of what was perhaps the dominant focal point of the Yangshao 'prehistoric' culture.[d] Creel[e] notes that the character *hsia* never appears on the oracle-bones with the meaning of a State, but concludes that there probably was some kind of integrated community or primitive State which had the name. Around the −6th century it was often used to denote the group of States

[a] Cf. Edkins (5).

[b] The extremely destructive attitude to the traditions has been somewhat modified in the most recent presentations of Ku Chieh-Kang (*2*).

[c] (5), p. 293.

[d] This is the opinion of Hsü Chung-Shu (*1*), p. 539. Some sinologists, however, are still willing to break a lance in favour of the full historicity of the Hsia, e.g. Erkes (2).

[e] (1), p. 130.

[1] 成湯 [2] 紂辛

which possessed full Chinese culture, as opposed to those which did not (like Hellenes and barbarians). As to the four books in the *Shu Ching* which were traditionally regarded as of Hsia date (and two of which, the *Yao Tien*[1] and the *Yü Kung*,[2] are of much scientific importance, for astronomy and geography[a] respectively) it is quite certain that they are of Chou, and probably not even of early Chou, provenance.[b] Creel[c] puts forward the suggestion that the conception of a Hsia State became concretised as a semi-conscious fiction forming part of the campaign of propaganda with which the Chou people prepared and then legitimised, or sanctified, their conquest of the Shang.

In this formative period of Chinese society, modern investigations (especially those of Eberhard, 1–3) have distinguished several separate components of the developing civilisation. Six such basic cultures have been provisionally identified: (1) the northern one (which had a predominating influence on the Yangshao and Lungshan people), 'proto-tungusic' in character; (2) a north-western one, influenced by the nomads and 'proto-turkic'; (3) a western one, 'proto-tibetan'; and (4–6) three southern or south-eastern ones, originally coastal, which brought in oceanic influences, and may be grouped together under the term 'Yüeh' (later the name of a State in the south-east); these have been called 'proto-thai'.[d]

Although these studies are still at an early stage, it seems possible to distinguish some of the culture-traits which these ancient groups contributed to Chinese culture as a whole. Thus the northern complex seems to have included shamanistic religion, bear-veneration and fox-myths, a matriarchal form of society,[e] and the use of pit-dwellings and bone arrow-heads. The culture of the Shang dynasty might be considered a fusion of this with a complex of elements coming from the south-east and south. What Eberhard calls the Yüeh culture had certain Indonesian affinities; it was maritime and fluvial, using long-boats, warships, and communal houses; it had boat-races (cf. the later Dragon Boat festival), dragon-myths, serpent-worship (*nāgas*), veneration of sacred mountains (most characteristic in later Chinese civilisation), dog-magic (cf. the later straw dog images) and bronze drums. The cross bow (in primitive form) was its weapon, bark cloth and tattooing were known, records were kept on knotted cords (*quipu*), agriculture was in burnt forest clearings (*milpa*), and there were spring and autumn festivals of mating (cf. Granet, 1). The southern complex brought wet-rice agriculture, irrigation and slope-terracing, the domestication of the water-buffalo, ancestor-worship, pig-sacrifices, fertility cults and poisoned weapons. Bamboo, iron and lacquer all came up from the south.

[a] Herrmann (1), Map 11.

[b] But, as will be explained in Section 20*e* below, the *Yao Tien* contains certain astronomical statements which may rest upon a tradition going back to the Shang or earlier.

[c] (1), p. 105.

[d] Herrmann (12) has attempted to chart roughly their geographical boundaries.

[e] Note the persistence of radical no. 38, 'woman', in the modern word for family name, *hsing*[3] (R. Wilhelm (1), pp. 34, 59).

[1] 堯典 [2] 禹貢 [3] 姓

Though the influence of the north-western and western complexes was felt throughout the second millennium, it would seem to have made its greatest contribution at the time of the Chou conquest of the Shang. The north-western style was primarily patriarchal nomadism, involving the worship of heavenly bodies (astral religion), horses and horse-sacrifices, tent-dwellings, tumulus-graves, earthenware drums and the levirate. The western complex ('proto-tibetan') has been hardest to analyse, but cremation, polyandry and the couvade[a] seem to have formed part of it.

What exactly may be the relation between these ancient culture-complexes and the tribal peoples who inhabit large areas of the western and south-western provinces today (the Miao,[1] Lolo[2] or Nosu,[3] Chiang,[4] Man,[5] etc., in Szechuan, Yunnan, Kweichow and Kuangsi),[b] and whose customs show clear survivals of many features from ancient Chinese life (dances at mating festivals, use of poisoned arrows,[c] etc.) remains a subject of great research possibilities.

(c) THE CHOU PERIOD, THE WARRING STATES AND THE FIRST UNIFICATION

Little is known of the origin of the Chou people, except that they were from the western regions (somewhere in the present provinces of Kansu and Shensi) and that they were less advanced culturally than the Shang, whom they admired.[d] Their conquest of the Shang area, which was roughly speaking the Yellow River valley, parts of the North China plain, and the lands between the Huang-Ho and the Huai River, took place at a time fairly close to the Aryan conquest of India. But though the Aryans were also bronze-age chariot-users, there is no evidence suggesting that the Chou came from anywhere west of China proper. They continued the Shang traditions of bronze-working, pottery and textiles, and took further steps in the development of the written language. Though possibly of pastoral antecedents, they quickly adopted the thoroughly agricultural character of the unfolding Chinese civilisation.[e] The new astronomical evidence arising from the study of eclipses mentioned on the oracle-bones (Tung Tso-Pin, *1*; Dubs, 26), indicates that the conquest took place in the latter half of the -11th century and not 100 years earlier as was traditionally supposed.

The outstanding feature of the Chou period was its systematisation of bronze-age proto-feudalism which had been sketched out under the Shang, until it became almost as fully developed as in the typical feudal period of Europe. The empire (as it had

[a] Cf. Graham (3).

[b] Cf. Graham (1, 2); Morse & Yen (1); Cook (1); Yang Ching-Chi (1); Chêng Tê-Khun & Liang Chhao-Thao (1). There are many important contributions on this subject by Academia Sinica scholars such as Fu Mou-Chi.

[c] Cf. Fêng Ta-Jan & Kilborn (1) (aconite).

[d] Shih Chang-Ju (*1*) has recently gone far to identify the Chou people with a red-pottery culture which had previously been known but not associated with them.

[e] Herrmann (1), Map 13.

1 苗 2 羅羅 3 羅蘇 4 羌 5 蠻

now become) was partitioned into fiefs, and these were held by the new aristocratic class, in a way somewhat reminiscent of the granting of domains after the Norman conquest of England. Much of the Shang population was deported to the dukedoms of Lu[1] and Chhi,[2] and one of the remaining Shang princes was invested with the fief of Sung[3] in order that the ancestral sacrifices of the former dynasty might be kept up.[a] The whole system rested on the work of the peasant-farmers in agricultural production, in which corvée labour on the lord's lands was a prominent feature.[b] Tribute was passed up from rank to rank in the feudal hierarchy, ultimately reaching the imperial court. As in the times of the Shang, the various feudal lords were expected to make journeys to the capital from time to time and to reside there for stated periods. A kind of chivalry developed.[c]

During the course of the first three centuries of the first millennium before our era, the Chou society began gradually to manifest a certain instability. Wars with the barbarians continually increased as the settled population inevitably sought expansion, and the Hunnish peoples to the north and north-west began those restive nomadic incursions which were to occupy the defensive military force of China for more than a thousand years. Moreover, the advancing cultural process did not lack internal opposition and, as Lattimore says,[d] many of the 'barbarians' against whom the feudal princedoms were constantly fighting (and with whom they often made alliances against each other when it suited them), such as the Ti,[4] the I[5] and the Jung,[6] were probably simply the less advanced groups of Chinese.

The form in which this instability showed itself was the collapse, in the −8th century, of the fiction that a tightly knit feudal empire existed. In −771 the emperor Yu[7] was killed by the army of one of the smaller princedoms, allied with barbarians;[e] and his successor Phing[8] had to move the capital from approximately Sian east to Loyang, leaving the western loess region to the great fief of Chhin,[9] now more properly called a State. During the next few centuries the domain actually ruled by the Chou Emperor was very small indeed, and some twenty-five feudal semi-independent States, acknowledging but a shadowy authority from Loyang, contended among themselves for the hegemony (see Table 6 and Fig. 12).[f] The first to win this leadership, which involved the presidency of the princes at the imperial capital, was the State of Chhi[10] in Shantung. This State had a peculiar quality of its own; it was the main source

[a] See Table 6.
[b] Cf. *Tso Chuan*, Duke Chuang, 30th year, Couvreur (1), vol. 1, p. 200.
[c] Cf. *Tso Chuan*, Duke Hsüan, 12th year, Couvreur (1), vol. 1, p. 611 and Duke Hsi, 13th year, Couvreur, p. 288.
[d] (1), p. 319.
[e] *TH*, vol. 1, p. 102.
It is of interest that six of the State names in the table were also names of later dynasties, and that seventeen of them later became common family names. The best map of the State boundaries is no doubt that in O. Franke (1) at the end of his first volume.

1 魯 2 齊 3 宋 4 狄 5 夷 6 戎 幽
8 平 9 秦 10 齊

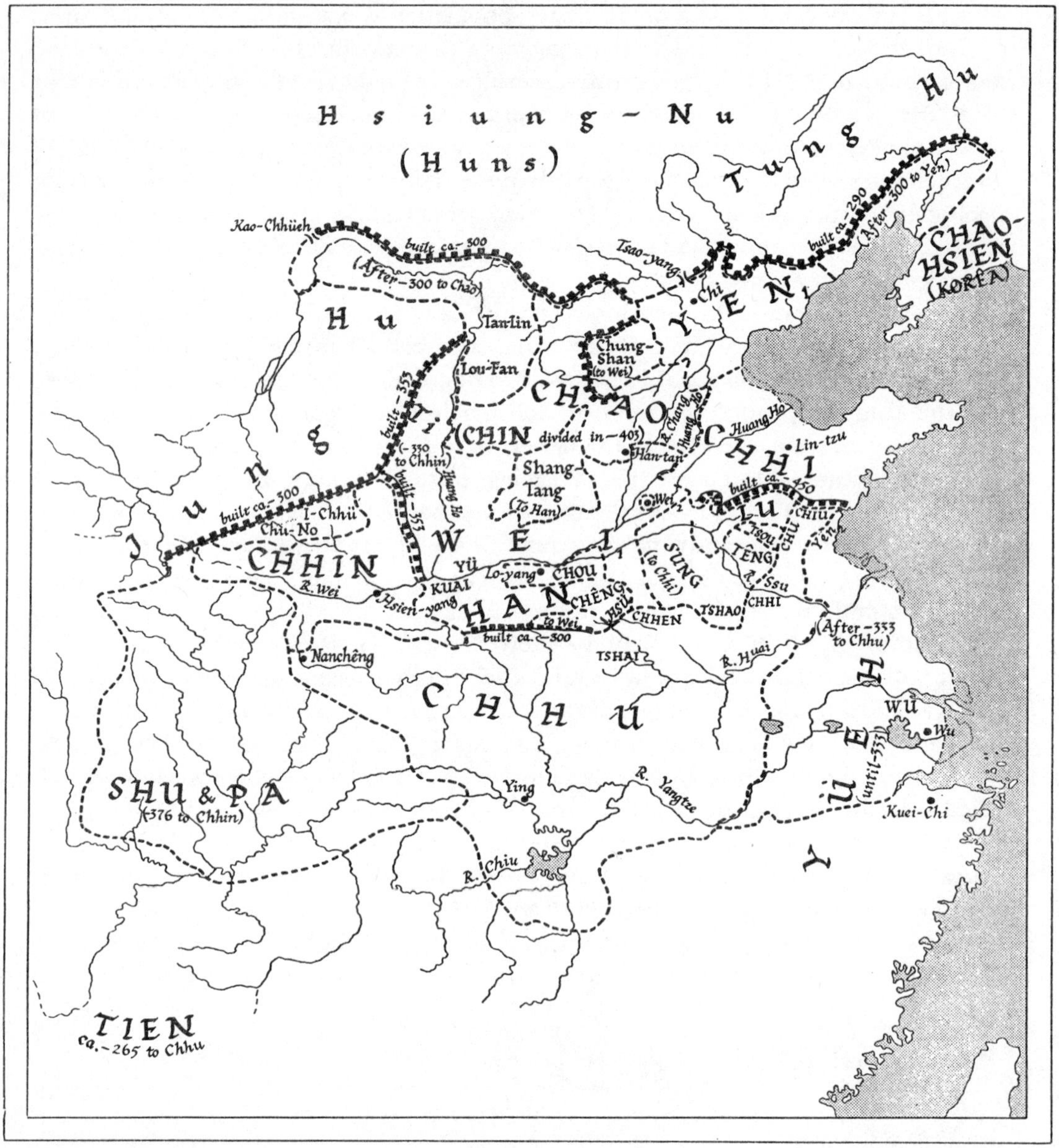

Fig. 12. Approximate boundaries of the Feudal States about the beginning of the −3rd century (after Herrmann, 1). In order to show the exact locations of the smaller feudal states a map on a much larger scale would be necessary. The walls shown are not all those which were built, but only the more important ones; for further details see Lattimore (1).

Besides that in Lattimore (1), p. 336; cf. maps in *TH*, vol. 1, no. 6, opp. p. 70, and no. 7, opp. p. 190.

of salt, made by evaporating sea-water at the coast, and acquired great wealth through the 'nationalisation' of this industry. It also had a leading position in the working of iron.

Though some authorities believe that iron became known in China as early as the −8th century (Janse (1) thinks *c.* −685), the generally accepted date for the first reference to it is −513,[a] though even this is considered a little too early by others (cf. Sect. 36). In any case there can be no doubt that its use began and quickly spread about the middle of the Chou period or somewhat before, and it seems therefore probable that it was one of the most important factors in disintegrating early Chou feudalism, and favouring the rise of independent States.[b] The fact that the first State to achieve the hegemony was also the first to be associated with iron-working, a process not available exclusively to the imperial power, may thus be rather significant. In course of time, the knowledge of the processes involved spread widely, rendering the States more uncontrollable than ever. As for Chhi, it always retained a definitely magical-scientific tradition, which we shall often have occasion to note later (cf. Sects. 13*c*, 20*f*, 26*b*, 33).

The hegemony passed afterwards to Sung, Chin,[1] Chhin and Chhu.[2] During the *Chhun Chhiu* period,[c] at which we have now arrived, the feudal hierarchy was organised in five grades[d] with titles which are usually translated duke, marquis, count, viscount and baron. The royal title of *wang*[3] was borne only by the Chou 'emperor'. The lesser nobility, not in any of the foregoing ranks, were called *chün-tzu*[4] as opposed to the plebs, the *min*[5] or 'little people' (*hsiao jên*[6]) of the classics.[e] It was presumably about this age that the custom grew up of dividing the mass of the population into the famous four groups *shih*, *nung*, *kung*, *shang*,[7] i.e. the lesser nobility or gentry, knights and scholars, then the peasant-farmers, then the artisans, and last of all, the merchants. The great significance of this order of classes (if the word is allowable) will have to be referred to again (Sect. 48). What may here be pointed out is that the last of these classes, the merchants, was given exactly the same name as that of the ancient dynasty.[f] H. Wilhelm[g] has plausibly suggested that in fact this was because the nobility of the *ancien régime*, though removed to particular locations such as Sung, and depressed in the Chou society, was not reduced completely to agricultural labour,

[a] *Tso Chuan*, Duke Chao, 29th year, Couvreur (1), vol. 3, p. 456.

[b] This was first pointed out by Kuo Mo-Jo (2).

[c] Hermann (1), Map 15.

[d] *Kung*,[8] *hou*,[9] *po*,[10] *tzu*,[11] *nan*.[12]

[e] This distinction later had a moral or ethical significance, *chün-tzu* gaining the meaning of 'gentleman', by which word it is often, unsatisfactorily, translated. Another expression for the ruling families was *pai hsing*[13] (the hundred names) as opposed to the common mass (*shu jen*[14]) or 'black-haired people' (*li min*[15]). In the course of ages the first of these lost its status, and in the form *lao pai hsing*[16] came to mean in our own time just all ordinary folk in general. The point was that in ancient times only members of the ruling class had names at all.

[f] *Shang*, perhaps 'the talkers' (K734).

[g] (1), p. 50.

1 晉 2 楚 3 王 4 君子 5 民 6 小人
7 士農工商 8 公 9 侯 10 伯
11 子 12 男 13 百姓 14 庶人 15 黎民 16 老百姓

Table 6. *The feudal States of the Chou period*

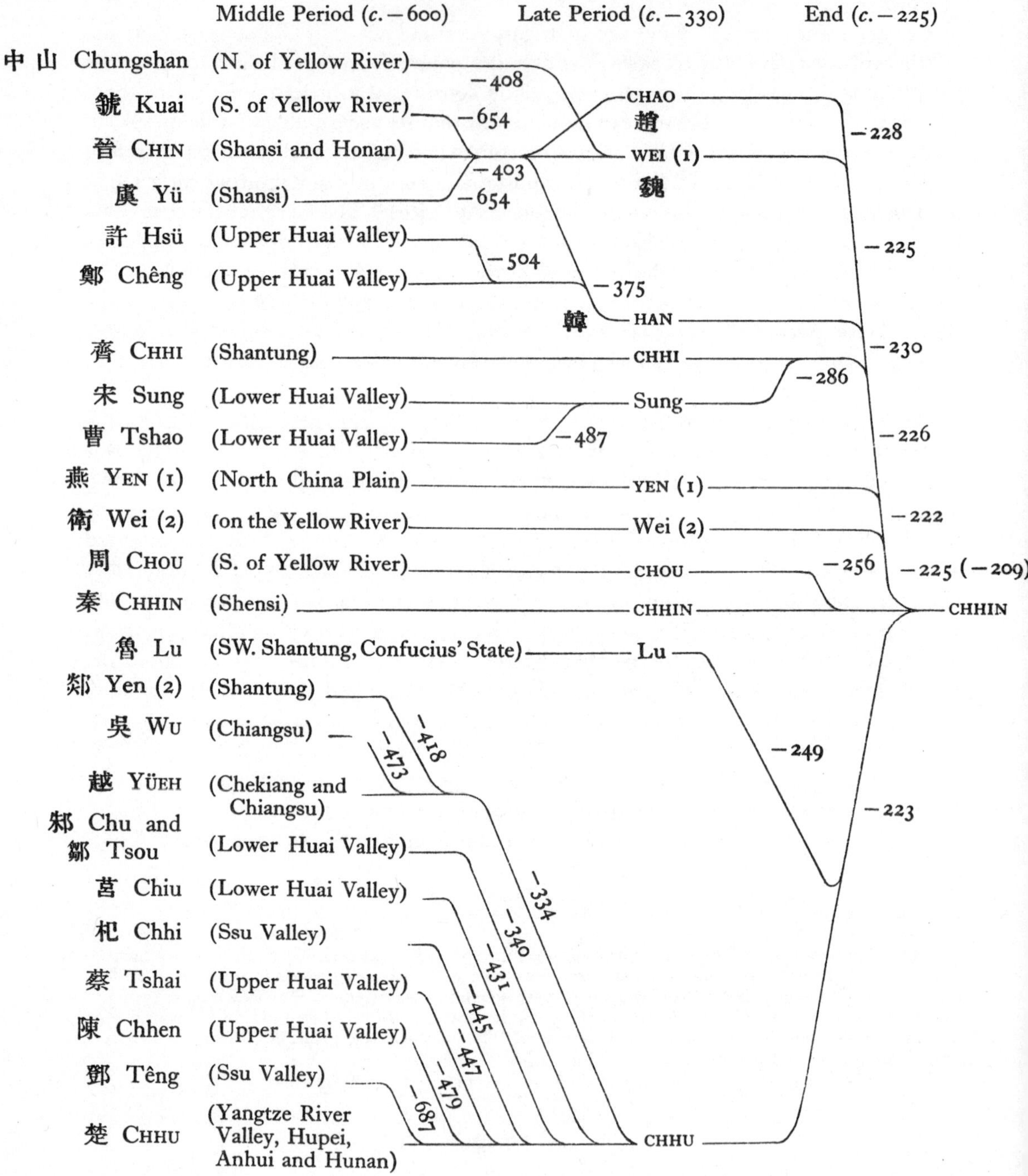

but found an outlet in commerce, scholarly activity, education and the embryonic beginnings of bureaucratism. As we shall see, he explains in this way some of the most interesting features of later Taoist thought.

By the −6th century we are indeed entering the greatest period of intellectual flowering of ancient China. The 'hundred schools' of philosophers were at their height between −500 and −250. Without anticipating the careful study which we must give them, from the standpoint of the history of scientific thought, we may refer to them here in their historical setting. Increasing population pressure, increasing conflicts with the barbarians and among the feudal States themselves, growing unrest among the people and the technological revolution caused by the coming of iron, all led to a demand for advisers on the part of the feudal lords, who often felt themselves at a loss in unfamiliar situations. Such at any rate are the only explanations which are available for the great rise at this time of philosophers (who have often been compared, not very happily, with Greek sophists) travelling from capital to capital with their disciples, and being prepared to take up positions as advisers and diplomats upon request.[a] From this time date the earliest Chinese books to which the names of individual authors are attached.[b]

Confucius traced his own family descent to the Shang. His short tenure of office in the State of Lu is so well known that it need only be mentioned—most of his life was spent in imparting his social teachings to his disciples. One might perhaps say that in his strong political support for the Chou emperor, constantly emphasised by his idolisation of the founders of the dynasty, he was fighting for the only unifying factor which he could see in the society of his day—as if someone in 1946 declared their support for the ideal of the United Nations. Besides the Confucians (the *ju chia*[1])[c] there were many other schools[d]—the Taoists (*tao chia*[2]), with a complex and subtle set of conceptions to which we shall have to give particular attention since it lies at the basis of all subsequent Chinese scientific thought—the Mohists (*mo chia*[3]), with their doctrine of universal love—the Logicians (*ming chia*[4]), arguing about hardness and whiteness—the Legalists (*fa chia*[5]), winning the favour of ruthless tyrants by their advocacy of draconic severity in administration—and others as well.

The beginnings of academies of scholars go back to this very early time in China. The most famous of them was the Academy of the Gate of Chi (Chi-Hsia[6]) at the capital of the State of Chhi;[7] here scholars from all the other states as well as Chhi itself were welcomed and provided with quarters and maintenance.[e] The more important of them held the rank of Great Prefect. Founded by King Hsüan[8] of Chhi about −318, this Academy united together many of the most brilliant scholars

[a] See Nagasawa (1), p. 30.
[b] Article on this point by Lo Kên-Tsê (*1*).
[c] Cf. Nagasawa (1), pp. 32 ff.
[d] Cf. e.g. Nagasawa (1), pp. 45 ff., 49 ff., 53 ff., 59 ff.
[e] Chavannes (1), vol. 5, pp. 258 ff.; Duyvendak (3), pp. 73 ff.; Dubs (7), pp. 22, 26, 27; Fêng Yu-Lan (1), p. 132.

1 儒家 2 道家 3 墨家 4 名家 5 法家 6 稷下
7 齊 8 宣

of the age towards the end of the −4th century and the first half of the −3rd. There was Tsou Yen,[1] the nature-philosopher and founder of the Yin-Yang school,[a] there were Taoists such as Thien Phien,[2] Shen Tao,[3] Phêng Mêng[4] and Yin Wên Tzu,[5] and perhaps even Chuang Tzu[6] himself; there was Sung Hsing[7] the Mohist; and Mencius[8] and Hsün Tzu,[9] Confucians.[b] It is interesting to notice the similarity in date between this Academy and the Academy founded by Plato at Athens which was already formed at the time of his death in −348, and continued for some centuries afterwards. The Stoic Academy founded by Zeno of Citium in the Painted Porch must have originated shortly after −300. So one may say that the Chi-Hsia Academy originated at a time just between that of the Peripatetics and that of the Stoics—another remarkable example of important events occurring almost simultaneously in Eastern Asia and Western Europe.

Side by side with these great intellectual developments, we find also many cultural advances, so many that this age has always been looked upon as China's 'classical' period. There were advances in craftsmanship and methods of production, the appearance of the animal-drawn plough and a great expansion of irrigation; the multiplication of market-places and the intensification of a money-economy tending to replace the ownership of land and corvée-labour as source of wealth and power. There were also military developments which, with the use of iron, made possible the rise of independent feudal States and sharpened their conflicts, particularly the perfection of the cross-bow. This invention seems to have been made earlier in China than anywhere else, in spite of China's relative lateness in acquiring bronze and iron. In the hands of horse-archers, who also had the old composite sinew-backed short bow, and used tactics consciously copied from the western and northern nomads, it put an end to the ancient chariot-fighting techniques which had been characteristic of the nobles of Shang and early Chou (cf. Sect. 30).

During the Warring States period,[c] a process of consolidation came about, the smaller and weaker States being continuously absorbed by the stronger ones. There were cases of fission, as when Chin split into Chao,[10] Wei,[11] and Han,[12] but from −500 onwards the general rule was intensification of power at a smaller and smaller number of focal points. Industrial concentration and the control of systems of hydraulic engineering played important parts in this process. By about −300 the two most important States were Chhu in the south (occupying eastern Szechuan, Chiangsi and Hunan) and Chhin (in Shensi) in the north-west. Chhu occupied a region, in and about the Yangtze valley, where water was available in plenty, while Chhin, based on the Wei valley and the loess country, was in great need of works of

[a] See Sect. 13*c* below. Cf. Fêng Yu-Lan (1), Vol. 1, pp. 159 ff.

[b] The Legalists do not appear to have participated in the Chi-Hsia Academy; Duyvendak (3) believes that their ideas originated at the court of King Hui of Wei, or in the State of Chhin itself, i.e. further west.

[c] Cf. Herrmann (1), Map 16.

1 騶衍　2 田駢　3 愼到　4 彭蒙　5 尹文子　6 莊子
7 宋銒　8 孟子　9 荀子　10 趙　11 魏　12 韓

irrigation. This it was which strengthened the central State power at the expense of the smaller feudal lords, and one may say that the process of replacement of feudalism by feudal bureaucratism had its beginning in the State of Chhin. These changes are associated with the names of King Hsiao[1] and his minister Kungsun Yang[2] (the Lord of Shang Yang), who, if the attribution is justified, was responsible for one of the principal writings of the Legalist school. It is at any rate certain that the doctrines of this group were put into practice in Chhin.[a] The realisation that the old feudalism was dying is shown in Hsiao's first decree, which threw open the buying and selling of land and the alienation of feudal fiefs. There followed the appointment of government officials, instead of the former donation of fiefs, to administer new lands won by the State. The whole population was militarised, a police and passport system introduced, and general coercion by draconic punishments became the order of the day.

The other feudal States were naturally much alarmed by the rise of so highly organised a power, and every kind of means was employed against it, by alliances and otherwise. One of the most famous episodes of this kind was the building of the Chêng-Kuo[3] canal, which connected the Ching[4] and Lo[5] rivers north of the Wei River and Sian (see Sect. 28*f*).[b] It is related in the *Shih Chi* (ch. 29) that Wei and Han sent a hydraulic engineer, Chêng Kuo, to suggest to Chhin the building of this canal, in order to keep its population occupied and reduce its military strength. But the result was that a vast area was irrigated (it can be seen even on a modern map); the Chhin rulers eventually realising that the additional wealth in grain would enable them to build up army reserves, and that they would be stronger strategically by the water-transport possibilities provided.[c]

The construction of large waterworks became a settled policy on the part of Chhin. A little later, after the occupation of the States of Shu[6] and Pa[7] (late-comer organisations in the Szechuan basin) in −316, a magnificent project was undertaken to irrigate the Chhêngtu plain, which is still functioning today. To Li Ping,[8] the Chhin Governor of Szechuan, and his achievement, we shall return in the proper place (Sect. 28*f*).

In −318 a confederation of States was overwhelmingly defeated in the field by Chhin,[d] and in −312 the same fate overtook the forces of Chhu.[e] Han and Wei were conquered in −293 and Chao in −260. In −256 the ancient Chou domains were entirely surrounded.[f] Then came a pause while internal adjustments were occurring in Chhin; King Chêng[9] came to the throne in −246 and made preparations for the final effort aided by two outstanding ministers, first Lü Pu-Wei[10] (originally a merchant, and leader of a group with scientific and technological interests[g] which

[a] Cf. *TH*, vol. 1, pp. 160 ff.
[b] Cf. Chi Chhao-Ting (1), p. 76.
[c] See below, p. 115.
[d] *TH*, vol. 1, p. 169.
[e] *TH*, vol. 1, p. 170.
[f] *TH*, vol. 1, p. 190. One of the best accounts in English of the whole process of unification is that by Lei Hai-Tsung (1), which merits reprinting in less inaccessible form.
[g] Cf. *TH*, vol. 1, pp. 193 ff.

1 孝 2 公孫鞅 3 鄭國 4 涇 5 洛
6 蜀 7 巴 8 李冰 9 政 10 呂不韋

produced the book *Lü Shih Chhun Chhiu*[1]); and after his fall, Li Ssu.[2] Then events followed rapidly. Han was formally annexed in −230, Chao in −228, Chhi in −226, and Wei in −225; and finally the last and greatest of the opponents, Chhu, in −222. Chêng then found himself at the head of a united China, and adopted the title Chhin Shih Huang Ti,[3] or the First Emperor of Chhin.

(*d*) COMPARATIVE RETROSPECT

Having arrived at this point it seems desirable to look back in order to form an idea of the comparative chronology of events in China and in the other parts of the civilised world.[a] As we have seen, history in China cannot be pushed back beyond about −1500. In comparison with this, the other river-valley civilisations of the ancient world are much older. The Neolithic begins in Egypt about −6000, but it is short, since copper became known there (much earlier than anywhere else) about −4500; in Europe it lasted three thousand years. European Neolithic culture of the north has some clear resemblances to Chinese, particularly in the use of spiral designs on pottery, which in both cases replace the geometrical and zoomorphic designs of southern Europe and the Middle East. It is provoking that so far no cave-paintings analogous to the famous Neolithic works of art of Western Europe and North Africa have come to light in China, since it would be just such paintings, if the Yangshao or earlier Chinese ever made them (and it is hard to believe that they did not), which might show us the earliest possible stages of the pictograms found so much more highly developed on the Shang oracle-bones.

By −3500 we are at the beginning of Chaldean Ur,[b] and two hundred years later at the beginning of the Egyptian Old Kingdom.[c] The Middle Kingdom is already over, and the Hyksos invasions in progress by −1700, i.e. just about the time of the beginning of the Shang. The period from about −3000 to −2000 sees also the flourishing of Akkad in Mesopotamia, the age of bronze in Egypt, Iran, Mesopotamia and the Indus cities such as Mohenjo-daro. Bronze reaches the eastern Mediterranean towards the end of the period, and Western Europe about −1900. Thus the Celtic tumulus-builders of the West European bronze age are the real contemporaries of the Shang bronze-founders at the other extreme end of the heartland. We shall later consider the suggestion that the similarity went far into social organisation. In any case it is sure that from the main foci of ancient civilisation in Egypt, the Fertile Crescent, and the Indus valley, arts and techniques spread both west and east.

[a] I have already referred to the helpful book of Furon (1); here should also be mentioned the brilliant expositions of Gordon Childe (1–6), the old though still useful book of Breasted (1), and the latest account of the Indus Valley civilisations by McKay (1).

[b] Cf. Woolley (1).

[c] Cf. Wallis Budge (1).

[1] 呂氏春秋 [2] 李斯 [3] 秦始皇帝

While the city of Anyang was in full life, about −1400, a non-Semitic but cuneiform-using people of Asia Minor, the Hittites,[a] with their capital near modern Ankara, first developed the manufacture of iron (though isolated pieces, probably of meteoric origin, had long been known, especially in Egypt). A couple of hundred years later the industry had spread to Nineveh, and at the turn of the millennium the Etruscans took a knowledge of it to Italy. Penetration both east and west was slower. The European iron age is divided into the Hallstatt[b] period, from −900 to −500, and the La Tène[b] age, from −500 to −100. To this latter age have been ascribed the familiar semi-legendary figures of our youth, Conchobar and Cuchulainn and Deirdre. One can see, therefore, that the appearance of iron among the Chinese feudal States, at about −600, was of late Hallstatt time, and indeed the latest appearance of iron in any of the great culture-areas. In spite of this fact, we shall see below (Sect. 36) how rapidly the Chinese surpassed all other parts of the world in iron technology.

The close coincidence in date between the appearance of many of the great ethical and religious leaders has often been remarked upon: Confucius, *c.* −550; Gautama (Buddhism), *c.* −560; Zoroaster (if a historical personage), *c.* −600; Mahāvīra (Jainism), *c.* −560, and so on. But the Chhun Chhiu period was also contemporary with many important political events, such as the taking of Nineveh by the Medes in −612, the fall of Babylon to Cyrus in −538, and the invasion of the Punjāb by Darius in −512, all examples of Iranian expansion.[c] At the beginning of the Warring States period, the Greeks checked Iranian expansion westwards (−480), and the middle of the −5th century saw the erection of the Athenian Parthenon. The concluding stages of the Warring States time are contemporary with many outstanding events, such as the conquests of Alexander the Great (*c.* −327), the foundation of the Maurya dynasty in India and the beginning of the reign of Asoka[d] (−300 and −274 respectively), and the Punic Wars in the Mediterranean (−250 to −150) which overlap with the first unification of China under Chhin Shih Huang Ti. But the beginning of the Roman Empire (−31) does not take place until well into the Han dynasty, to which we must now turn, after a few words on the short-lived triumph of the Chhin.

[a] Cf. Delaporte (1), vol. 1, p. 59; vol. 2, p. 186.
[b] Site-names; cf. Hubert (1, 2).
[c] Cf. Huart & Delaporte (1).
[d] V. A. Smith (1); Masson-Oursel, Willman-Grabowska & Stern (1).

6. HISTORICAL INTRODUCTION. THE EMPIRE OF ALL UNDER HEAVEN

(a) THE CHHIN DYNASTY

As soon as the new unified empire was established, strong measures were taken to circumscribe the old feudalism and to set up that bureaucratic type of government which, prefigured in the Chhin domains, was to characterise the whole of subsequent Chinese history.[a] Holders of large feudal estates were expropriated, in the interests of control by high officials, and in so far as a nobility of birth continued, its representatives were forced to reside at the capital (Hsienyang[1] near modern Sian in Shensi). The whole country was divided into thirty-six, and later forty-one, prefectures, with military governors and civil administrators.[b] The farmers were given more rights over their land than they had had in feudal times, but were subjected to clearly specified taxes. Everything was standardised, from weights and measures to the gauge of carts and chariots. But merchants were discriminated against by means of sumptuary and other laws. A network of tree-lined roads was begun, and some of the walls which had been built at various times by the feudal States (see Fig. 12 and below in Sect. 28*c*)[c] were, in the north, joined together to form a continuous defence, the Great Wall (Wan Li Chhang Chhêng[2]), the building of which has remained a focal point in Chinese folksong ever since.[d]

Lattimore (2) acutely says that the Great Wall was intended to check the drift of Chinese groups towards coalescence with nomadic life, or the formation of mixed economies, at least as much as to keep the nomads out. It was realised that any fusion would be likely to react back later in the form of 'tribal' inward military pressure. How justified this premonition was can be seen from the many subsequent centuries during which large parts of North China were held as independent States by barbarian or semi-barbarian nomadic houses able to call upon Chinese technicians and peasants as well as the resources of their own pastoral background (cf. the Northern Wei, Liao and other dynasties, pp. 119, 120, 133 below).[e]

There was, in fact, a fairly clear line of cleavage between the territories and peoples which could advantageously be included within the centralised agrarian empire, and

[a] Cf. *TH*, vol. 1, pp. 209 ff.

[b] Herrmann (1), Maps 18, 19.

[c] Cf. Lattimore (1, 2); Puini (1).

[d] Cf. the *Ballad of Mêng Chiang Nü*,[3] on which see Hsiao Yü (1); Needham & Liao (1); Nagasawa (1), p. 221; Ku Chieh-Kang (1) (tr. Hummel, pp. 124 ff.).

[e] A striking example of how far technology could be involved in this is the invention of the automatic grain-milling wagon by Chieh Fei and Wei Mêng-Pien for the Hunnish Later Chao dynasty, *c.* +340 (see below, Sect. 27*c*).

1 咸陽 2 萬里長城 3 孟姜女

those which could not. Thus neither the Chhin nor the Han could long retain control of the Ordos Desert. Hence the course taken by the Wall depended on the possibilities of agricultural productivity. Dividing the steppe from the sown, it was designed to keep the peasant population in no less than to keep the nomad horsemen out. This subject may be pursued further in the penetrating studies of Lattimore (1) on the general relations between Chinese agricultural and Nomadic pastoral life and culture. As a piece of military technique the Wall seems to have fulfilled its purpose. Large bodies of horsemen could only pass through by the capture of one or more gates, or by demolishing the structure at some point or building ramps up to it, and all such procedures would take time, permitting the arrival of Chinese reinforcements. Conversely, emigrating families would be subject to control at the gates where the Wall crossed routes in common use.[a]

Driven to find still further employment for the armies which he had formed, the first emperor pushed further to the south than any element of the Chinese *oikoumene* had done before, and in a campaign remarkable for its geographical extent, conquered the coastal province of Fukien (then as now very forested and mountainous), the two Kuang provinces (Kuangsi and Kuangtung), and even Tongking in what is now Indo-China. This feat was made possible by the construction of a canal crossing the watershed between the Yangtze and the West River (cf. Sect. 28*f*).

Guided by his minister, Li Ssu,[1] Chhin Shih Huang Ti went on to standardise the written language.[b] According to the account universally accepted for centuries in China, he also ordered the destruction, by burning, of all books except the imperial archives, and works on medicine, divination and agriculture, considering that the literature of the day was too much impregnated by feudal ideas.[c] It is also said that he put to death a large number of the literati. But there seems some reason to doubt these stories, which are paralleled by others told of similar hero-kings in other parts of the world.[d] On the other hand, there is no need to question the emperor's interest in magic and alchemy, which is well authenticated, and to which we shall have to return (Sects. 10 and 33). He worked indefatigably, handling '120 pounds of reports' daily, and travelled a great deal; his palace was described as one of the wonders of the world. It is interesting to note[e] that the processes of unification by communication

[a] The best descriptions of the Great Wall which I have seen are those of Clapp (1) for the eastern parts between Shanhaikuan on the sea and Ninghsia, and of Geil (3) for the western part from Ninghsia to the 'Tibetan' end at Chiayükuan, the gate on the Old Silk Road. On this western part there is also information in Cable & French (1); Langdon Warner (1); and Needham & Needham (1). Cf. Figs. 13, 14, 15; and for other illustrations Warwick (1). An excellent map is found in the first of these articles; it should be compared with Herrmann (1), Map 17, to show its relation to the rest of Asia.

[b] Li Ssu's biography (ch. 87 of the *Shih Chi*) was translated long ago by Pfizmaier (30), and anew by Bodde (1), who has made a modern study of him.

[c] The list of exempted subjects (*Shih Chi*, ch. 6, p. 23*b*; ch. 87, p. 7*a*) is, literally, 'medicine, pharmaceutics, divination by scapulimancy and milfoil, agriculture, and forestry' (Chavannes (1), vol. 2, p. 173; cf. Bodde (1), pp. 24, 162).

[d] This was the opinion of Gustav Haloun.

[e] With Bishop (2), p. 315.

[1] 李斯

systems was characteristic of at least two other great empires arising about the same time, namely, those of Persian Darius and Indian Chandragupta Maurya (−320). But only the second of these approached the Chinese State in one basic particular, namely, the importance attached to works of hydraulic engineering.

In −210 the Great Unifier[a] died, and was succeeded by his son, who proved a quite ineffectual ruler. Almost immediately revolts broke out, and after eight years of constant fighting, not so much with the adherents of Chhin as between rival claimants to the throne, Liu Pang,[1] an army officer from the east, founded the dynasty of the Han. With a short interregnum it was to last four centuries.

At the beginning the revolutionaries fought under the slogan 'Back to Feudalism', and the shadow of events then very recent is seen in the fact that some of their detachments called themselves the Avenging Army of Chhu.[b] Some of the former feudal States even set up administrations again for a year or two, but this was not the pattern of the future, and the real decision rested with the leader who could outdo all his competitors in staying-power. Liu Pang's own story illustrated the self-negating character of the draconic laws of Chhin, for having been in charge of a group of convicts who had escaped, and being himself therefore doomed to death, he deserted and placed himself at the head of one of those groups of 'bandits' whose fortune it proved many times in subsequent centuries to found new dynasties.[c]

(*b*) THE HAN DYNASTIES

However great the desire may have been in some quarters to return to the pure feudalism of the Chou, it soon became clear that this was impossible, although throughout the dynasty it was advocated by backward-looking scholars and those who had something to gain from it. The original settlement of Liu Pang, after he had ascended the throne as Han Kao Tsu,[2] was a compromise. A considerable number of smaller feudal tenures were confirmed, but in its broad lines the State system perpetuated the Chhin rather than the Chou system. All sorts of measures were taken against the holders of feudal estates or small 'kingships' (*hou, wang*[3]); a succession system was introduced which tended to divide them up at each inheritance,[d] and there grew up a practice of having officials of the central government reside at the seats of hereditary vassal princes (rather like the 'residents' at rajah's courts in British Indian times).[e] It became the set policy of the dynasty to enfeeble the vassal kingdoms by taking away some territory whenever a kinglet committed any mis-step.[f] Others

[a] Cf. Bodde (1).
[b] *TH*, vol. 1, p. 229.
[c] *TH*, vol. 1, p. 232. Ch. 33 of the *Chhien Han Shu*, concerning these events, has been translated by Pfizmaier (38).
[d] Cf. *TH*, vol. 1, p. 384.
[e] But of course they were not the agents of a foreign power.
[f] Dubs (2), vol. 1, pp. 295, 297; striking examples in Wilbur (1), pp. 326, 331, 355, 370 ff., 393, 419 and 444.

[1] 劉邦 [2] 漢高祖 [3] 侯王

PLATE IV

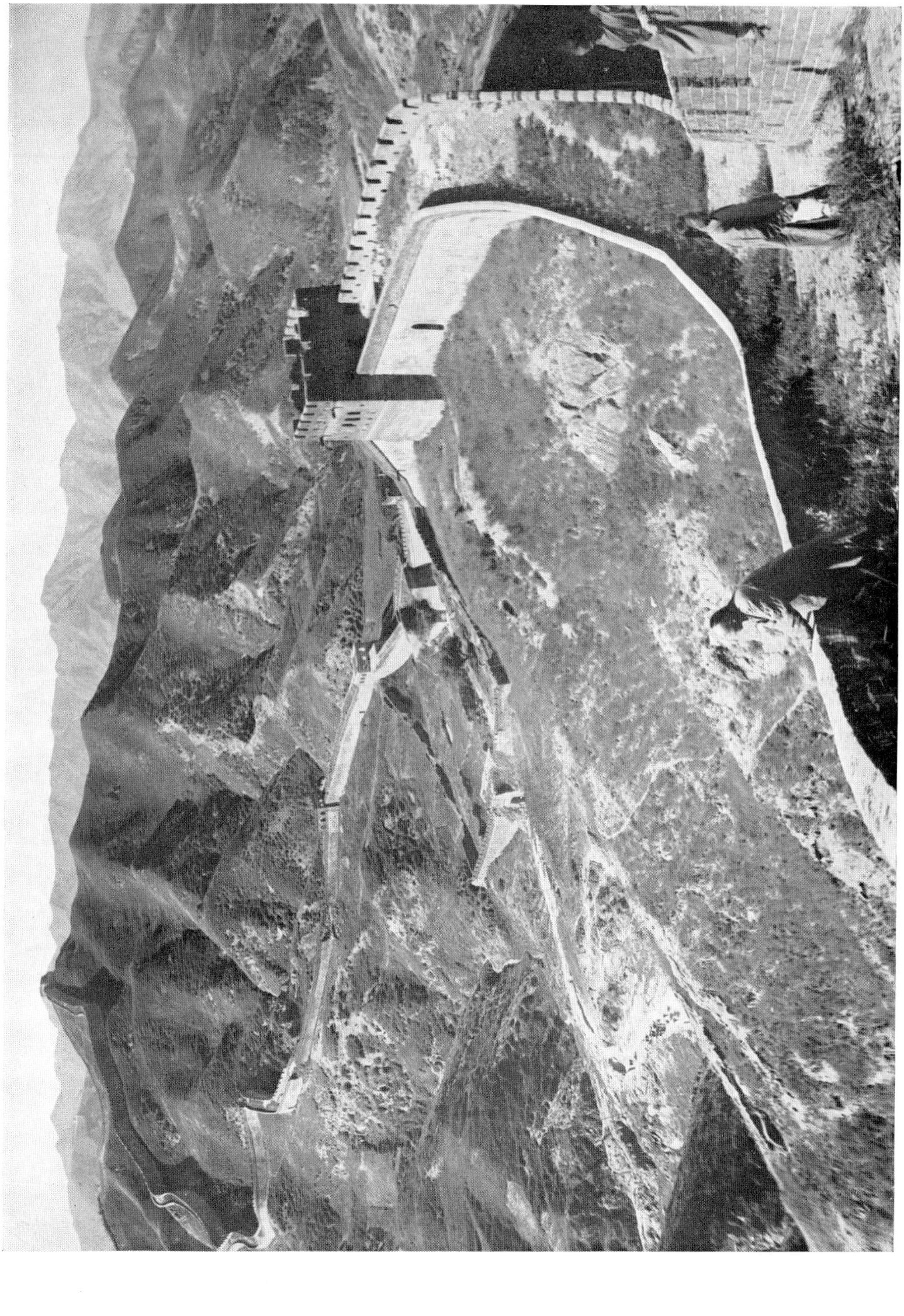

Fig. 13. The Great Wall near the Nankow Pass, north of Peking (photograph of about 1910: Ponting).

PLATE V

Fig. 14. The Fort of Chiayükuan at the western end of the Great Wall (orig. photo., 1943).

Fig. 15. Ramparts of a Han dynasty fort of stamped earth in the San-wei-shan mountains guarding the valley which runs down to Chhien-Fo-Tung (the Thousand Buddha Caves) near Tunhuang (photo. Alley).

(*kuan nei hou*,[1] lords 'within the passes') had to live on rents at the capital. Watched and supervised by the officials, the princes had become so unimportant by the time of the Hsin interregnum that they could then be reduced to private life with little difficulty.

From the new capital at Chhang-an (modern Sian),[a] built a few miles from the destroyed Chhin capital at Hsienyang, thirteen provinces were organised, along practically the same lines as those existing today.[b] Their administration required a great growth of the bureaucracy, and the need for administrators was met by a process which gave the Confucian school its permanent hold on Chinese society. 'The formation of the official class[c] of literati', says H. Wilhelm,[d] 'is one of the most dramatic chapters of Chinese history, and signifies the last and most decisive step in the integration of Chinese society. It was only possible by means of a renaissance of Confucianism, and it was because of this that Confucius came to occupy his place as the "uncrowned ruler" of all China.' There are many revealing episodes of this rapid development from feudalism to feudal bureaucratism. Thus the *Thung Chien Kang Mu* says, of the year −196:

(The chamberlain) Lu Chia,[2] constantly quoted the Odes and the Annals (*Shih Ching* and *Shu Ching*) to the Emperor, who ended by becoming exasperated. 'I conquered the empire on horseback', he cried, 'what is the good of these Annals and Odes?' Lu Chia replied 'That is true, but it is not on horseback that you will be able to govern it. The (ancient) emperors Thang and Wu got it by violence, but governed it following (the people's will). War and peace are two aspects of an eternal art....If the Chhin, having become the masters of the empire, had governed it in humanity and righteousness, if they had imitated the ancient sages, you would not have got it.' The Emperor changed colour, and said 'Show me then what it was that lost the empire for the Chhin, and how it was that I got it, and what it was that won or lost kingdoms of old'. So Lu Chia wrote a book dealing generally with the causes of the rise and fall of states, in 12 chapters, which he read one after the other to the Emperor, who never failed to praise them. The book was called *Hsin Yü*[3] (New Discourses).[e]

Compare the civilising efforts of Shusun Thung, in −201:

The Emperor had abolished the complex and difficult rites of Chhin...but the result was that when the officers drank together, they disputed about precedence, got drunk, shouted and banged their swords on the columns of the halls. The Emperor was disgusted. Shusun

[a] Herrmann (1), Map 21.
[b] Herrmann (1), Maps 22, 23.
[c] He uses the word 'class' here not in a technical sense, since the literati or scholar-gentry were never a class *inheriting* power and privilege in the sense of a feudal aristocracy. There was a fluidity of rise and fall into and out of it which varied in completeness from dynasty to dynasty.
[d] (1), p. 84.
[e] Ch. 3, p. 46*b*, tr. auct.; adjuv. Wieger, *TH*, vol. 1, p. 299. The passage comes originally from *Chhien Han Shu*, ch. 43, p. 6*b*. We still possess the *Hsin Yü*, see Forke (12), p. 6; tr. von Gabain (1).

[1] 關內侯 [2] 陸賈 [3] 新語

Thung[1] said to him 'Scholars may not be able to conquer an empire, but they can help to conserve it. I suggest that you convoke all the literati of Lu and instruct them to draw up an imperial code of rites.'...(After the first trial of it) the Emperor said, 'This day for the first time I see what imperial majesty means.'[a]

It seems fairly clear that the Confucianist school was able to step into the position of the new bureaucratic orthodoxy because the Legalists had become overwhelmingly identified with the excessively tyrannical rule of the Chhin state and later of the Chhin dynasty. In −206 the Chhin laws were all repealed, and simpler and more humane principles introduced.[b] Not all the practices of Chhin authoritarianism were dispensed with, however, since the first emperor at least felt the need of some of them; thus, though passports (*chuan*[2]) were abolished in −168,[c] we read of periodical 'great searches' (*ta sou*[3]) after closure of city gates, for strangers, criminals, beggars, etc.[d]

What is not so obvious is why the Confucians dominated so as to wipe out even the memory of other schools which had been equally interested in social problems, such as the Mohists. Perhaps these had been completely destroyed already by the authoritarian Chhin Legalists; only the Taoists were irrepressible, as we shall see.

An interesting account of the victory of Han Confucianism in the founding of the bureaucracy has been given by Dubs (3).[e] The examination system (of which more later, Sect. 48) may be said to have begun when Kao Tsu, at the instance of his chancellor Hsiao Ho,[4] asked his provincial administrators to recommend persons with excellent reputations and manifest virtue for official positions. Shortly after the accession of the Emperor Wu, in −141,

the Marquises of Wei-Chhi (Tou Ying) and of Wu-An (Thien Fên) became his lieutenant-chancellors, and made Confucianism flourish. When moreover (Tung) Chung-Shu[5] wrote (his famous) replies to the (examination) questions (set by the Emperor Wu, he advocated) promoting and making glorious (the teaching of) Confucius, and repressing and degrading (the advocates of) the hundred (other schools of) philosophy. The establishment of offices for a (government) university and schools, and the recommendation of (persons with) Abundant Talents, and of Filially Pious and Incorrupt (persons to the imperial government) by the provinces and commanderies, all arose from the proposal of (Tung) Chung-Shu.[f]

Tung Chung-Shu was a scholar who was always afterwards looked upon as one of the most important founders of bureaucratic practices and procedure; his most famous work was the *Chhun Chhiu Fan Lu*[6] (String of Pearls on the Spring and

[a] *Thung Chien Kang Mu*, ch. 3, p. 25*b*; tr. Wieger, *TH*, vol. 1, p. 281; eng. auct. The passage is based on *Chhien Han Shu*, ch. 43, pp. 15*b* ff.
[b] *Chhien Han Shu*, ch. 1A, tr. Dubs (2), vol. 1, p. 58.
[c] *Chhien Han Shu*, ch. 4, tr. Dubs (2), vol. 1, p. 252.
[d] *Chhien Han Shu*, ch. 6, tr. Dubs (2), vol. 2, p. 104.
[e] Cf. Nagasawa (1), pp. 9 ff., 105 ff.
[f] *Chhien Han Shu*, ch. 56, tr. Dubs (2), vol. 2, p. 341.

[1] 叔孫通 [2] 傳 [3] 大搜 [4] 蕭何 [5] 董仲舒 [6] 春秋繁露

Autumn Annals), still extant.[a] This was significant, for as the Han time went on, great attention was given to the surviving historical writings of the Chhun Chhiu (*Tso Chuan*) period. Assemblies of scholars,[b] which afterwards became famous, were held to determine a kind of constitutional case law and to decide upon precedents from the ancient writings; hence the meaning of the schools of textual criticism which now appeared. The first of these meetings, held in the Shih-Chhü[1] (stone canal) Pavilion of the palace, from which it takes its name, in −51, has been compared in importance with the Council of Nicaea in the West (+325).[c] The second, in +79, was held in the White Tiger Lodge (*Pai Hu Kuan*[2]), and the record of the discussions has been preserved until now in the *Pai Hu Thung Tê Lun*[3] of Pan Ku. Another notable assembly of Han times, that in which the nationalisation of the salt and iron industries was discussed, even more important for our interests, will be the subject of closer attention later (Sect. 48). But long before this date, the Legalists had been liquidated, as witness the action taken in −141:

> The lieutenant-chancellor, (Wei) Wan, memorialised 'Of those capable and good (persons) who were recommended, some have applied themselves to (and are conversant with) the sayings of Shen (Pu-Hai), of Shang (Yang), of Han Fei, of Su Chhin, and of Chang I. (Such persons) cause confusion in the government of the State. I beg that they all be dismissed.' The memorial was approved.[d]

The Han emperors now understood that central power, indifferent to any possible coalition of feudal lords, and able both to guard or extend the frontiers, and to carry through the largest public works projects, depended on a competent and learned civil service, recruitment to which could have nothing to do with birth. The 'carrière ouverte aux talents' was in a sense a Chhin and Han discovery. But they understood it almost too well. Eunuchs began in the Han to play an important part in politics, always condemned by the scholar-gentry, but always renewing their influence after every setback, especially at times when the family of the empress or the dowager attained power and importance. What must be one of the earliest references to eunuchs occurs in the *Tso Chuan* for the year −535,[e] but it was only under the Han that they became a dominant factor. In the early stages of the developing bureaucracy it was quite natural that provincial governors or viceroys should have a tendency to try to make their offices hereditary by founding great feudal houses (as happened exactly in the case of the Carolingian Counts where what were originally public offices

[a] Cf. O. Franke (3); Wu Khang (1); Shryock (1). His replies to the emperor's questions are in the passage from *Chhien Han Shu*, ch. 88, translated by Shryock (1), pp. 49 ff. Ch. 56, also concerning the matter, has been translated by Pfizmaier (45).

[b] Dubs (2), vol. 2, p. 271; Nagasawa (1), p. 127.

[c] The fragmentary report which still exists (*Shih Chhü Li Lun*) has been translated by Tsêng Chu-Sên (1).

[d] *Chhien Han Shu*, ch. 6, tr. Dubs (2), vol. 2, p. 28.

[e] *Tso Chuan*, Duke Chao, 6th year, Couvreur (1), vol. 3, p. 121.

[1] 石渠閣 [2] 白虎觀 [3] 白虎通德論

became imperceptibly hereditary fiefs).[a] Against this it was convenient indeed to have at the centre a staff of a loyalty which was unimpeachable because they could not, in fact, found feudal families. Eunuchs throughout history played a considerable part in the politics of the court, and, as has rightly been said, the fact that history was written mostly by their enemies, and that many of those who served the State well were not written about at all, should moderate the gloomy view of their activities adopted by all the official historians.[b]

At the time of the great Emperor Wu, we see the flowering of this system.[c] Previous emperors, though absolute monarchs, were under the influence of the Taoist doctrine that they could govern best by doing nothing, so that their official acts were determined by the group of officials whom they chose as their executives. But the active and ambitious Han Wu Ti was not content with such a position, and after −131 took the government into his own hands. Unfortunately, heading the bureaucracy in person removed an important check to misgovernment, since officials, however highly placed, could be criticised, but hardly the emperor himself. Power was thus driven underground, since the Masters of Writing (*shang-shu*[1]) instead of the Lieutenant-Chancellors (*chhêng-hsiang*[2]), inevitably screened all memoranda before they reached the emperor. The chancellors became figureheads, and the most important position was held by the Intendant of Affairs of the Masters of Writing (*chih shang-shu shih*[3]). Later, power shifted still further underground, to the Palace Writers (*chung-shu*[4]), who were eunuchs, and so able to reach the emperor when in the inner apartments, as other scholars could not. Moreover, the very triumph of Confucianism, with its doctrine of the duty of favouring one's own family and relatives, brought about the custom of distributing the highest posts to the family of the empress. All this led to those bloody palace revolutions which recurred so frequently at changes of reign in all later times.

Now that the bureaucratic system was becoming well established, it followed that some systematic attention must be paid to its recruitment, and hence that schools and colleges must be set up. In −124 the Po Shih Kuan,[5] or, as Dubs (3) terms it, the Imperial University, was set up, with a chair for each of the classical books.[d] This supplied candidates for the government official posts for many years. But provincial education had begun spontaneously some time before, around −145, as the result of the initiative of a governor of Szechuan, Wên Ong.[6] The *Chhien Han Shu*[e] says:

> On taking up his post, he discovered that the district was uncivilised, the culture resembling that of the Man barbarians, so he tried to educate and improve the people.... In Chhêngtu

[a] Bloch (1), vol. 2, pp. 83, 182. In China, owing to the power of the central bureaucracy, just the opposite took place, titles such as marquis (*hou*) becoming ultimately non-hereditary honours.

[b] One of the first of these 'shock-troops of bureaucratism' was Shih Hsien,[7] who rose to great power about −47. The *Chhien Han Shu* (ch. 93, p. 5*a*) specifically says that he was favoured because he had no 'party' or clique of relations and descendants (*wu wai tang*[8]).

[c] Dubs (2), vol. 2, pp. 8, 10, 145, 292, 294. [d] *Chhien Han Shu*, ch. 88. [e] Ch. 89, p. 2*b* ff.

1 尙書 2 丞相 3 知尙書事 4 中書 5 博士官
6 文翁 7 石顯 8 無外黨

he established a department of education (*hsüeh kuan*[1]) (administered by men whom he had himself trained and sent to the capital for further study), inviting boys of outlying districts to come and study under them. The best students became candidates for official positions, while those of lesser ability received an honorary title (*hsiao ti li thien*,[2] lit. Filially Instructed Farmer). On Wên Ong's official visits throughout the district he took these students with him, and honoured them in various ways, such as allowing them the privilege of entering his official residence through a special gate. As a result the people respected the students and desired to become scholars themselves, the wealthy even paying for the privilege. Culture therefore improved, and the scholars were equal to those of the former States of Chhi and Lu. The emperor Wu (later) established schools in the provinces, but the system of government education really began with Wên Ong. He died in Szechuan. The officials and the people erected a temple in his memory, and sacrificed to him at the seasons of the year. Even now (after 200 years) the people of Szechuan love culture and education, and this is due to the work of Wên Ong.[a]

For further information the reader is referred to E. Biot (2), whose book remained for a century the only study in a European language on the history of education in China, until the recent and valuable work of Galt.

The 54-year reign of Han Wu Ti (−140 to −87) was one of the most important epochs of Chinese history. He had to try to bring the economic situation, which had been getting more and more unbalanced under his predecessors, into some kind of order. The merchants, goaded and bewildered by erratic anti-mercantile edicts, had speculated heavily and driven prices so high that the coinage no longer sufficed; then to remedy this the right of private minting was given to certain families, leading to an extremely localised capital accumulation.[b] Now the emperor took the most capable of these merchants and financiers into the civil service, especially in relation to the salt and iron monopoly. Experiments in currency were made, including the first attempt at paper money, namely, the skin of certain white deer which were found only in the emperor's hunting reserves; these were issued to notables coming to the court, for compulsory purchase against very large sums.[c] The conception of the 'ever-normal granary' was also introduced. But throughout the period taxation was heavy and constantly increased, largely for the financing of the wars against the Huns.[d]

The period was also remarkable for the contact which took place with other countries, some far away. The journey of Chang Chhien,[3] which began in −138, took him on a diplomatic mission to Bactria, and involved a ten-year detention by the Huns. It was one of the most remarkable explorations of antiquity, not only because

[a] Tr. Shryock (1), p. 68. Wên does not sound very like a family name, and there is a tradition that his family name was Shu.[4]

[b] Cf. *TH*, vol. 1, p. 339.

[c] *TH*, vol. 1, p. 428; and further, Swann (1).

[d] These continued, with few intermissions, throughout the Han period (see de Groot (1) and McGovern (1)). There is a translation of the chapter on the Huns in the *Chhien Han Shu* (ch. 94) by Wylie (2) but, as Hirth (2), p. 90, rightly warns, it was done at the very end of his life and cannot be used without checking against the Chinese text. Chs. 54 and 55 of the *Chhien Han Shu*, on the great Chinese generals of the time, have been translated by Pfizmaier (42, 44).

[1] 學官 [2] 孝弟力田 [3] 張騫 [4] 舒

of its long duration and the distance covered[a] (his informants may have come from as far as the Persian Gulf overland), but also because of the great variety of plants and other natural products which he brought back with him.[b] The epic of Chang Chhien will be frequently referred to in this book. It was of cardinal importance for the westward expansion of the Chinese empire, and the establishment of what has come to be known as the Old Silk Road. In geopolitical terms, as Chavannes (12) and many others have pointed out, the chain of oases along the foot of the Nan Shan constituted a 'fault' or line of least cohesion between the Hunnish and Mongol nomads on the one hand and the Tibetan tribes on the other. It was therefore a natural line of communication between the Chinese and Iranian culture areas, and in −120 the fortress-cities of Ganchow, Liangchow and Chiu-chhüan were firmly occupied by the Han. Besides all this, there were several maritime 'embassies' from the Western countries including some Romans or Roman Syrians, during Han times, as we shall see.[c]

The Han emperors, but particularly Han Wu Ti, are criticised by conventional historians for their great interest in 'superstitious practices', and it was these, no doubt, which contributed spice to the legends which grew up later, and which came to be embodied in books such as the *Han Wu Ti Nei Chuan*[1] and the *Han Wu Ku Shih*.[2] The Emperor Wu extended and enlarged the imperial sacrifices and official magico-religious rites, and spent a great deal of time in trying to cultivate relations with supernatural beings. He was thus brought into close contact with the shamanist strain of Taoism, and the accounts of his relationships with magicians such as Luan Ta,[3] Li Shao-Chün and Shao Ong[4] are unforgettable. His attempts to attract the supernatural beings, however, invariably ended in failure, for as Dubs says, 'he was too acute to be easily fooled, yet he could not repress the feeling that some of the magicians' practices might not be entirely fraudulent'.[d] In fact, they were not, for realising as we do now the intimate connections in earlier ages between magic and science, there is little doubt that the emperor's magicians were unearthing a certain proportion, perhaps small, of true and valuable natural observations, as in alchemy, magnetism, pharmaceutical botany, and so forth (see on, Sects. 10, 26*i*, 33, 45).[e]

It was in connection with this complex of ideas that the Chinese first reached out to the east. Chhin Shih Huang Ti had sent Hsü Fu[5] to sea on a peaceful expedition in order to make contact with the supernatural beings who were supposed to dwell in the islands of the Pacific.[f] Han Wu Ti followed this up in a more warlike

[a] Nearly 40° of longitude. See pp. 173 ff. below.

[b] The story is well told by Hirth (2); cf. *TH*, vol. 1, p. 409.

[c] Pp. 192 ff. below.

[d] (2), vol. 2, p. 19.

[e] There is a certain parallel between these 'gentlemen possessing magical recipes' (*fang shih*) and the sophists of Greece, who, opposing τέχνη to φύσις, offered to teach people all kinds of arts, not only disputation, but also writing, medicine, household economy, boat designing, and the like (cf. Lovejoy & Boas (1), p. 194).

[f] Yetts (2) has pointed out the parallel between these expeditions and the Carthaginian efforts to reach the 'Fortunate Isles' which, in the form of the Canaries and Madeira, had an equally sound factual basis. Cf. Sect. 29 below.

1 漢武帝內傳 2 漢武故事 3 欒大 4 少翁 5 徐市

manner by penetrating most of Korea and establishing a colonial government of four provinces there. This centre of civilisation had a great influence on the slowly developing culture of Japan.[a] In −130 another expeditionary force opened up Kweichow,[b] and in −111 the kingdom of Nan Yüeh,[1] the centre of which was Canton, was annexed.[c] The Hunnish wars, too, were almost continuous.[d]

The Han Dynasty was cut in half from +9 to +23 by the so-called interregnum of Wang Mang,[2] the first and last Hsin[3] emperor. It was marked by a series of far-reaching reforms, the significance of which is controversial to this day.[e] They have been called socialist, but as we shall see on closer examination later (Sect. 48) they were essentially an attempt to strengthen the bureaucratic State. The agrarian reform declared the entire land of the country to be State property, redistributed large holdings among the peasant-farmers, and taxed uncultivated fields. Male slaves were at first freed, but as this could not be enforced, their owners were subjected to heavy taxation. State monopolies were extended. The calling-in of gold coins in exchange for bronze ones resulted in enormous accessions to the treasury, and a drain on the world gold circulation;[f] Dubs (4) has calculated that at the end of Wang Mang's reign about 5,000,000 ounces had been accumulated[g]—more than the total supply of medieval Europe. The ever-normal granary system, whereby the government purchased grain when prices were low, and released it again when prices rose, was tried once more. With a willing and honest civil service, as Goodrich says,[h] Wang Mang might have succeeded; as it was, the officials enriched themselves, merchants and financiers were driven to desperation, the mass of the population grew restive, and finally open revolt broke out which led to the collapse of the reign and the death of the reformer. Notable at this point was the activity of an element which has played a great role in Chinese history, namely, the secret societies (in this case the Red Eyebrows). These were usually of Taoist inspiration and must therefore be returned to in that connection.

The historian of science is likely to have a weakness for Wang Mang, for apart from the seeming rationality of his reforms, he was undoubtedly interested in technology and the science, if so it could be called, of his day. It was under his auspices

[a] Sansom (1), pp. 18 ff.

[b] *TH*, vol. 1, p. 420; details in Rosthorn (1).

[c] There is a translation by Wylie (3) of ch. 95 of the *Shih Chi* which deals with the conquest of Min, Nan Yüeh and Korea, but it was made at the very end of his life and is not reliable without checking against the text. The same applies to the translation of the same chapter in the *Chhien Han Shu* by Pfizmaier (51).

[d] Detailed accounts in Lattimore (1) and McGovern (1).

[e] Cf. the paper of Sargent (1) and the reply by Dubs (1). The biography in the *Chhien Han Shu* has been translated by Stange (1).

[f] It has been pointed out that it was just at this time that Tiberius (+14 to +37) prohibited the wearing of silk because Roman gold had to be exported to pay for it.

[g] Chinese gold of this period probably came from Siberia; Herodotus (1, 215) said that the Massagetae (Scythians) made their horse-trappings and belt-buckles from the metal.

[h] (1), p. 42.

1 南越 2 王莽 3 新

that the first assembly of scientific experts in Chinese history was called together. In +4, when he was chief minister to the Emperor Phing Ti,

A convocation (of assembly) was sent out to all persons throughout the empire who were learned in the lost classics and ancient records, in astronomy (*thien wên*[1]) and calendrical science (*li*[2]), in mathematics (*suan*[3]) and the acoustics of the standard musical tones (*chung lü*[4]), in philology and history, in magical or medical techniques (*fang shu*[5]), in the botany of woody plants and herbs (*pên tshao*[6]), and in the five classics (including the) Confucian Analects (*Lun Yü*[7]), the Filial Piety Classic (*Hsiao Ching*[8]) and the Literary Expositor (*Erh Ya*[9]). These doctors were given (as a special mark of honour) an authorisation to add a second horse to each of their chariots. More than a thousand of them assembled at the capital.[a]

It is to be regretted that no record of their deliberations has reached us. Later, in +19, pressed by the Huns, and having conscripted one man in thirty of the whole population, Wang Mang

enlisted the help of those who possessed strange arts.... One said he knew how to cross waters without the aid of boats or oars, so that even troops of a hundred or a thousand horse could ride over; another said that it was not necessary to maintain army grain supplies as he had a medicinal substance which would dispense with food and hunger; a third claimed to be able to fly a thousand *li* in one day and suggested that he should be sent to bring back intelligence of the movements of the Huns. Wang Mang had all these alleged methods tested without ceremony, and found that none of them were usable.[b]

Wang Mang, too, may have a connection with the origin of the magnetic compass (see Sect. 26*i*).

Upon Wang Mang's death, Liu Hsiu,[10] a cousin of the former Han emperors, emerged triumphantly from a short period of confusion, and inaugurated the Later or Eastern dynasty of the same name, which, after moving its capital from Chhang-an to Loyang, essentially consolidated the practices and policies of the Former Han. The wars with the Huns had come to a temporary halt in −51, when their empire was divided into two parts, one of which, the southern, accepted the position of tributary state to China. But fighting continued during the whole of the first century of the Later Han. The famous general Pan Chhao,[11] Governor-General of Central Asia for many years,[c] made his influence felt as far as the Caspian Sea. The silk trade through the Parthian lands, which alone separated the regions of Chinese and Roman influence, continued; and there were further mercantile contacts by sea, especially by the Persian Gulf route, after about +120, with Arabia and Syria.

[a] *Chhien Han Shu*, ch. 12, p. 9*a*; tr. auct.

[b] *Thung Chien Kang Mu*, ch. 8, p. 66*a*; tr. auct. adjuv. Wieger, *TH*, vol. 1, p. 618; originally from *Chhien Han Shu*, ch. 99*c*, p. 5*b*.

[c] Cf. his biography, translated from *Hou Han Shu*, ch. 77, by Chavannes (16). See also his story, well told, in McGovern (1), pp. 264ff.

1 天文 2 曆 3 算 4 鐘律 5 方術 6 本草
7 論語 8 孝經 9 爾雅 10 劉秀 11 班超

The Han time (especially the Later Han) was one of the relatively important periods as regards the history of science in China, so that references to events in it will recur in later chapters. Much was done in astronomical science and calendar improvement, there was an outstanding invention in the earth sciences, the beginnings of systematic botany and zoology were laid down, and a definitely sceptical and rationalist current of thought was abroad which found its most perfect expression in the philosopher Wang Chhung[1] (Sect. 14*c*).[a] Closely associated with Taoist materialist mysticism, alchemy was continuously patronised by the court and the high officials, and the period saw the appearance of the first book on it in any language (cf. Sect. 33). At least two imperial princes took part in this intellectual life. Prince Tê[b] of Ho-Chien[2] was a scholar and bibliophile;[c] he it was who preserved for us the technologically important *Khao Kung Chi*[3] (Artificers' Record) section of the 'Record of the Rites of Chou',[d] the *Chou Li*.[4] More famous, even legendary, was Liu An,[5] Prince of Huai-Nan,[6] who was deeply interested in the magical and proto-scientific aspects of Taoism, and gave his name to one of the most important monuments of ancient Chinese scientific thought, the *Huai Nan Tzu*,[7] a book probably compiled by the group of naturalists who surrounded him.[e] But the prince having become involved in some treasonable activities against Han Wu Ti, or rather having been pushed into them by his enemies, committed suicide or disappeared in −122.[f] The Confucian orthodoxy of the bureaucrats described him as having been 'skilled in perverse teachings' (*chih yü hsieh shuo*[8]).

In technology the age was marked by the invention of paper,[g] numerous ceramic developments such as the beginning of glazes and proto-porcelain (as it is called), and a level of textile techniques which was not approached by Persia or Europe till several centuries afterwards. The architectural processes, for example the making of decorated bricks and tiles, greatly advanced. A large number of natural products new to the Chinese were imported, such as improved horse breeds, alfalfa and the grape-vine from the west; and oranges, lemons, areca nuts and *lichis* from the south and south-west. Jade came from Khotan and perhaps Burma. Then there was a systematic development of bibliography, the first book lists, embodied in the *i wên chih*[9] of the Han histories, were compiled by experts in magic, medicine, military science, history, philosophy, divination and astronomy; they list some 700 works written on wooden or bamboo tablets and on silk.[h] Manuscripts of wood, and even a complete volume of

[a] Cf. Nagasawa (1), p. 135.

[b] Posthumous title Hsien,[10] by which name he is more usually known.

[c] *TH*, vol. 1, p. 381.

[d] Cf. Nagasawa (1), p. 122. The *Chou Li* is almost certainly a Han compilation, though it may well contain many fragments of earlier writings.

[e] Cf. *TH*, vol. 1, p. 385.

[f] *Chhien Han Shu*, ch. 6, Dubs (2), vol. 2, p. 58. Ch. 44, which concerns this princedom, was translated by Pfizmaier (39).

[g] By Tshai Lun,[11] d. +114.

[h] Wu Khang (2).

1 王充 2 河間德王 3 考工記 4 周禮 5 劉安
6 淮南王 7 淮南子 8 忕於邪說 9 藝文志 10 獻王 11 蔡倫

seventy-five tablets bound with the original thongs, were found by Stein (1–5) in his wonderful series of excavations in Chinese Central Asia, where the study of the rubbish-heaps outside ancient Han forts (Fig. 15) revealed all kinds of objects preserved without decay by the aridity of the climate of the Gobi and Sinkiang. There are lists of weapon stores, battalion orders, fragments of writings on astrology, medicine, military technique and divination, primers, calendars for years such as −63 and −39 (cf. Chavannes, 12).

About +65 Buddhism appeared in China for the first time, and a century later the first sūtras were being translated into Chinese at Loyang.[a]

Towards the end of the dynasty, palace revolutions and disturbances became increasingly frequent, and an agrarian crisis led to a peasant revolt in +184.[b] The suppression of it put certain army generals into a position of such power that a shogunate might have developed, but for this the conditions were unsuitable, and in +220 even the appearance of a central government could no longer be maintained. For the next half-century China was divided into the Three Kingdoms (San Kuo) from which the period takes its name.

(c) THE SAN KUO (THREE KINGDOMS) AND THE KEY ECONOMIC AREAS

During this period the three States were in a condition of permanent hostility to one another (Fig. 16). Wei occupied the north and north-west, being based essentially on the Yellow River valley, and with the capital at Loyang. Wu had the south and south-east, including the Yangtze valley but also the two Kuang provinces; its capital was located first at Wu-chhang (near modern Hankow) but later at Nanking. The third, Shu (or Shu Han, since it considered itself the legitimate successor of the Han), was based on the Szechuan basin, with Chhêngtu as its capital, but also controlled the hills of Kweichow and part of Yunnan.[c] The ultimate victory gained by Wei is generally ascribed to the cautious policy adopted of using military agricultural colonies,[d] enlarging their own irrigated supply bases, and starving out their enemies instead of meeting them in pitched battles. Shu was reduced in +264 and Wu in +280, but the House of Wei did not continue, since it was one of their generals, Ssuma Yen,[1] who founded the new unified dynasty of Chin.

The battles and manoeuvres in this struggle became legendary, and inspired, for example, one of China's most famous novels,[e] as well as many plays. Tshao Tshao,[2]

[a] *TH*, vol. 1, p. 689. Here the basic papers are those of O. Franke (5) and H. Maspero (5, 19, 20). The traditional account of the coming of Buddhism is a fable, but there are good reasons for retaining the traditional date (see Sect. 15*a* below).

[b] This time the 'Yellow Turban' secret society played a similar part to the 'Red Eyebrows' of Wang Mang's time (*TH*, vol. 1, p. 773).

[c] Prefectures in Herrmann (1), Map 25.

[d] On these, see Biot (18).

[e] The *San Kuo Chih Yen I*[3] by Lo Kuan-Chung[4] of the Yuan. Cf. p. 40 above.

[1] 司馬炎 [2] 曹操 [3] 三國志演義 [4] 羅貫中

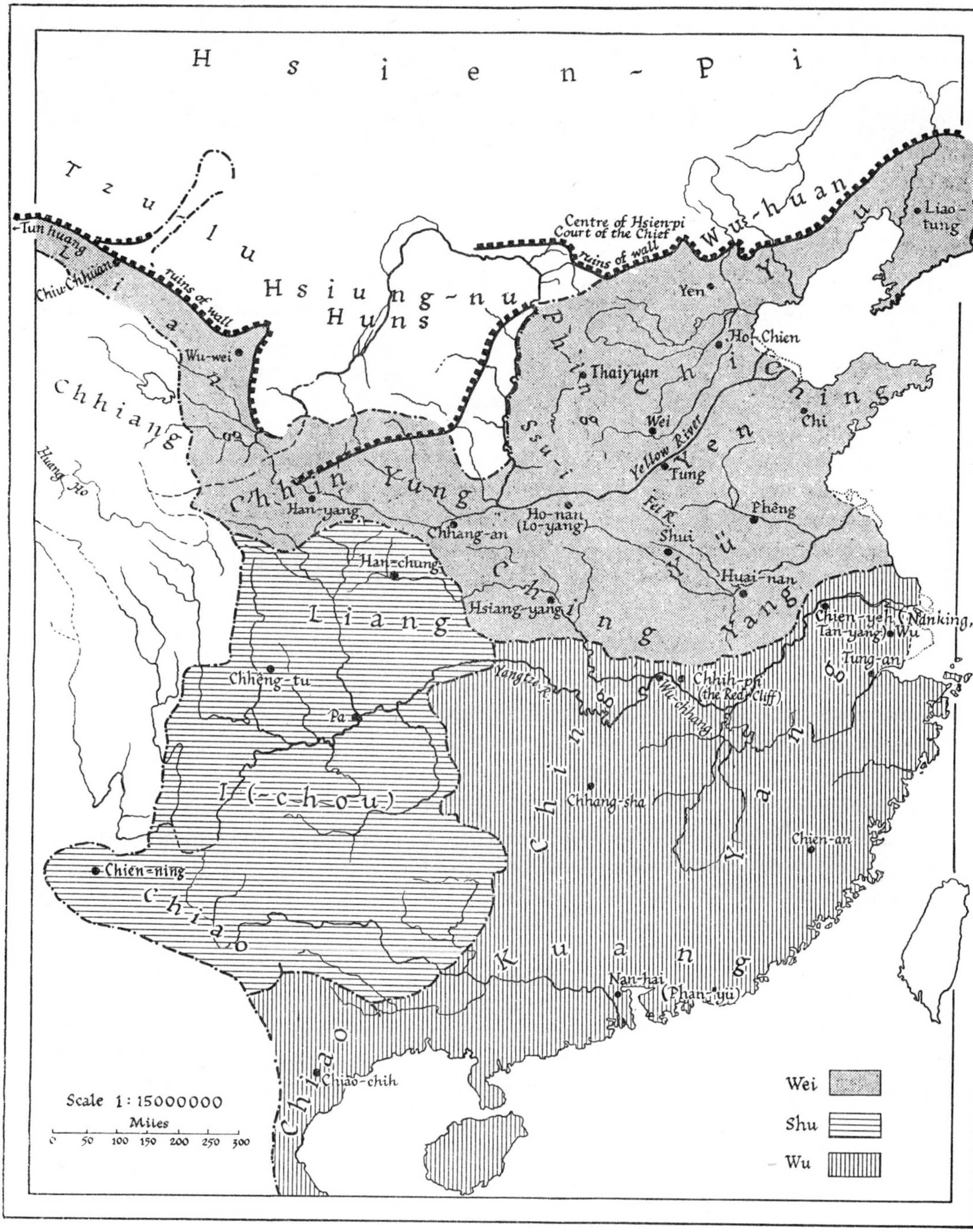

Fig. 16. China in the Three Kingdoms period (+220 to +265), after Herrmann (1).

the Wei leader, became the very type and model of a brave, but ruthless and cunning ruler.[a] Chuko Liang,[1] the leader of the Shu, an equally important general, was more of a diplomat and strategist, and seems to have been associated with certain technological inventions. Another Shu Han soldier was Kuan Yü,[2] who, though killed at the beginning of the period, made such an impression on his contemporaries that he was afterwards deified as the God of War.[b] Among the established facts concerning these wars is the development of naval combats, as in the Battle of the Red Cliff (Chhih-Pi), when in +208 the Wei fleet was completely destroyed near Hankow.[c]

The division of the country into three units has, however, a significance far more important for us than the military and dynastic events of the time. The political boundaries of the States (Fig. 16) may be compared with a map showing the boundaries of what Chi Chhao-Ting, in a book which is perhaps the most brilliant contribution so far in any Western language to the mechanics of Chinese history, has called the 'Key Economic Areas' (Fig. 36, folding map).

> Unlike other later periods of division [he says], which were complicated by the simultaneous occurrence of barbarian invasions, the period of the Three Kingdoms was a typical case of division generated by the internal forces of Chinese society. The material and fundamental factor responsible for such division was the rise of rival economic areas, whose productivity and location enabled them to serve as bases for a sustained challenge of the authority of the overlord who commanded the central or main key economic area. In this case it was the increasing maturity of Shu, and the adolescent exuberance of Wu...which produced the balance of power represented by the San Kuo.[d]

This conception recalls to us the basic fact that Chinese feudal-bureaucratic society rested on an intensive agriculture requiring irrigation, and an accumulation of tribute grain at the source of power. Since both of these required important works of hydraulic engineering, water-control was, from the beginning, essentially a function of the central State authority, and later we shall ask to what extent this conditioned the whole course of Chinese history (Sects. 47, 48). Political power had thus a very close connection with regional geography. In this way can be explained the recurrent dominance of one region over others; the achievement of political unity in large areas apparently including marked diversity, and the course taken by the wars and popular revolutions which usually preceded and accompanied changes of reign or dynasty. One has to realise that the development of commerce and transportation never succeeded in overcoming the regionalism of the purely agricultural economy—there were only the tens of thousands of practically self-sufficient villages on the one hand, and the provincial and national capitals collecting grain-tax from them on the other. This

[a] Actually like Huai Nan Tzu, he and his sons favoured Taoist magicians, who gathered at the Wei court.

[b] See Doré (1), part 1, vol. 6, pp. 54 ff.

[c] *TH*, vol. 1, pp. 826, 836.

[d] (1), p. 96.

[1] 諸葛亮 [2] 關羽

grain went into a reserve, which could be released in times of shortage, but which had the more important function of supplying the armies necessary to quell rebellions, or to furnish expeditionary forces, or to fight foreign wars. In the *Hsin Shu*[1] of Chia I[2] (one of Han Wên Ti's advisers), he says:

> Although the Han dynasty has been in existence for 40 years, both public and private reserves are in a pitiable condition. In case of famine in a territory of two or three thousand *li*, where could relief be procured? And if there are emergencies on the border, how could the provisions for several tens of thousands of soldiers be supplied?[a]

Unity and centralisation of State power simply meant, as Chi says, the control of an economic area where agricultural productivity and facilities of transport permitted a supply of grain-tribute so superior to that of any other areas, that the group which controlled the key area controlled China. But in the San Kuo period there were in fact three areas of comparable resources. It therefore openly reveals to us, as if by a dissection, the geopolitical muscular system normally hidden by the subcutaneous fat and skin of constitutional and military history as it is generally written.[b]

Let us now compare the map in Fig. 36 with the Table of Dynasties (p. 78). As we have seen, the Shang and Chou ages were essentially Yellow River valley cultures, and their history was one of expansion within that northern area, extending at the end of the Chou to include the Huai valley within it, but only in very undeveloped form the greater part of the second, the east-central Yangtze valley area. The first unification in the Chhin and Han dynasties was definitely based on a superior development (begun, as we have seen, by the Chhin kingdom) of the northern area, the Yellow River, and its upper tributaries, the Ching,[3] Wei[4] and Fên[5] rivers. When this unity fell apart at the end of the Han it was because the lower Yangtze valley (east-central or Wu area) had largely caught up with the northern area and produced enough to be able to afford an independent centralisation. Simultaneously the Szechuan basin (western or Shu area), owing, among other things, to the extensive irrigation works already mentioned (p. 97), had also now become capable of independence.

To continue the story, anticipating the following pages, we may say that this centrifugal tendency continued until the end of the +6th century, since the short-lived unification of the Chin dynasty proved unavailing, and it soon had to retire to the east-central area, leaving the north to the domination of foreign houses. But by the time of the Sui and Thang dynasties (+589 to +907) the Yangtze valley had increased its agricultural productivity and transport facilities so much beyond that of the north that it was able to become itself the main key area, and with the building of the Grand Canal the capital could be placed once again in the north with some

[a] Ch. 24; tr. Chi. Chia I's dissertation on the mistakes of the Chhin dynasty has been translated by Margouliès (1), p. 56.

[b] Of course, as Lattimore (3) has emphasised, the difficulty which most Chinese dynasties had to face was that the key economic areas and the key strategic areas were very often not the same.

[1] 新書 [2] 賈誼 [3] 涇 [4] 渭 [5] 汾

assurance of stability. A further seesaw takes place, however, during the Five Dynasties period, and after a renewal of centralisation in the Northern Sung, the northern area is again taken over by barbarians (the Liao and Chin, +907 to +1280), while the Southern Sung rule over the east-central area. Eventually the overwhelming Mongol might brings about a final unification, and by the end of the +14th century transport facilities have so much improved that whether under the Chinese dynasty of the Ming, or the foreign Manchu dynasty of Chhing, the country does not splinter again.

So much for the future. But from this vantage-point of the divisions of the San Kuo period we can look back to the past and reinterpret certain fundamental events which have already been described. The conquest of the Shang by the Chou was a spilling over of power from the Shensi basin, so that the lower Yellow River lands were brought under the domination of the people who controlled the loess domain which we spoke of geographically as Kuanchung (especially the Wei valley). This was a readjustment within the northern key economic area. But the epic struggle some seven centuries later in which the State of Chhu was defeated by the State of Chhin may be regarded in this light as the first rivalry between the northern (Yellow River) and the east-central (Yangtze) economic areas. As Bishop (5) has well set forth, the Chhin people, wheat-eating chariot-fighters, probably under considerable nomadic and Iranian influence, faced the Chhu people of Oceanic, south-east Asian and Indian affiliations, whose staple cereal was rice and who understood naval warfare using large canoes. The strategic upper Han valley was at first annexed by Chhu, but the Chhin people turned them out of it. Thus 'the troubles', says Lattimore (2), 'which attended the fall of the first inclusive imperial dynasty had in part the character of a reaction of the South against the North'. We may remember that the first Han armies called themselves 'the Avenging Army of Chhu'. To a large extent the Han settlement was a compromise. Just as after the English civil war the restored monarchy was by no means the same as the feudal and autocratic monarchy against which the parliamentary armies had fought, so when the first Han emperor attained power, there was to be a centralised empire, but with due regard to the Yangtze provinces, and with the liquidation of the hated Legalist school on which Chhin had relied.

Now that we have come to think of Shu (Szechuan) as an independent kingdom,[a] it is interesting to note, with Liang Chhi-Chhao (1), that 'whenever there were disturbances in China, Szechuan was held by an independent ruler, and it was always the last to lose its independence'. This happened seven times in Chinese history. After Wang Mang's time it had sixty-one years of independence; it was one of the San Kuo for forty-two years, it went its own way again for several decades when the Chin empire was dismembered, and then during the Five Dynasties period between Thang and Sung it had two dynasties of its own in sixty-five years. There were similar episodes both at the beginning and end of the Ming, and during the Chhing. Yunnan, though independent for many centuries until the Sung, usually followed whatever Szechuan did.

[a] Cf. Chêng Tê-Khun's (1) short history.

Fig. 36 may now be dismissed by a reference to the very late-developing southern economic area of the two Kuang provinces, and to the special character of certain regions such as Shansi in the north, defensively strong but never really economically self-sufficient; and Fukien in the south-east, a country of mountains, forests and sailors' ports. None of these affected the main outlines of history.

We return to the confused campaigns of the San Kuo. Chi has pointed out that hydraulic works played a considerable part in those decades of strategy. Wu's chief problem was shortage of man-power needed to develop the virgin fertility of natural resources in the upper Yangtze basin below the gorges. In the middle of the period the Wu State completed the important Chüyung[1] canal in Chiangsu and made the artificial lake at Chhihshan[2] for irrigation near Tan-yang,[3] not far from modern Nanking. The evidence for Shu is not so clear, but it is thought that works were carried out when Shu captured the southern bank of the upper Wei valley in +234. The Wei people, on the other hand, paid the greatest attention to waterworks, constructing between +204 and +233 three large reservoirs and two trunk canals in the Huai river system, and six important canals in the provinces of Shantung, Honan, Hopei, Shensi and Shansi. When the Huai valley was fully incorporated in the northern area the victory of Wei was inevitable.

In this period of disruption and war devastation it was natural that people should turn to other-worldly religions as a refuge, and Buddhism was already there to fill the need. We shall examine later its effect upon Chinese thought, but it certainly provided a consoling popular faith which could be deepened by those devotionally inclined, an elaborate theology for intellectual exercise, and also a cosmological philosophy which had been somewhat lacking from indigenous systems. There was much activity in translating the sūtras in Nanking in the +3rd century, and one of the effects of the spread of the new religion was greatly to increase foreign contacts. Scholar-monks came as missionaries and expounders, mostly from India and Ceylon, but also from Western and Central Asia (Parthia, Sogdia, Gandhāra, Kashgaria, Kucha, Turfan, etc.) and even from Java and Cambodia. The influence resulting was certainly not confined to religion, but included science and probably technology too.

About the same time there occurred the great transformation whereby Taoism, uniting the doctrines of the Taoist school of ancient philosophers with the mass of magical-scientific accretions which had crystallised around the nucleus of primitive shamanism, changed into an organised religion which competed strongly, and often successfully, with Buddhism. The first steps in its systematisation were taken, significantly enough, in the north, where its first 'pope' resided in the Wei capital (see on, Sect. 10*j*).

In comparison with these trends Confucianism was in decline, though it received some official support from the Wei State. Examinations were no longer held, or else ceased for the time to be the main entry to official life. Nevertheless, it was at this time that the historical classics (*Shu Ching*, *Chhun Chhiu*, *Tso Chuan*) were

1 句容 2 赤山 3 丹陽

carved in stone at Loyang, a measure which contributed greatly to the preservation of the exact texts. The unsettled state of society led to intense discussion of social problems.[a]

The foreign contacts referred to above in connection with Buddhism were by no means the only important ones at this time. Roman and Syrian merchant-'ambassadors' who before the end of the Han would have been sent to Chhang-an now came through Tongking to Nanking, the Wu capital, which also received tribute gifts from all the newly Hinduised countries and islands of south-east Asia. Correspondingly, the Wei court maintained relations as far west as Ferghana, and with the Huns of the Ördös and Alashan regions, and the Hsien-pi[1] of eastern Mongolia; it was also in contact with Korea and Japan. It was quite natural therefore that one science at least should flourish during these distracted times, namely, geography, and this in fact it did, bearing China's greatest cartographer, Phei Hsiu,[2] whom we shall meet again (Sect. 22*d*). So also, doubtless due to the stimulation of meeting with new products, the botany and mineralogy of the southern regions were extensively studied, as in the books *Nan Fang Tshao Mu Chuang*[3] (Plants and Trees of the Southern Regions) by Chi Han,[4] in which occurs what must be the first reference in any literature to the entomological control of plant pests, the *Nan Chou I Wu Chih*[5] of Wan Chen,[6] the *Nan I I Wu Chih*[7] (Strange Things of the South) of Yang Fu,[8] etc. Among the new customs of the +3rd century was the use of tea, the first reference to which is of +273,[b] though it took some centuries to spread in the north. In technology, we find water-mills becoming widespread at this time (see on, Sect. 27*f*) and there was also the invention of the wheelbarrow, which is connected with Chuko Liang (see on, Sect. 27*c*). It would not be unplausible to attribute the appearance of these two labour-saving devices to the depopulation caused by the wars, and the intense need for man-power felt by each of the three States for the strengthening of their respective economic areas. The first had arisen more or less contemporaneously in China and Asia Minor about the end of the −1st century, and its earlier beginnings are still obscure; the second, though so simple, was an entirely new development and (almost unbelievably) was not used in Europe until ten centuries later.

[a] Cf. the analysis by Balazs (1, 2) of the writings of a half-dozen of the most typical thinkers of the time. There was, as he says, a revival of the extremist schools of thought of the Warring States period, which had not been attractive during the long Pax Romana of the Han.

[b] Pelliot (1).

1 鮮卑 2 裴秀 3 南方草木狀 4 稽含 5 南州異物志
6 萬震 7 南裔異物志 8 楊孚

(*d*) THE CHIN DYNASTY AND ITS SUCCESSORS (WEI, LIU SUNG AND LIANG)

No sooner had the Chin house of Ssuma once more unified the country by a re-establishment of the primacy of the northern economic area, than it was faced by very great pressure from a variety of northern semi-barbaric peoples. These Turkic, Mongol, Hunnish and Tungusic tribes had been fighting among themselves for years, and in +304 the Huns and then the Hsien-pi were called in to help in Chinese internal struggles. But such spirits once raised could not be laid again. By +316 both Chhang-an and Loyang had been lost, and the Chin had to withdraw south of the Yangtze, setting up their capital at Nanking, i.e. in the less advanced east-central economic area.[a] Between +304 and +535 no less than seventeen 'dynasties' contended with each other in the north—of these four were Hunnish, four Turkic (Tho-pa[1]), six Mongol (Hsien-pi[2]), and only three ruled by houses of Chinese stock.[b] Nevertheless, throughout this time the 'barbarians' were sinified much more than the northern Chinese were barbarised. Nomadic dress was doubtless widely adopted, but in general the Chinese agriculture and administration continued and barbarian customs were adapted to it; intermarriage was universal and encouraged, and even the polysyllabic names of the barbarian chieftains were exchanged for Chinese ones.

We may later speak again of what has been called the extraordinary integrative and absorptive power of Chinese civilisation,[c] a power which no invader before modern times was able to withstand. It must surely have been connected with the highly characteristic nature of Chinese agriculture and administration. Whoever might rule, it was always the scholar-gentry who had to be called in to administer. They alone possessed the necessary mastery of the written language, of official procedures, and of essential techniques such as hydraulic engineering. H. Wilhelm[d] draws a very interesting parallel with the influx of Germanic tribes into the late Roman Empire. There, on the contrary, the invaders could neither maintain their own cultural traditions nor adopt fully those of the conquered imperium. Wilhelm suggests that the cultural stratum of the Romans was laid too thinly over the masses of the working people to be able to withstand such a strain; while, on the other hand, even in Han China, the culture had driven deep roots into the vast population of peasant farmers, whose standards were in fact the classical Confucian or Taoist valuations. From my own experiences in China I am much inclined to agree with him. It must, moreover, be remembered that China had no large slave element in the population, as had Rome, but that is another story, which must be told in its place.

[a] Lei Hai-Tsung (2) considers that the Battle of the Fei[3] River, fought in +383, in which the Chin defeated a large barbarian army, was one of the most decisive in Chinese history, since it permitted the peaceful further development of purely Chinese culture, which afterwards the barbarians adopted. The Fei River is one of the tributaries of the Huai, in Northern Anhui.

[b] Herrmann (1), Map 29. Much the best account of these States and dynasties is given by Eberhard (9).

[c] Cf. Wittfogel (1).

[d] (1), p. 100.

1 拓拔 2 鮮卑 3 淝

The longest lasting of these dynasties was that of the Northern Wei, which, with its capital at Ta-thung, based itself on Shansi (see Fig. 17). It eventually controlled all the north with the exception of Shantung, and extended as far west as Tunhuang.[a] Buddhism was taken up with particular enthusiasm by the Wei people, as may be seen in the +4th and +5th-century fresco paintings of the Caves of the Thousand Buddhas (Chhien-Fo-Tung[1]) at Tunhuang,[2] where the Wei work is the most beautiful of all the styles (cf. Figs. 19, 20);[b] and also in the famous carvings of the Yünkang[3] caves near Ta-thung.[4] These are the two sites which correspond, one may say, with India's Ajantā and Ellora respectively.[c] In other parts of China, Buddhist philosophy was much developed in this period, and new sects were formed. Fa-Hsien,[5] the great Buddhist pilgrim, set out for India in +399 and returned to Chhang-an in +414 after having visited Central Asia and almost every part of India.[d] Chinese medicine owed something to Buddhist importations.

At the same time the Taoists were active, probably receiving accessions of strength from the troubled state of society. Military, rather than administrative, talents were in demand, and it was difficult for philosophers to be effective in political life. Hence perhaps arose a favourable atmosphere for speculations about Nature. Thus the Taoists produced in the early +4th century one of their greatest naturalists and alchemists Pao Phu Tzu[6] (Ko Hung[7]). In historical geography, this age saw the first of what afterwards developed into a vast class of writings, the so-called 'gazetteers', *ang chih*[8] or *hsien chih*,[9] i.e. local topography and records, which, as Wylie says,[e] 'are probably unrivalled in any nation for extent and systematic comprehensiveness'. This was the *Hua Yang Kuo Chih*[10] of Chhang Chhü,[11] written in +347; it describes the region of south Shensi and north Szechuan, giving accounts of the building of the Shu capital at Chhêngtu, biographies of local notables, local customs, monuments, birds and other animals, and commodities such as copper, iron, salt, honey, drugs, bamboo, tea, etc. 'For centuries', says Goodrich,[f] 'one after another of these local records were compiled, sometimes as a labour of love by a retired scholar of the district, sometimes by official order to provide all the information about the local governments for which there came a succession of men from the outside.' More

[a] Herrmann (1), Maps 30, 31.

[b] See Grousset (1), pp. 316ff. The first cave was consecrated in +366; cf. *TH*, vol. 1, p. 1009. Unfortunately the numbering of these caves is in a state of confusion since each has now three or four different ones painted on it. I have adhered to the numbers in large Chinese script black on white (to the best of my ability, that is to say, for the notes which I made when at Chhien-Fo-Tung in 1943 were not intended to serve scholarly purposes, and other writers refer to the caves according to one or other of the different systems). Standardisation is urgently needed.

[c] There is a rough correspondence also in time, for the best work at Ajantā is of the +5th century, while the Kailāsa, Ellora's crowning work, is of the +8th, i.e. Thang, when so much was done at Tunhuang. In 1953 the discovery was announced of a third group of caves, as important as Tunhuang or Yünkang, at Maichi near Thienshui (Kansu).

[d] Cf. *TH*, vol. 1, p. 1047.

[e] (1), p. 35.

[f] (1), p. 92.

1 千佛洞 2 敦煌 3 雲岡 4 大同 5 法顯
6 抱朴子 7 葛洪 8 方志 9 縣志 10 華陽國志
11 常璩

Fig. 17. China about +440 (after Herrmann, 1).

than 6500 of these books are known, but few are earlier than the Thang. Larger geographical works were also written now, however, such as Li Tao-Yuan's[1] famous Commentary on the Waterways Classic (*Shui Ching Chu*,[2] about +510). Nor were mathematicians lacking, such as the eminent Tsu Chhung-Chih.[3]

Finally, the Northern Wei split into an eastern and a western portion, each of which was taken over, after +550, by Chinese successor states, the Northern Chhi and the Northern Chou respectively.[a] Meanwhile, during the previous century, the Chin in the south had been replaced by the Liu Sung (Sung, of the house of Liu), and this again gave place to three further dynasties of a duration of some forty years each. The second of these, the Liang, was notable for its devotion to Buddhism, one of its emperors becoming a monk in Nanking.[b] At last a man appeared who knew how to unite once more the economic areas, namely Yang Chien.[4] He ruthlessly destroyed the Northern Chou in +581, gaining thereby the Chhi domains which had fallen to the Chou in +577, and then, crossing the Yangtse, made an end of the Chhen dynasty in the south in +589.[c] By +610 the new Sui emperor had under his control every part of the continent from Annam and Formosa in the south to Tashkent and Hami in Central Asia. Though his dynasty did not last long, it was a preparation for one which did.

(*e*) THE SUI DYNASTY

With the exception of the short period of unification under the Former or Western (really northern or north-western) Chin, China had now been in division and confusion for 330 years (some 60 under the San Kuo, and some 270 under the Northern and Southern Empires, as they came to be called). If the general interpretation of Chinese history already outlined were right, one would expect that the first action of a really strong ruling group which unified the country would be to take unprecedented action to improve the connections between the northern and southern (or rather east-central) economic areas.[d] This was precisely what the second Sui emperor, Yang Ti,[5] did, for he completely overhauled the previously fragmentary water transportation systems between the Yellow River and the Yangtze and built the main links in that great waterway which came to be known as the Grand Canal (Yün Ho[6]). In his time the east-central Yangtze valley had become the leading key economic area, but the political centre of gravity still lay in the north, and both Chhang-An and Loyang were

[a] Herrmann (1), Map 33.

[b] *TH*, vol. 2, p. 1212.

[c] See the translation of seven chapters of the *Chhen Shu* by Pfizmaier (59), and eleven chapters of the *Pei Chhi Shu* by Pfizmaier (60).

[d] Vidal de la Blache (1, p. 249) has acutely observed that when thirteen centuries later 'China adopted the locomotive, care was taken to lay the fundamental line of the system between the North and the South', Peking-Hankow-Canton, i.e. medievally speaking between Cathay and Manzi, or geologically speaking, the whole length of the Neo-Cathaysian Geosyncline, having only to surmount the Nanling to reach the last late southern key economic area.

[1] 酈道元 [2] 水經注 [3] 祖沖之 [4] 楊堅 [5] 煬帝 [6] 運河

State capitals, the former achieving great brilliance during the succeeding dynasty. 'This anomalous situation', says Chi,[a] 'rendered the development and maintenance of a transport system linking the productive south with the political north a political necessity. It was provided by the Grand Canal, which engaged the attention of the best minds of China for more than ten centuries, and demanded countless millions of lives and a large portion of the wealth of the country for its improvement and maintenance.' As a Ming writer, Yü Shen-Hsing,[1] put it, Sui Yang Ti, by his extravagance in public works construction 'shortened the life of his dynasty by a number of years but benefited posterity to ten thousand generations. He ruled without benevolence, but his rule is to be credited with enduring accomplishments.'[b]

Without anticipating the more detailed description of the Sui waterworks, we may note here that the first Sui emperor (Wên Ti[2]) concentrated on the union of the Huai Valley with the Yangtze, greatly improving a former small canal between Shanyang and the south. A Sung book, *Ta-Yeh Tsa Chi*,[3] thus described it:

> From Shanyang (near Huaian) to the Yangtze River, the water surface of the canal was forty paces wide. Roads were constructed along both banks and planted with elms and willows. For over 2000 *li* from the eastern capital (Loyang) to Chiangtu (modern Yangchow) shadows of trees overlapped each other. An imperial resting place was built between every two post stations, and from the capital at Chhang-an to Chiangtu there were more than forty such pavilions.[c]

This canal passed right across the traditional battlefields between the north and the south. Wên Ti's successor, Yang Ti, improved the waterways north of the Huai River, and also completed the section between the Yangtze and Hangchow, much further to the south. The *Khai Ho Chi*[4] (Record of the Opening of the Canal), an anonymous Sui work,[d] throws a light on what the concentration of resources meant in forced labour to the people. Some 5,500,000 workers, including all commoners in certain areas between the ages of fifteen and fifty, assembled, and worked under the guard of 50,000 police. Every fifth family was required to contribute one person to participate in the supply and preparation of food. Those who could not or would not fulfil the demands made on them were 'punished by flogging and neck-weights'; some had to sell their children. Over two million men were said to have been 'lost' (*chê*[5]).[e] While the great cruelty involved in such mass operations in an age lacking what we should consider adequate records and communication facilities can easily be appreciated, it is a little difficult to follow the usual interpretation of the figures as

[a] (1), p. 113.
[b] Quoted from Fu Tsê-Hung's[6] Golden Mirror of the Flowing Waters (*Hsing Shui Chin Chien*[7]), ch. 92.
[c] Tr. Chi (1), p. 117, mod. Goodrich (1), p. 114.
[d] Which has been preserved in the same chapter of Fu Tsê-Hung's book already quoted.
[e] The word does not necessarily mean 'died'. They might have run away to distant provinces.

[1] 于愼行 [2] 文帝 [3] 大業雜記 [4] 開河記 [5] 折
[6] 傅澤洪 [7] 行水金鑑

implying enormous loss of life in medieval Chinese engineering projects, even allowing for a probable absence of adequate safety factors in the calculations. Perhaps there were grave epidemics caused by the unusual crowding of human masses (though the record does not hint at it), perhaps the food supplies ran short, and no doubt the protection of the workers' living-quarters from the elements was totally insufficient. But this was the way things went throughout Chinese history; condemned to eotechnic methods, the principle of 'a million men with teaspoons'[a] was the only one which could be adopted.

The reign of the Sui[b] was too short to have much effect in cultural matters, but the emperors consciously sought a renaissance of Confucianism. Before it could take effect, the immense cost and disturbances of the public works, the extravagance of military expeditions to Central Asia and especially Korea, and other causes of complaint, led to widespread unrest. Revolution broke out when Yang Ti was himself surrounded by Turkic tribes at the siege of Yenmên. Li Yuan[1] and his son Li Shih-Min[2] raised the flag of a new dynasty in +617. Their family was related both to the Sui and to the Turks; its forces took Chhang-an, and in the following year the first emperor of the Thang ascended the throne.[c]

Some have seen a certain parallel between Chhin and Han on the one hand and Sui and Thang on the other. Without pushing the analogy too far, it would seem that something of a parallel exists, in that the Legalist authoritarianism of Chhin, after unifying the country, went too far, and had to be modified by the milder government of the Han. So also the Sui, in the matter of the transport system, did what had to be done, but they did it too fast and too ruthlessly, with the result that a long period of stability could not be attained under their aegis. It remained for the house of Li to establish a regime which most historians, both Chinese and Western, have regarded as China's Golden Age.

(*f*) THE THANG DYNASTY

Building on the foundations laid by the Sui, the Thang emperors were able to enlarge the boundaries of China's influence and territorial possessions to an extent not reached since the Han. The invasions which the Turkic tribes attempted during the early years were not only successfully beaten back, but the war was carried into the nomad lands, and by +648 they accepted the sovereignty of the Chinese 'khan'. At this time also Chinese influence penetrated Tibet, a country which had not till now been unified, but whose king (the great Srong-btsan Sgam-po) accepted a Chinese princess as wife, and with her much civilising influence. For example, it is about this time that such inventions as water-mills and iron-chain suspension-bridges begin to be found in Tibet. By +660 practically the whole of Korea and Manchuria had been

[a] The phrase is Mr Ritchie Calder's.
[b] Sui prefectures in Herrmann (1), Map 36.
[c] Cf. the translations of the *Hsin Thang Shu* by Pfizmaier (66, 73).

[1] 李淵 [2] 李世民

brought under Thang rule. It was at this time that the Chinese suzerainty over Sinkiang ('the new dominion') was definitely achieved, two new provinces being created, one to the north of the Thien Shan mountains and one to the south.[a]

Maximum expansion was reached about +750, after which a slow decline and contraction set in[b] which ultimately left a dangerous legacy to the following dynasty, the Sung. Unfortunate diplomatic incidents with the prince of Tashkent led to an Islamic counter-attack, and in +751 at the Battle of the Talas River (which many consider should be regarded as one of the decisive battles of world history) the Chinese army was completely defeated. As a result Turkestan (Sinkiang) was lost to China, and the whole region, which had been a stronghold of Buddhism from its earliest times, Islamised. About the same time Mongolia became independent as the home of the Uighurs, and the Thai principalities in the south-west, occupying all of modern Yunnan and the forested mountains south and west of it, rebelled and formed a separate dynasty (Nan Chao[1]), which lasted a long time. In the north-east the situation was no better; the Chhi-tan[2] Tartars (who ultimately founded the Liao dynasties, lasting from +907 to +1168) were beginning to set up strong bases in Manchuria; and before long the previously quasi-independent state of Silla[c] in Korea absorbed the Chinese protectorates so that the whole of that country was lost to China. Nor did the good relations with the Tibetans continue; indeed, these folk became such a menace both to west China and to the dominions of the Caliph in Khurāsān and Muslim Central Asia that in +787 envoys from Hārūn al-Rashīd to the court of Chhang-an were able to arrange an Arab-Chinese alliance against them.[d]

Among the alternating periods of receptiveness and xenophobia which have characterised Chinese (as well as European) history, the Thang period was certainly one in which foreigners of every kind were welcome at the capital. Chhang-an, no less than Baghdad, became a meeting-place of international fame.[e] Arabs, Syrians and Persians came there from the west to meet Koreans, Japanese, Tibetans and Tonkinese and to discuss religion and literature with Chinese scholars in the elegant pavilions of the great city in the Wei Valley.[f] Khotanese painters such as Yüchhih Po-Chih-Na and his son made successful careers. It became a commonplace[g] for wealthy Chinese to employ Central Asians as grooms and camel-drivers, Indians as jugglers, Bactrians and Syrians as singers and actors. Fig. 22 shows a terra-cotta tomb figure[h] of this period depicting a Persian or Arabian, perhaps a major-domo requesting instructions,

[a] Thang prefectures in Herrmann (1), Map 40.

[b] Beginning with the exhausting rebellion of An Lu-Shan[3] about +756; cf. *TH*, vol. 2, p. 1429. For possible economic causes of this civil war cf. Waley (12), p. 52. Cf. Pulleyblank (1).

[c] Chinese Hsin-Lo.[4]

[d] *TH*, vol. 2, p. 1465. Map of relations with the Caliphate in Herrmann (1), 38/39.

[e] This tendency had begun already during the Sui, when there was Japanese music at the court of Wên Ti (Goodrich & Chhü, 1).

[f] Schafer (2) has analysed the material about Persian merchants in Thang dynasty tales.

[g] As Fitzgerald (1), p. 367, has well emphasised. Cf. Lips (1).

[h] In the collection of Dr Dorothy Needham; purchased at Sian.

1 南詔 2 契丹 3 安祿山 4 新羅

or a merchant offering to display gem-stones. The Chinese themselves also voyaged far;[a] the classical example is the pilgrim monk Hsüan-Chuang,[1] who was away in India gathering Buddhist sūtras from +629 to +645.[b] After initial setbacks at the beginning of the dynasty Buddhism saw a period of great expansion, and flourished in close association with the best artists of the period, as witness the Thang frescoes at Chhien-Fo-Tung[c] (Figs. 21, 23). These reflect the internationalism of the period by showing monks and lay people sometimes with brown or even red hair, and blue or green eyes, as well as occidental features. The rather excessive proliferation of temples and of monks and nuns within them was, however, ultimately checked, and in +845 the Confucian bureaucracy, probably having realised that a State within a State was growing up which might ultimately challenge the accepted foundations of Chinese society, succeeded in bringing about the destruction of 4600 temples, the secularisation of 260,000 monks and nuns, the abolition of 40,000 shrines, the confiscation of millions of acres of arable land, and the manumission of 150,000 slaves.[d]

There was one direction, however, in which the influence of Thang Buddhism was extremely fruitful, namely, its stimulus to the invention of printing. For some reason or other the Buddhists had always felt the necessity for unending repetitions of sacred names, sūtras, pictures of Bodhisattvas, holy ejaculations, etc. This is obvious, for example, to the most casual visitor to Buddhist cave-temples or other shrines. Later we shall briefly unravel the story of what is known concerning the invention of printing,[e] but it is surely significant that the earliest block-printing known is that of a Buddhist charm of +770. As Goodrich[f] well puts it, 'the time was ripe for such an invention. The Chinese had long used paper and ink; they knew how to make seals out of metals, stone, or clay; it was usual for them to take rubbings of printed inscriptions on bronze or stone. There was furthermore a demand—text-books were required by thousands after the civil examination system was instituted, and charms for prayer formulae and for warding off evils and diseases were desired in Buddhist and Taoist circles.' The most probable date for the first experiments, the results of which have not come down to us, is the middle of the +6th century.[g]

The Thang emperors by no means favoured Buddhism or Taoism at the expense of Confucian orthodoxy. Though systematic written examinations had been begun in the Sui,[h] they did not become stabilised and definitive until the Thang—oral and unorganised examinations had of course been currently used since the Han, as we have seen. Orders that Confucian temples (*wên miao*[2]) should be set up in every city, and that the civil officials should make sacrifices in them, were issued in +630. An

[a] Cf. Mills (3).
[b] Cordier (1, vol. 1, pp. 551 ff.) has a good chapter on the pilgrims in general.
[c] See Pelliot (25); Stein & Binyon (1); Shor & Shor (1); Bohlin (1); Wright (1).
[d] *TH*, vol. 2, p. 1491.
[e] See Sect. 32.
[f] (1), p. 137.
[g] A small printed notice 'Beware of the dog' has survived from as early as +594 (see Schindler, 4).
[h] Cf. Shryock, (1).

[1] 玄奘 [2] 文廟

Fig. 18. General view of the Thousand Buddha Caves (Chhien-Fo-Tung) near Tunhuang in Kansu (orig. photo.).

Fig. 19. Central shrine of a cave-temple of Northern Wei date (*c.* +5th century) at Chhien-Fo-Tung, Tunhuang (cave no. 248, 249; orig. photo.). Predominant colours, grey, dark red, maroon, purple, cream and bright blue.

Fig. 20. Fresco paintings in a cave-temple of Northern Wei date (*c.* +5th century) at Chhien-Fo-Tung, Tunhuang (cave no. 83; orig. photo.). A constellation god in the right-hand top corner.

Fig. 21. General view of a cave-temple of Thang date (*c.* +8th century) at Chhien-Fo-Tung, Tunhuang (cave no. 255; orig. photo.). Predominant colours, pale red, pale green; with darker blue, green and brown in ceilings.

ordinance of +647 set forth the Confucian sages whose tablets they were to contain. The systematisation of examinations led to the re-organisation of the university (Kuo Tzu Chien[1]), which about the middle of the century had some 5000 students, as many as Cambridge today.[a] Besides this, an Imperial Academy (the Han-Lin Yuan,[2] the 'Forest of Pencils') was founded in +754, and continued until the very end of the Chhing.[b] If we may consider it the forerunner of the Academia Sinica of today, the latter is older than any existing European Academy by nearly a millennium.

Another direction in which literary studies developed was the codification of law. In general Chinese thinkers were from the earliest times averse to such standardisation, wishing to avoid the abstract and believing that every case should be judged on its merits. Later we must investigate how far this absence of codified jurisprudence affected the development of scientific thought (Sect. 18). Certain it is that, though China has few figures comparable with Justinian's jurists and the members of the School of Beirut, Thang scholars produced a new code which dealt mostly with criminal law. Civil law had been previously, and remained, very little used in China, partly because a genius for compromise had long been characteristic of private and commercial life.

Literary activity as such, however, was the particular glory of the Thang period. The poets Li Pai,[3] Tu Fu[4] and Pai Chü-I[5] are world-famous, but there was a host of others, of whom Wang Wei,[6] Wei Ying-Wu[7] and Mêng Hao-Jan[8] are among the greatest.[c] Painters and sculptors achieved levels never before reached. Models of Chinese prose for all time were set by the great Confucian Han Yü,[9] known for his courageous polemics against Buddhism,[d] and an outstanding prose work was the *Thung Tien*[10] encyclopaedia of political history by Tu Yu.[11] It is interesting that many of the poets and prose-writers of the Thang were deeply affected by Taoism.[e]

For the historian of science, however, the period is not so interesting as that which follows it, the Sung. The two dynasties have entirely different atmospheres, the Thang humanistic, and the Sung much more scientific and technological. Nevertheless, there is no doubt that the Thang Taoists were very active in alchemy, and (as we shall see, Sect. 33) some chemical texts of that time are still available to us, though one has the impression that there were great losses when the Taoist Canon (the *Tao Tsang*[12]) was 'purged' in the Sung, on account of the sexual techniques for

[a] The rudiments of the university went back beyond the Chin, when it was called Kuo Tzu Hsüeh.[13] Cf. *TH*, vol. 2, p. 1320, and p. 106 above.

[b] It should be understood that this was not an academy for research only, though individual members of it doubtless, like the imperial librarians, spent much of their time in literary research. It had charge of all the literary activities of the imperial court, and its members had to draft edicts, prepare State papers, propose forms for prayers at State sacrifices, and so on.

[c] Many fine translations are available, notably those of Waley (2, 3) and Chhu Ta-Kao (1). By far the best book known to me on Chinese medieval poetry is that by Lu Chih-Wei (1). Full historical details in Nagasawa (1), pp. 192 ff.

[d] Tr. *TH*, vol. 2, p. 1473; Margouliès (1), p. 198.

[e] Cf. Belpaire (1) and Li Chhang-Chih (*1*).

[1] 國子監 [2] 翰林院 [3] 李白 [4] 杜甫 [5] 白居易
[6] 王維 [7] 韋應物 [8] 孟浩然 [9] 韓愈 [10] 通典
[11] 杜佑 [12] 道藏 [13] 國子學

attaining longevity discussed in it. The greatest problem here for the history of science is the real effect of the numerous books which appeared in or before the Thang with titles showing that they were compiled from Indian or West Asian sources, and it is a hard one, since all these texts were afterwards lost and references to them are scarce.

The official history of the Sui dynasty, completed in +636 by Wei Chêng,[1] contains in its bibliographical catalogue[a] a number of books all beginning with the words 'Po-lo-mên' or Brahmin.[b] Thus we have *Po-lo-mên Thien Wên Ching*[2] (Brahmin Astronomical Manual), *Po-lo-mên Suan Fa*[3] (Brahmin Mathematics), *Po-lo-mên Yin Yang Suan Li*[4] (Brahmin Calendrical Methods), *Po-lo-mên Yao Fang*[5] (Brahmin Pharmaceutics), etc. The same bibliography also lists a *Hsi-Yü Ming I so chi Yao Fang*[6] (The Best Prescriptions collected by the most famous Physicians of the Western Countries). It would make a good research problem to try to work out what the effects of these books are likely to have been; we may find some hints as we go on.

It is also significant that a small number of Buddhist scholars achieved an abiding fame for their work in science at this time. The monk I-Hsing[7] particularly comes to mind; noted as an astronomer and mathematician (cf. Sects. 20, 26), he calculated rather precisely the magnitude of the sidereal year fraction. It is probable that the magnetic needle was first used, for geomancy, in this period.

With regard to the famous physicians of the Western countries referred to above, one must not forget the influx of religions new to the Chinese at this time. Zoroastrianism arrived early in the +6th century from Persia, and a hundred years later was a recognised cult under government control.[c] Then came Nestorian Christianity carried by Syrian monks about +600; it seems doubtful whether it succeeded in getting any great following among the Chinese, though its church-temples were quite numerous.[d] The famous stele called the 'Nestorian Stone', erected in +781, is still standing in the museum next door to the Confucian temple at Sian. At the end of the +7th century arrived Manichaeism[e] from Persia. It made its fortune at first by converting the Uighur people, among whom were also many Nestorians, but when the Uighur State declined, it declined also, and by +843 it was dead in China. There are records that Sogdian Manichaean priests made contributions to astronomy.[f]

[a] There is no translation of this, but its counterpart in the *Hsin Thang Shu* has been put into German (very badly, it must be said) by Pfizmaier (70, 71, 72).

[b] Altogether twelve are listed in Yin-Tê Index No. 10 (the index to the bibliographies in all the official histories), including three on astronomy, three on mathematics, and three on pharmaceutics. Cf. Wieger (3), p. 180; Bretschneider (1), vol. 1, p. 93 (who adds another three); Mikami (1), p. 58; Sarton (1), vol. 1, p. 450.

[c] The last Sassanid king and his son were given refuge by the Thang emperor in +674. See below p. 214. Cf. *TH*, vol 2, p. 1347; Chhen Yuan (*1*); Huart & Delaporte (1), p. 361.

[d] Saeki (1); A. C. Moule (1).

[e] Burkitt (1); cf. *TH*, vol. 2, pp. 1383, 1486; Chhen Yuan (*2*).

[f] Goodrich (1), p. 129; Yule (2), vol. 1, p. 63; one such person arrived in China in +719.

[1] 魏徵 [2] 婆羅門天文經 [3] 婆羅門算法 [4] 婆羅門陰陽算曆
[5] 婆羅門藥方 [6] 西域名醫所集要方 [7] 一行

Fig. 22. Thang tomb-figure in terra-cotta depicting a Persian or Arabian merchant or major-domo (collection of Dr Dorothy Needham). Height $8\frac{1}{2}$ inches.

Fig. 23. A military guardian deity (*lokapāla*) in one of the cave-temples of Thang date (*c.* +8th century) at Chhien-Fo-Tung, Tunhuang (cave no. 12 or 305; orig. photo.). On the back wall fresco, a scene of battle between good spirits and demons; on the side wall, the Western Paradise. Statues in reinforced plaster, about life size.

Judaism and Islam[a] also touched the borders of China in the Thang, but did not take root until much later.[b]

One would expect to find considerable technological influences in connection with all this coming and going, but China seems still to have given much more than she received. Besides printing, mentioned above, the mastery of the art of porcelain manufacture now added a new achievement to Chinese technology. The discovery that a feldspathic mineral can be made fusible and used as a glaze may have been accidental, but it led to the discovery of a method of incorporating the feldspar in the body of the stoneware itself, and so to true porcelain.[c] By the beginning of the +9th century the art had made great strides forward, and the jade-green celadons were being produced. As export articles the beautiful objects were in great demand, and quickly spread to the west. Broken pieces of Thang porcelain have been found in the +9th-century Mesopotamian city of Samarra, and in the rubbish-heaps of al-Fusṭāṭ (Old Cairo) in Egypt. The appreciation of the Arab merchant and traveller Sulaimān in +851 has often been quoted, and we shall cite it in the proper place.

In such an age of international intercourse, cartography could not but flourish, so it is not surprising to find at least one eminent name, that of Chia Tan,[1] whose map of +801, the *Hai Nei Hua I Thu*,[2] was long famous.[d]

In economics, the State continued to fulfil its traditional role, and many new water-works were undertaken, as well as the repair of old ones. It continued to press heavily on the peasant-farmers, but even more so on the merchant groups, however wealthy the new external trade was making them. The merchants would have wished to use their capital accumulations in manipulating food stocks (since the farmers could never regulate market prices) as well as investing in land, but they were greatly inhibited by the failure of an independent banking system to arise.[e] The individual merchant or his guild continued to be the only source of credit. Confucian officialdom was always in fear of a ruination of the agricultural basis of production by the speculative activities of merchants, and now the Thang returned to the Han expedient of 'compensation offices' or 'ever-normal granaries'. Government buying during gluts prevented prices falling, and conversely release of seed grain at the right time of year prevented scarcity; the government storehouses were also, at any rate partially, effective against minor famines. The State continued nervous, however, of substantial capital accumulation by merchants, whether derived from hand industry or trade, and thus must be explained the nationalisation of the tea export during the Thang. The amassing of wealth was not frowned upon if done by the officials of the bureaucracy, and it was even expected that they should make up their salaries by lending money upon interest,

[a] Cf. *TH*, vol. 2, p. 1355.
[b] Map of religious sites in Herrmann (1), 45. General discussion by Pelliot (7).
[c] The earliest achievement of this seems to have occurred before +400, when a bluish-green porcellanous ware was being produced at Wênchow in Chekiang (see Sect. 35).
[d] Cf. *TH*, vol. 2, p. 1466; and Sect. 22 below.
[e] This did not happen until the rise of the Shansi bankers in the Ming.

[1] 賈耽 [2] 海內華夷圖

but in the end this only strengthened the merchants, who showed much more experience and skill in handling funds than the officials.

Towards the end of the dynasty, peripheral weakening allowed the rise of numerous semi-feudal semi-autonomous states with hereditary rulers which the bureaucratic central government was hardly able to control. After +875 rebellions became general in Shantung, in Shansi and in Kuangtung, and by +906 there could no longer be any question of a Thang emperor. The empire fell apart for the third time for some fifty years, during which five dynasties succeeded one another in unstable equilibrium.

(*g*) TIME OF THE FIVE DYNASTIES AND THE TEN INDEPENDENT STATES

Chinese history recognises the term 'Five Dynasties' (Wu Tai[1]) as an understatement, since at least ten other independent states co-existed with them. But the shifting mists of the +10th century can be illuminated by the light of the concept of key economic areas—China had in fact simply reverted to its original cellular unaggregated state.

All the dynasties which historians afterwards agreed, somewhat arbitrarily, to consider as having carried on the 'apostolic succession' of the Son of Heaven had their capitals in the easterly zone of the northern economic area. The first, the Later Liang[2] (+907 to +923), settled at Khaifêng,[3] a city which had not hitherto been a capital, but they did not control by any means all of the area they were in. A separate State, Chin,[4] held the North China Plain, and to the west on the Honan side of the Yellow River valley was another, Ching-Nan.[5] Further to the north the Chhi-tan Tartars (actually a Mongol people, descendants of the Hsien-pi) were establishing their long-lasting Liao dynasty. To the west, the cradle of Chinese civilisation in the Wei valley was in barbarian hands.

Similarly, the east-central economic area of the Yangtze valley was also split up. North of the river was the State of Huai-Nan[6] (strange echo of the scientifically minded prince of the Han); south of it was Wu-Yüeh[7] (an equally anachronistic title),[a] and higher up the valley was independent Hu-Nan.[8] Even in the Far West the economic area was split, Szechuan being independent as Shu[9] and the Nan Chao continuing in Yunnan.[b] To complete the picture, the Southern Economic Area, which appears for the first time, was independent under the title Ling-Nan.[10] And this by no means completes the list of petty States.[c]

Badgered about by difficulties to right and left, the imperials maintained their location, oscillating up and down the Yellow River to Loyang and back as conditions

[a] Description by Chavannes (2).
[b] The history of this western dynasty has been translated into French by Sainson (1).
[c] Cf. *TH*, vol. 2, Map 21, opp. p. 1528; see also Herrmann (1), Map 41.

[1] 五代 [2] 後梁 [3] 開封 [4] 晉 [5] 荆南 [6] 淮南
[7] 吳越 [8] 湖南 [9] 蜀 [10] 嶺南

permitted. Later Liang was succeeded after sixteen years by Later Thang,[1] which lasted thirteen (+923 to +936), Later Chin,[2] lasting ten (+936 to +946), the second Later Han,[3] unworthy successor of its namesake, of only four years' duration (+946 to +950), and finally the second Later Chou,[4] nine years long and even more absurdly named (+951 to +960). All except the first and last of these were the creation of Turkic Sha-tho families. By this time the sands had run out and a new unifier appeared.

What is not quite clear is why it was that at this Third Partition period the fragmentation was so severe as to dissect even the economic areas, especially remembering that nearly a thousand years of hydraulic engineering work had been put in on the water supplies for irrigation and the waterways for grain transport. It is insufficient, with Cordier, to compare the numerous ruling houses of this time with medieval European robber barons, attractive though the analogy might superficially be. More convincing is the fact, brought out by Chi, that there had been great neglect of the waterworks during the latter part of the Thang, hardly any having been undertaken after +750. He also believes that in spite of the great advances in communications which had been made in the east-central area, it still formed a series of geographical units very loosely bound, thus permitting extreme fragmentation to occur. But it seems unlikely that all the factors in this process have yet been analysed, and one may have a suspicion that new forms of warfare, especially the use of gunpowder, the first reference to which may take us back to the beginning of the partition, i.e. +919, may have had a temporarily disintegrating effect (see Sects. 30, 34).

The most curious thing about this time of confusion was that printing took great steps forward. A beginning was made of cutting the Confucian classics on wood blocks in +932, and the work was finished in +953; after that time, the books printed from them in 130 volumes, were widely diffused throughout the country. Already before the end of the Thang, Taoist books had become fairly common in Szechuan, where they were seen by an imperial official in +883 at Chhêngtu (Sect. 32). It was because Szechuan had been under imperial control for a few years from +929 that knowledge of printing had been brought from Chhêngtu to Loyang. The process went steadily on in spite of all political disturbances. Both Buddhists and Taoists printed their respective canons (or patrologies) immediately after the establishment of the Sung, the former being finished in +983 and the latter in +1019.

It is interesting to reflect that this fragmentation represented a state of affairs which in Europe would have been thought perfectly natural. But the Chinese, in spite of the vastness of their country, had a tradition of unity. This alone indeed might not have been enough to have kept them together, but in conjunction with the algebraic quality of their language, where meaning was independent of pronunciation, and the written character was universal, it meant that a tendency to unity must dominate. Only the man was wanting, and he appeared in the person of Chao Khuang-Yin,[5] who in +960 was raised to power by the army after the manner of the later Roman emperors. He and his house kept it much longer than most of them.

1 後唐 2 後晉 3 後漢 4 後周 5 趙匡胤

(*h*) THE SUNG DYNASTY, AND THE LIAO AND CHIN (TARTAR) DYNASTIES

No sooner had the new Sung empire been definitely established than the emperor, Thai Tsu,[1] called together the army commanders who had helped him in his rise to power, and there followed a scene which is one of the most interesting in Chinese history.

The emperor invited them all to a banquet, and when the company had drunk deeply and were in cheerful mood, he said, 'I do not sleep peacefully at night.' 'For what reason?' enquired Shih Shou-Hsin[2] and the other generals. 'It is not hard to understand,' replied the emperor. 'Which of you does not covet my throne?' The generals made deep bows and all protested, 'Why does your Majesty speak thus? The mandate of Heaven is now established. Who still has treacherous aims?' The emperor replied, 'I do not doubt your loyalty, but if one day one of you is roused at dawn and forced to don a yellow robe, even if unwilling, how should he avoid being obliged to overthrow the Sung (just as I against my will was forced to overthrow the Chou?)' All protested that none of them was sufficiently talented for such a thing to be thought of, and asked for his counsel. The emperor said, 'The life of man is short. Happiness is to have the wealth and means to enjoy life, and then to be able to leave the same prosperity to one's descendants. If you, my officers, will renounce your military authority, retire to the provinces, and choose there the best lands and the most delightful dwelling-places, there to pass the rest of your lives in pleasure and peace until you die of old age, would this not be better than to live a life of peril and uncertainty? So that no shadow of suspicion shall remain between prince and ministers, we will ally our families with marriages, and thus, ruler and subject linked in friendship and amity, we will enjoy tranquillity.'... The following day, the army commanders all offered their resignations, reporting (imaginary) maladies, and withdrew to the country districts, where the emperor, giving them splendid gifts, appointed them to high official positions.[a]

Thus was ushered in a period during which the empire, though not fortunate in its military exploits and hemmed in by more or less barbarian States, rose to a height of culture and science previously undreamed of. Carter[b] has well described it as a time of maturity. Lyric poetry gave way to learned prose, religious faith to philosophical speculation and scientific description,[c] while in technology, the Sung period put to practical use what the Thang had conceived.

Before discussing these cultural and scientific achievements, a brief outline of the political situation from +960 to +1280 must be given. During most of this time the Liao kingdom in Manchuria, Mongolia, and the northern part of the North China plain, under nine emperors, continued (after long-drawn-out hostilities ending in

[a] *Hsü Thung Chien Kang Mu*, ch. 1, p. 24*b*; tr. Wieger, *TH*, vol. 2, p. 1559; eng. Fitzgerald. The passage is based on *Sung Shih*, ch. 250, p. 2*a*.

[b] (1*a*), p. 55.

[c] But the ornamentation of the Tunhuang caves went on (Fig. 24).

[1] 太祖 [2] 石守信

Fig. 24. A Buddha or Boddhisattva in a cave-temple of Sung date (*c.* +12th century) at Chhien-Fo-Tung, Tunhuang (cave no. 164; orig. photo.). Predominant colours, green, black, blue, white, gold.

+1005) on fairly friendly relations with the Sung empire, sweetened by the payment of an annual consignment of gold and silk. This 'danegeld', a strange reversal of the usual direction of tribute from neighbours to the imperial court, set the style for systematic payments to barbarians which lasted throughout the Sung time. In the end the Chhi-tan[1] Tartars of the Liao (who, as we have seen, were really descendants of the Mongol Hsien-pi) became so sinified that the customs of their territory were indistinguishable from those of China proper. Their rule in North China, which often extended as far down as the Yellow River, came to an end in +1125, when, after a decade of fighting, they were driven away to the west by the Jurchen,[2] a nomadic Tungusic people from the Amur in the far north, and the forefathers of the Manchus who were to conquer all China five hundred years later. The remnants of the Liao were welcomed in Central Asia by the Uighurs, their former vassals, and established a new State in Sinkiang and Turkestan known as the West Liao (Qarā-Khiṭāi), lasting until +1211, when it was swept away in the general cleaning of the slate by Chinghiz Khan.[a] The Liao State, even in its prime, was essentially uncultivated, and offers little of interest from the point of view of the history of science and technology. Its only scholarly achievements were the preparation of good versions of Buddhist texts, and the introduction of an alphabet using characters derived from Chinese in +925.

Unlike the Chhi-tan Liao, the Jurchen, who took the name Chin[3] or Golden for their dynasty, were on permanently bad terms with the Sung and fighting never really stopped until they also were swept away by the Mongols under Chinghiz in +1234. After +1142 heavy danegeld was paid by the Sung empire to them. Nevertheless, and in spite of appeals by Chin rulers to the contrary, this State also became almost entirely sinified. Whoever occupied the northern key economic area and took the profit from it, Chinese officials administered it and Chinese peasant farmers tilled it.

The third of the northern principalities was a particularly interesting one, the Hsi-Hsia[4] dynasty of the Tanguts, a Tibetan tribe, which set itself up athwart the Old Silk Road, occupying the modern provinces of Kansu and Ninghsia, with the capital at Ninghsia. Their first emperor began his reign in +1032 and the State lasted, benefiting from Sung danegeld, until the time of Chinghiz, when in +1227 it was utterly destroyed. It must have been the scene of great literary activity, for many thousands of Tangut writings have survived, written in the extremely complicated Hsi-Hsia script, which looks not unlike Chinese.[b] Confucian, Taoist and Buddhist texts were translated into Hsi-Hsia, and there was a Hsi-Hsia learned academy.

[a] It was the foundation of the Western Liao which gave rise to the legend of Prester John in Europe, first mentioned in +1145 (Cordier (1), vol. 2, p. 372). Both the Liao kingdoms have been the subject of detailed study in the large monograph of Wittfogel, Fêng Chia-Shêng *et al.* (1).

[b] Cf. Laufer (2); Wang Ching-Ju (*1*).

1 契丹 2 女眞 3 金 4 西夏

Meanwhile great activity was going on in the Sung, the heart of which was the east-central economic area. The Sung capital was first at Khaifêng,[a] but in +1135 the Chin armies made a disastrous raid on it which resulted in the capture of the emperor and almost all the government officials—what remained fled southward with one of the princes and set up the Southern Sung empire, as it is called, at Hangchow.[b] From that time on the frontier was along the Huai rather than the Yellow River.[c] Nevertheless, neither this disaster nor the continuous wars prevented the Sung from making improvements in communications. After the time of fragmentation described in the previous section it would be expected that the Sung should have paid great attention to water conservancy engineering, and so they did, no less than 496 projects being brought to fruition under the Sung as against ninety-one under the Thang.[d] Their activity in this field was no doubt motivated partly by the necessity to find employment and food for the large accessions of population which fled from the Chin Tartars in the north. So great was the pressure that there was a big development of marsh and lake drainage,[e] but this in turn raised almost as many problems as it solved.

Whenever one follows up any specific piece of scientific or technological history in Chinese literature, it is always at the Sung dynasty that one finds the major focal point. This is as true for the applied as for the pure sciences. Improvements in hydraulic engineering, as well as its intensified use, are to be noted, such as lock-gates and new surveying instruments. In bridge-building the use of the ingenious transverse shear-wall and the caisson were introduced. In architecture the classical Chinese compilation on the subject, the *Ying Tsao Fa Shih*[1] of Li Chieh,[2] appeared about +1100. Shipbuilding took great strides, the sternpost rudder, which had probably been invented some time before, became general, and sea-going junks of considerable size were built.[f] Treadmill-driven paddle-wheel ships, again not new, became more general. Most important of all, chemical science, pursued in the obscure laboratories of Thang Taoist temples, had really found something at last, and gunpowder was made available for the armies—the Chin-Sung wars were its first proving-ground. Later we shall follow with attention the story of the discovery and application of the first explosive known to man; here it must suffice to say that by +1000 at least, explosive grenades and bombs were being fired from catapults. By +1040 the great work *Wu Ching Tsung Yao*[3] stabilised the Chinese name for gun-powder[g] and registered an extremely rapid development of projectile weapons,

[a] Map in Herrmann (1), 48.
[b] Cf. Ferguson (4).
[c] Maps of Sung, Chin, Liao and Hsi-Hsia in Herrmann (1), 42, 43, 46, 47.
[d] A special study has been devoted to this by Chhüan Han-Shêng (*1*).
[e] Chi (1), p. 134.
[f] Brief general account by Mills (3). See Sect. 29 below.
[g] *Huo yao*[4] ('fire chemical').

[1] 營造法式 [2] 李誡 [3] 武經總要 [4] 火藥

poisonous and signalling smokes, flame-throwers, and other novel inventions. These were used in naval as well as in land warfare.

If chemistry was thus applied in destructive war, the biological sciences also flourished, and were applied for the benefit of man. Many famous physicians appeared in the Sung and old systems such as those of pharmaceutics and acupuncture were codified, while relatively new discoveries such as variolation (precursor of vaccination) were diffused. About +1111 an Imperial Medical Encyclopaedia was compiled by twelve of the most eminent medical men of the time (the *Shêng Chi Tsung Lu*[1]). The books of pharmaceutical botany reached an unprecedentedly high standard, and several recensions of the *Pên Tshao* or Codex were produced, notably not only in the Sung but also in the northern Chin country. Thus the illustrations in certain editions of the *Ta-Kuan Ching Shih Chêng Lei Pên Tshao*[2] of the +12th and +13th centuries were better than those of European botanical books of the +15th and early +16th. Particularly characteristic of the period, however, are the numerous botanical and zoological monographs, of which Han Yen-Chih's[3] *Chü Lu*[4] (Orange Record) of +1178 may be considered a type-specimen; it deals in detail with all aspects of citrus horticulture, and was the first book on the subject in any language—but besides this there were monographs on bamboos, *lichis*, aromatic plants, cucurbits, and flowering trees, as well as Crustacea, birds and fishes.

This class of literature goes over into another, that of the 'miscellaneous notes and records' (*pi chi*, pen records, or *pi than*, pen conversations) in which numerous scientific observations are to be found. Here the type-specimen is the *Mêng Chhi Pi Than*[5] (Dream Pool Essays) of Shen Kua,[6] perhaps the most interesting character in all Chinese scientific history. Shen Kua was born in +1030 and had the usual career of a scholar in government service; he was at various times ambassador to the Hsi-Hsia and other countries, military commander, director of hydraulic works, and chancellor of the Han-Lin Academy. But wherever he went on his numerous travels, and however much occupied with his official duties, he never failed to note down all that was of scientific or technical interest. His *Mêng Chhi Pi Than*, the date of which is about +1086, is one of the first books to describe the magnetic compass, and later we shall take it as a fixed point in the analysis of this subject; but it also contains much astronomy and mathematics, together with notices of fossils, the making of relief maps and other matters of cartographic interest, descriptions of metallurgical processes, and a high proportion of biological observations. The scientific parts form rather more than half the book.

It is really surprising that no translation of the *Mêng Chhi Pi Than* into any Western language has ever been undertaken. In view of its importance as a landmark in the history of science in China I shall pause here to give a tabular analysis of the material contained in it. In later sections numerous extracts from the book will find their place as the story proceeds. The book has twenty-six chapters, with an additional four as

1 聖濟總錄 2 大觀經史證類本草 3 韓彥直 4 橘錄
5 夢溪筆談 6 沈括

appendices, and each of these is divided up into from fifteen to thirty paragraphs or entries. The distribution of these paragraphs is roughly as follows:

	No. of paragraphs	
Official life and the imperial court	60	
Academic and examination matters	10	
Literary and artistic	70	
Law and police	11	
Military	25	
Miscellaneous stories and anecdotes	72	
Divination, magic and folklore	22	
Total of humanistic material		270
On the *I Ching*, Yin and Yang, and 5 elements	7	
Mathematics	11	
Astronomy and calendar	19	
Meteorology	18	
Geology and mineralogy	17	
Geography and cartography	15	
Physics	6	
Chemistry	3	
Engineering, metallurgy and technology	18	
Irrigation and hydraulic engineering	6	
Architecture	6	
Biological sciences, botany and zoology	52	
Agricultural arts	6	
Medicine and pharmaceutics	23	
Total of natural sciences		207
Anthropology	6	
Archaeology	21	
Philology	36	
Music	44	
Total of humanistic sciences		107
		584

Thus of science, taken in its broadest sense, almost three-fifths of the book is composed.

A little-known, and very short, book has survived from the pen of Shen Kua, which gives us a fascinating glimpse of the travelling methods of this omnivorous observer; it is the *Wang Huai Lu*.[1] In it he describes the best way to make an official inspection-carriage comfortable, how to arrange that from the lacquered interior one shall get a good view of the scenery or other objects of interest, and what are the most

[1] 忘懷錄

essential things to take with one in one's travelling cases. Here he mentions a raincoat, a chest of medicines, plenty of spare clothes and combs, a box of preserved food supplies and tea, another containing paper, ink, scissors, a rhyming dictionary and a lute. Candles, knives, chessmen and a folding chessboard are not to be forgotten, and a box should be prepared to receive books which may be bought en route, together with some insecticide powder[a] for keeping worms away from them. Lastly, he advises that 'mud-boots'[b] should be taken, as there are some places which cannot be reached without them.

Shen Kua was a friend of the famous poet Su Tung-Pho,[1] with whom he collaborated in a book on medical prescriptions, the *Su Shen Liang Fang*.[2]

Shen's interest in mathematics has been mentioned; the Sung produced some of China's greatest mathematicians of any age, particularly Chhin Chiu-Shao,[3] and Li Yeh[4] and Yang Hui.[5] Possibly due to stimulation by Arab influence (though there is no clear evidence even for a transmission of ideas), the Sung was the period of the great algebraists, and China was the country where during these centuries this branch of mathematics reached its highest level.[c]

Nor did humanistic studies[d] lag behind. The prose of Ouyang Hsiu[6] continued the tradition of Han Yü, and Ssuma Kuang[7] in +1084 produced the first complete history of China up to his time, the *Tzu Chih Thung Chien*,[8] which was rewritten and condensed by the Neo-Confucians into the *Thung Chien Kang Mu*, finished about +1190. A notable chronological encyclopaedia was prepared by a group under the editorship of Li Fang;[9] the *Thai-Phing Yü Lan*[10] of +983, which gives systematic quotations from ancient and medieval authors arranged under subject-headings of every kind in 1000 chapters, and is still a work of first-rate value for consultation today. Then there was a geographical encyclopaedia, *Thai-Phing Huan Yü Chi*[11] in 200 chapters, compiled by Yüeh Shih.[12] Chêng Chhiao's *Thung Chih* has already been mentioned (pp. 36, 74); it was carried further in the famous *Wên Hsien Thung Khao*[13] of Ma Tuan-Lin,[14] finished in +1254. The tremendous volume of intellectual effort needed for these monuments of scholarship was no doubt the product of a general rise in the level of education which took place in this dynasty, manifested, for example, by the crystallisation of private schools or academies (*shu yuan*[15]), financially aided by benefactors, around distinguished scholars.

Another aspect of the movement was the rise of the Neo-Confucian school of philosophers, whose greatest exponent was Chu Hsi[16] and whose relation to the progress of scientific thought we must later examine.[e] Theirs was an empirical rationalism,

[a] *Yün tshao*;[17] we shall meet it again in Sect. 42.
[b] Cf. Sect. 41.
[c] See Sect. 19 below.
[d] Cf. Nagasawa (1), p. 227.
[e] See Sect. 16 below.

1 蘇東坡 2 蘇沈良方 3 秦九韶 4 李冶 5 楊輝
6 歐陽修 7 司馬光 8 資治通鑑 9 李昉 10 太平御覽
11 太平寰宇記 12 樂史 13 文獻通考 14 馬端臨 15 書院
16 朱熹 17 芸草

a kind of scientific humanism, perhaps a reaction to Buddhist theology which had shown up the absence of cosmological explanations in traditional Confucianism.[a] Taoism, however, was also very much alive during the Sung;[b] the printing of its canon has already been mentioned, and the majority of existing Chinese medieval alchemical texts, some of which are well illustrated with diagrams of apparatus, date from this time. All the schools of thought were now in close contact with one another, and it is significant that one of the most important commentaries on the oldest Taoist alchemical book is from the pen of Chu Hsi himself (cf. Sects. 13, 33).

It remains to say a word or two about the economic life of the period. It produced the second of the two great reformers of Chinese history, Wang An-Shih.[1] Unlike Wang Mang, he never attained, or desired, the imperial yellow, but like him was interested in practical matters, technology and science; he himself referred to his study of books on botany, medicine, agriculture and weaving. After +1069, when he became minister, he introduced a long series of reforms which, partly owing to the storm of opposition which they aroused, became famous ever afterwards. His methods were primarily financial. He first reorganised the finance ministry and by administrative rationalisation, economies at court and measures to stop peculation, managed to save as much as 40 per cent of the national budget. He then proposed to abolish the ancient system of transporting tax grain in kind to the capital, substituting for this a system of government warehouses in all large cities, from which the produce could be sold on the spot and the taxes remitted to the central government in money. At the same time state money-lending arrangements were made—another attack on the merchants as source of credit—so that farmers could obtain advances on the security of growing crops, at rates of interest less than those commercially obtaining. New land surveys were carried out, and the taxes based upon them. Commutation of forced labour by money payments was made possible (cf. the decline of corvée in late European feudalism). By the *pao chia*[2] system, every ten families formed a group, all the members of which were held responsible for any misdeed committed by any one of them, and this was coupled with a conscription law whereby each of such groups had to furnish a stated number of men for the army. Large property owners had similarly to furnish horses. By this means the size and power of the standing army were minimised. Further anti-mercantile laws bore heavily by taxation on hoarding of commodities, and restricted the production of luxury goods.

Opposition of extraordinary intensity then unleashed itself, and Wang An-Shih had against him most of the famous scholars and officials of his time.[c] It can only be

[a] Cf. Nagasawa (1), pp. 266 ff.

[b] The third Sung emperor engaged in elaborate Taoist mystifications in order to increase the imperial prestige; see the amusing translation in *TH*, vol. 2, pp. 1572–82. I agree entirely with Fitzgerald (1), p. 384, as to the justifiability of his political motives for this.

[c] The annals use the term *tang*[3] for the groups of disputants; this is the word which in modern times came to mean a political party. For an account of the personalities engaged see *TH*, vol. 2, p. 1595, and Fitzgerald (1), pp. 391 ff. Later the Reforming Party became associated with Taoism, which was a not unnatural development in view of the conservative nature of Confucianism.

[1] 王安石 [2] 保甲 [3] 黨

explained by saying that the measures of Wang An-Shih (which in any case demonstrate the originality of the Sung time in economics as in so many other fields) went against the grain of Chinese feudal bureaucratism, though intended essentially to strengthen and reinforce it. It could not, in fact, sustain the imposition of a money economy. Had Wang An-Shih been able to win the support of the masses of the agricultural population, he might have been able to carry out his programme, but they, while glad of the liberation from forced labour, resented the conscription methods, and especially the *pao chia* system, under which, given the tyranny of the petty police, the innocent would often suffer for the guilty. The officials and the scholar-gentry, adjusted for more than a thousand years to the fundamental conceptions of the use of the people in the construction of waterways on which tax grain would be transported, felt an abyss opening beneath their feet. If the peasants were no longer to be commanded in this way, if what seemed a dangerous reliance on unstable currency (paper money had become general from +970 after an initial trial under the Five Dynasties about +910) were to take the place of the good old long-hauls of grain in kind, and worst of all, if quantitative accounting were to restrict or close the field for peripheral 'squeeze' and the modest enrichment of functionaries which everyone expected—the end of the world had come.[a]

Wang An-Shih died in retirement and defeat in +1086 (just as Shen Kua was finishing his essays), though under other ministers, such as Tshai Ching,[1] his policies were largely carried through until the collapse of the dynasty. Wang An-Shih merits great praise for the reforms which he introduced into the examination system, though their effects were not very long-lasting. While they were in force, candidates laid aside the classics and the poets, and studied history, geography, economics, law and medicine.[b]

What seemed to be the end of the world did actually come in the middle of the +13th century.[c] At the beginning of it (+1204) Chinghiz was proclaimed khan of all the northern nomadic Mongols, and shortly afterwards attacked the Chin Tartars, winning Peking in +1215. Hsi-Hsia fell in +1227 and the city of Khaifêng in +1233. For the next forty-five years the Mongols fought the Sung, finding in them the technically best equipped and the subtlest fighters of any that they had had to face in Europe or Asia. But by +1276 the Mongols had overrun practically all the Sung territory,[d] and three years later the dynasty was extinguished when the last surviving Sung prince perished in a sea battle.[e]

[a] The government's difficulties were of course greatly increased by the drain of metal caused by the 'danegeld' paid to the northern neighbour-countries. Currency also left China in large amounts to Japan and the south seas in exchange for imports.

[b] This was not an entirely new idea. In +958 Shuang Chi,[2] who had been an official of the Later Chou dynasty, had gone to Korea and advised the Koreans to introduce mathematics, geography and medicine into their examination system. Cf. Balazs (4) on Li Kou.

[c] *TH*, vol. 2, pp. 1653, 1665, 1669 ff.

[d] At the other end of Asia, Baghdad had fallen to the Mongols in +1258.

[e] *TH*, vol. 2, p. 1697. For the general history of the Mongol Empire see Grousset (2) and Bouvat (1).

1 蔡京 2 雙冀

(*i*) THE YUAN (MONGOL) DYNASTY

The Mongols ruled China for just over a century. The great Temujin[1] (Chinghiz Khan) never ruled China himself, for he died in the Hsi-Hsia campaign early in the +13th century (+1227). At first, under Ogotai and Mangu, the interest of the Mongols lay mostly in the West, but gradually the possibilities of China attracted them more, and Setsen Khubilai Khan concentrated attention on East Asia, becoming the first Yuan emperor as Shih Tsu[2] in +1280. The last of the line, Toghan Timur (Shun Ti[3]), fell in +1367.

This was the greatest clash in Asian history between the nomadic culture of the steppe and the civilisation of intensive agriculture.[a] History preserves a record of the astonishment with which the Mongols first saw what Chinese agriculture was like.

> When Chinghiz invaded the western countries, he did not have in his stores a single measure of rice or a single yard of silk. When (they came to the first Chinese provinces) his advisers said 'Although you have now conquered the men of Han, they are no use to us; it would be better to kill them all and turn the lands back to pasture so that we can feed our beasts on it.' But Yehlü Chhu-Tshai said 'Now that you have conquered everywhere under Heaven and all the riches of the four seas, you can have everything you want, but you have not yet organised it. You should set up taxation on land and merchants, and should make profits on wine, salt, iron, and the produce of the mountains and marshes. In this way in a single year you will obtain 500,000 ounces of silver, 80,000 rolls of silk and 400,000 piculs of grain. How can you say that the Chinese people are no use to you?'... So Chinghiz agreed that it should be done.[b]

Whereupon Yehlü Chhu-Tshai,[4] who was himself a descendant of the royal house of the Liao, organised the beginnings of a Yuan civil service from Chinese scholars, and took every opportunity of propagating Confucian doctrine at the Yuan court. Both Yehlü Chhu-Tshai[c] and a Chinese, Kuo Shou-Ching,[5] who also took high office under the dynasty, were capable men of science and set up one of the most important astronomical observatories of the age at Peking (see Sect. 20).

China under the Yuan became better known to Europe than at any previous or subsequent time until the twentieth century. This was because the region under Mongol control extended for the full breadth of the heartland; it was the first and the last time in history that the whole area north of the Himalayas from Shanhaikuan to Budapest and from Canton to Basra was under one political authority. The roads across Central Asia were busier and safer than ever before or since, and the court

[a] Cf. Lattimore (1); Vladimirtsov (1, 2); R. Fox (1).

[b] *Hsü Thung Chien Kang Mu*, ch. 19, p. 27*b*; tr. auct. adjuv. Wieger, *TH*, vol. 2, p. 1656. The passage is based on *Yuan Shih*, ch. 146, p. 4*a*, where the use of the north as a war supply base for the attack on the south is more distinctly stated.

[c] Rémusat (9) wrote a short life of him; see also Grousset (1), vol. 2, p. 427.

[1] 鐵木眞 [2] 世祖 [3] 順帝 [4] 耶律楚材 [5] 郭守敬

of the Khan was full of Europeans and Muslims who had some skill or craft to practise as well as of ambassadors from Tibet, Russia or Armenia.[a] Although, as we have seen, the Mongols recruited considerable numbers of Chinese scholars into their bureaucracy, the highest posts were generally reserved for foreigners, and if a Mongol was not available the emperor was quite prepared to appoint someone of any non-Chinese origin; hence the career which Marco Polo made for himself without much difficulty between the years +1271 and +1297. In spite of what the bilious Wieger says,[b] there is no reason to think that it was not an honourable one, and the probability is that it included, besides other posts, high office in the salt bureau.

Not only outside China proper but also within it, roads were improved and government post-stations established along them.[c] Particular attention was paid to the section of the Grand Canal north of the Yellow River, and several new lengths were dug in that region, since the capital was fixed at Peking.[d] But other districts were not neglected.[e] The Yunnan region, which was now incorporated in the empire (and served as a base for a temporarily successful Mongol campaign in Burma), was much improved by hydraulic works under the governorship of Sa'īd Ajall Shams al-Dīn, a Muslim perhaps from Bokhara.[f] Doubtless it was in the Yuan time, and through such influences, that Yunnan acquired its large minority population of Muslims.

It was quite in accord with the spirit of the age that geography should flourish, and so we find one of the greatest of Chinese geographers, Chu Ssu-Pên,[1] producing his great atlas, the *Yü Thu*[2] between +1311 and +1320. Khubilai even sent expeditions to ascertain the true source of the Yellow River. In the other direction two attempts at an invasion of Japan were made, the most important being that of +1281, but no success was achieved and the Japanese afterwards looked back to the typhoon which had scattered the Mongol fleet in the same kind of way in which Englishmen remembered the great armada of Philip II.[g]

This was the time of the second, or Franciscan, penetration of China by Christianity. But it had in the end no greater success than the earlier Nestorian wave,[h] probably because its communications with its home bases were cut by the Ming policy before it had had time to take root.[i] Islam remained, since Islamic countries bordered directly on north-west China. Taoism suffered badly under the Yuan, its books being

[a] Cf. Chhen Yuan (3). Cordier has three interesting chapters on this subject (1), vol. 2, pp. 369 ff.

[b] *TH*, vol. 2, p. 1693. The best edition of Marco Polo is that edited by Moule & Pelliot, but the notes in the edition of Yule & Cordier (Yule, 1) are of permanent value. An excellent account of Polo's life in China is that of Beazley (1), vol. 3, pp. 15 ff.

[c] Yuan prefectures in Herrmann (1), Map 52.

[d] Chi (1), p. 139.

[e] Impey (1) has given an interesting description of the summer capital of Khubilai Khan at Shangtu (Xanadu), north of the Wall.

[f] When I first went to Yunnan in 1942 I was much impressed by these Yuan waterworks in the Kunming basin, with their lines of trees on the embankments.

[g] Cf. Mills (3).

[h] Nestorianism also flourished in China under the Mongols (*TH*, vol. 2, p. 1715).

[i] It is associated with the names of John of Plano Carpini and John of Monte Corvino (*TH*, vol. 2, pp. 1669, 1723) and Odoric of Pordenone (*TH*, vol. 2, p. 1731), all of whom left narratives (Yule, 2).

[1] 朱思本 [2] 輿圖

burnt in +1258 and +1281 successively—it is not quite clear why this happened, except that perhaps it was a reaction on the part of the Buddhists some of whose temples Taoists had occupied, and whose doctrines were after all those of Lamaism which had now penetrated China and which was destined to convert officially the whole of the Mongol nation. Taoism, driven underground, rapidly took on the character of a national cult associated with the struggle against foreign domination, a role entirely congenial to it in view of its long revolutionary affiliations, which we shall later examine (Sect. 10).

Meanwhile the more the Chinese scholar-gentry penetrated back into the bureaucracy the more insistent became the demand for re-establishment of the State examinations, which had been broken off in the middle of the +13th century. This was done in +1315. It is generally considered that it was owing to the interruption of the examinations, and the unemployment which at times reigned among the scholars, that so many talented writers turned their attention to drama and novel-writing,[a] which had not previously been considered orthodox in Chinese culture, but which never subsequently altogether lost the position they now attained. In the later Yuan, Confucianism regained much ground. The university was re-established, the Confucian temple in the capital reconstructed, and new imperial honours decreed to the sage.

About the middle of the +14th century, the days of the dynasty were seen to be numbered. On the whole, it had not been able to organise a loyal civil service, and the finances were in great disorder. Secret societies dedicated to the expulsion of the Mongols became active,[b] and the national movement was sufficiently strong to capture Nanking in +1356 under the leadership of Chu Yuan-Chang,[1] a man of grotesque appearance but great ability[c] who had at one time been a monk. The army of the Ming was strong enough a decade later to take Peking itself,[d] and by +1382 the last Mongol rule was overthrown by the conquest of Yunnan.

[a] Details in Nagasawa (1), pp. 255 ff., 302 ff.

[b] The Red Turbans this time (*TH*, vol. 2, p. 1734).

[c] His portrait hangs today in the hall of the beautifully situated tomb dedicated to him on the southern slope of the Purple Mountain at Nanking.

[d] These were the campaigns in which metal-barrel cannon were employed for the first time on a considerable scale, though the Mongols probably used them half a century earlier. The earliest dated Chinese iron bombard is contemporary with the Battle of Crécy (+1346), and thus some fifty years older than the earliest dated European cannon. The casting of metal-barrel guns in China cannot therefore (as has sometimes been thought) date only from the coming of the Portuguese. As we shall later see in detail (Sect. 34), they were the last step in a logical succession of experimental devices which can be traced back at least to the +11th century. It would be interesting to know what part this technological development played in the success of Chu Yuan-Chang.

[1] 朱元璋

(j) THE MING DYNASTY AND THE CHHING (MANCHU) DYNASTY

The period began by the establishment of the capital in the east-central economic area, at Nanking. A new law code was promulgated, irrigation works undertaken, and the civil service thoroughly overhauled.[a] But equilibrium between the north and the south was not yet attained. The first emperor, who took the reign title Hung-Wu,[1] led an army into Mongolia and burnt the Mongol capital at Karakoron (the centre of the world a hundred years before; Fig. 5), pursuing the fugitives as far as the Yablonoi mountains, the furthest northern point ever reached by a Chinese force. Similarly, Manchuria was firmly annexed, and systematic Chinese colonisation by settlement there began for the first time. But Hung-Wu's grandson successor became involved in a civil war with his uncle Chu Ti,[2] who held the northern economic area, and a devastating struggle ended with the victory of the latter, in +1403. Once more therefore the capital shifted to Peking, and under the golden Yung-Lo[3] reign-period which followed, acquired architecturally much of its present-day appearance. The development of each of the two great economic areas being now approximately equal, and the Grand Canal being in operation, it did not matter a great deal where the political centre was placed.

Although in fact it was placed in the north, the Ming dynasty did not orientate itself mainly to the north and north-west, but rather to the far south. There was a contraction of territory in Central Asia, the Ming 'Gate of China' being placed at Chiayükuan[4] at the western or Tibetan end of the Great Wall, rather than further west at Yümên,[5] where it had been in Han and Thang. Hami remained an outpost. On the contrary, the Ming was the greatest age of maritime exploration in Chinese history.[b] In +1405 the eunuch admiral Chêng Ho[6] left with a fleet of sixty-three ocean-going junks which visited many parts of the south seas and forcibly brought back the kings of Palembang and Ceylon to do homage at the imperial court. During the next thirty years seven such expeditions set forth, returning each time with abundant information concerning geography and sea routes as well as large quantities of the produce of the isles and India. Such animals as ostriches, zebras and giraffes were now seen for the first time in China. States as far west as the Persian Gulf sent tribute. The reasons for these expeditions are not known; they may have been intended to counterbalance the foreign trade which had now dried up over the land routes, or to increase the grandeur of the imperial court, or even, as the official annals said, to seek out the emperor's predecessor and nephew (who, in fact, had

[a] Including the restoration of the Ministry of Public Works (Kung Pu[7]).

[b] Detailed accounts by Pelliot (2) and Duyvendak (8, 10). Duyvendak (1) has described a MS of sailing directions preserved in the Bodleian Library. See also Mayers (3); Mills (3).

1 洪武 2 朱棣 3 永樂 4 嘉峪關 5 玉門 6 鄭和
7 工部

disappeared underground as a Buddhist monk and was found many years later in a succeeding reign). In any case they stopped as suddenly as they had begun, again for reasons which are now obscure. Whether or not some feud between the eunuchs and the Confucian bureaucrats was involved, the upshot was that the command of the Indian Ocean was left to the Arabs and to the Portuguese. Worse results followed nearer home; the coast was severely preyed upon by Japanese pirates, and Annam became independent in +1431.

Europeans now began to approach the shores of China,[a] the Portuguese coming for the first time in 1514,[b] the Dutch in 1622, and the English[c] in 1637. In 1567 and again in 1619 the Russians tried unsuccessfully from Siberia to open relations with the Ming court.[d] The Spanish occupation of the Philippines in 1565 brought about a great increase of Chinese trade; this was the origin of the introduction of the Mexican silver dollar into Chinese commerce.[e]

'China's domestic history', says Goodrich,[f] 'during the Ming, is one of restoration of her native culture. Defences, paved highways, bridges, temples and shrines, stupas, tombs, memorial arches and rock gardens were built in great profusion.' The walls of some 500 cities were entirely reconstructed. The Grand Canal was repaired and deepened throughout. For all these and similar enterprises a large bureaucracy was, according to Chinese traditions, necessary. An estimate for +1469 seems to show that there were over 100,000 civil and 80,000 military officials in the whole empire. Examinations were conducted every year and were still theoretically open to all candidates, though according to various customary procedures it was not possible to fulfil the obligations of taking a degree unless a considerable sum of money was available. During the Ming there was, however, a great development of the use of eunuchs in the high ranks of the administration—the admiral Chêng Ho has already been mentioned—and this led to severe internal struggles with the Confucian scholars, both sides enriching themselves with the spoils of office. At this time, therefore, the interesting institution of the Censorate was very active. An office or ministry in the government, it had the duty of reporting to the emperor all cases of misgovernment in the provinces, but as its functionaries were not protected by any special immunity, the exercise of the duty to remonstrate often required great moral courage and was quite likely to end in the ruin of the censor concerned.

In the later Ming, the censors adopted a method of unanimous proceedings, and if one member of the bureau was put to death for his protest memorandum, another would step into his place and repeat the attack. But in spite of all the heroism of these scholars the eunuchs gradually got the upper hand. Rooted out from the bureaucracy the scholars formed organisations which were half academies, half political parties,

[a] *TH*, vol. 2, p. 1761.

[b] Cordier is particularly good on this (1), vol. 3, pp. 96 ff. Map in Herrmann (1), 53.

[c] Capt. John Weddell.

[d] *TH*, vol. 3, p. 1790.

[e] The use of this continued until the very end of the Chhing dynasty in 1911. The only other alternative to strings of copper cash was the tael or shoe-shaped silver ingot.

[f] (1), p. 193.

such as the Tung-Lin[1] academy at Wu-hsi. Most of the best men of learning in Ming times were members of one or other of these associations.[a]

An enormous amount of scholarly work was done in the +15th and +16th centuries. The Chinese encyclopaedic movement reached its climax in the Yung-Lo period, when in +1403 the *Yung-Lo Ta Tien*[2] was commissioned. Although in 11,095 chapters, the work was so well organised that it took only four years to complete, more than two thousand scholars being employed.[b] The collection was really a kind of *tshung-shu*, whole works being written straight into it, and for this purpose rare books were collected from all over the empire and copied before being returned to their owners. Unfortunately, the result was esteemed too large to print, so that only two further copies were made, and the main collection perished in the destruction of the Yuan Ming Yuan[3] palace during the Boxer Rebellion of 1901.[c] Some 370 surviving volumes are scattered in libraries all over the world;[d] in Cambridge we carefully preserve two, which deal with mathematics. In philosophy the age was dominated by Wang Yang-Ming,[4] who moved away from the scientific humanism of the Neo-Confucians to a rather anti-scientific idealism.[e]

But though this may have been harmful to the natural sciences, which tended to be left to such despised geniuses as Wang Khuei[5] the zoologist (cf. Sect. 39), the scientific method as such made great advances in the Ming, and in a region where it might not have been expected, namely, phonetics and philology. The importance of the inductive method as a tool for research was first made plain by the Chinese, as Hummel (1) points out, in the field of phonetics. One of the preoccupations of Chinese scholarship had been, since the Sung, the recovery of the rhyming sounds of ancient poetry, as in the Book of Odes (*Shih Ching*). Scholars knew that ancient verses could not be read suitably with the prevailing pronunciation, but were unable to say with confidence what the ancient pronunciations had been. The migrations and disturbances of nearly two thousand years had brought about great changes in the sounds of words though the meanings had remained stable. Chhen Ti[6] may be said, therefore, to have founded the science of phonology when in +1606 he published his*Mao Shih Ku Yin Khao*[7] (Investigations on the Sounds in Mao's Version of the Odes), using the inductive method. Examples adduced to show how a word was pronounced were assembled under two heads, *pên chêng*,[8] 'internal evidence', and

[a] Besides these political struggles, there were also ideological ones in this period. Orthodox Confucianism had now greatly ossified, and the unhappy fate of such a scholar as Li Chih,[9] who was attracted by Buddhism and combated the conventional ideas of the time, shows that originality could then be dangerous (his memory has been revived by Wu Yü (*1*); O. Franke (4); and Hummel (4)). Later, in the Chhing dynasty, a severe literary censorship was established, partly in order to root out books which had been written by supporters of the Ming (Goodrich, 2).

[b] Mayers (2); L. Giles (1).

[c] Palaces had already been looted in 1860 by foreign troops; *TH*, vol. 2, p. 1829.

[d] Census by Yuan Thung-Li; see also Sarton (1), vol. 3, pp. 830, 1851. Description by O. Franke (9) and Kuo Po-Kung (*1*).

[e] Nagasawa (1), p. 274; see below, Sect. 17.

[1] 東林黨 [2] 永樂大典 [3] 圓明園 [4] 王陽明 [5] 王逵
[6] 陳第 [7] 毛詩右音考 [8] 本證 [9] 李贄

phang chêng,[1] 'external evidence'. 'We can claim', says Hummel, 'for Chinese literary scholarship of the beginning of the +17th century a systematic application of the inductive method and a use of the very terminology which we associate with that method in the West.' His work was continued by another famous scholar,[a] Ku Yen-Wu,[2] who declined to serve under the Manchus,[b] while an equally great man, Yen Jo-Chhü,[3] founded the science of philology in China by courageously attacking the authenticity of much of the *Shu Ching* and so making himself the forerunner of the modern school of sceptical humanists already referred to (p. 87).[c]

In some luminous pages Hu Shih[d] has contrasted this renaissance of the humanities with that other scientific movement away in Europe which was happening about the same time.[e]

> Four years before Ku Yen-Wu was born [he says], Galileo had invented his telescope and was using it to revolutionise the science of astronomy, and Kepler was publishing his studies of Mars and his new laws of the movements of the planets. When Ku Yen-Wu worked on his philological material and reconstructed his archaic pronunciations, Harvey had published his great work on the circulation of the blood, and Galileo his two great works on astronomy and the new science. Eleven years before Yen Jo-Chhü began his critical study of the *Book of History*, Torricelli had completed his great experiment on the pressure of air. Shortly after, Boyle announced the results of his experiments in chemistry, and formulated the law that bears his name. The year before Ku Yen-Wu completed his epoch-making *Five Books*[f] on philological studies, Newton had worked out his calculus and his analysis of white light. In +1680 Ku wrote his preface to the final texts of his philological works; in +1687, Newton published his *Principia*.

The striking similarity in the methods of work used, Hu Shih goes on to say, shows up in greater contrast the difference between the subjects examined. The West worked with stars, balls, levers, inclined planes and chemical substances—China with books, words and documentary evidence. But the important point is not so much, as Hu Shih says, that Chinese humanistic science created only more book-learning while Western natural science created a new world; it is rather that whatever inhibiting factors we may later find to have been at work in Chinese society, there was evidently nothing

[a] Nagasawa (1), pp. 307 ff.

[b] Like Wang Chhuan-Shan,[4] who combated Wang Yang-Ming's idealism and expounded a materialist theory of history (Nagasawa (1), pp. 311 ff.). H. Wilhelm (8) has drawn attention to the important effect which the refusal of scholars to serve in the Manchu administration had in furthering the humanistic sciences.

[c] Nagasawa (1), p. 318.

[d] (1), p. 70.

[e] The full exposition of this is in Chinese; see Sect. 14*i* below.

[f] These were the *Yin Lun*[5] (Study of Ancient Pronunciations), *Shih Pên Yin*[6] (Dictionary of the Original Sounds of the Book of Odes), *I Yin*[7] (Dictionary of the Original Sounds of the Book of Changes), *Thang Yün Chêng*[8] (Thang Dynasty Rhyme Sounds), and *Ku Yin Piao*[9] (Catalogue of Ancient Pronunciations).

[1] 旁證 [2] 顧炎武 [3] 閻若璩 [4] 王船山 [5] 音論
[6] 詩本音 [7] 易音 [8] 唐韻正 [9] 古音表

in the Chinese mind to prevent the development of a body of knowledge answering to the most rigorous canons of evidence, accuracy and logical systematisation.

Another field in which it was active was geography, for during the Ming a flood of *hsien chih*[1] (local historical geographies) poured forth.[a] The greatest traveller was Hsü Hsia-Kho,[2] who spent his life exploring the vast areas of west and south-west China which were still practically unknown. His greatest discoveries were the true sources of the West River and the Yangtze, and the fact that the Mekong and the Salween are two quite different rivers.

As in the Han, imperial princes took part in the flowing tide of learning. At the beginning of the dynasty, Chou Ting Wang,[3] the fifth son of the first emperor, aided by his son Chou Hsien Wang,[4] took a great interest in botany and wrote the *Chiu Huang Pên Tshao*[5] (Herbal for the Prevention of Famine).[b] During the last twenty years of the +14th century the elder prince maintained a large botanical garden near Khaifêng, where his estates were, and cultivated with particular attention all such plants as were thought to be of value as food suitable for use in emergencies and famine periods. The first edition is of +1406; its woodcuts thus antedated the first European woodcuts of plants, contained in Conrad of Megenburg's book, by some seventy years. Then the seventeenth son of the emperor Hung-Wu was also scientifically active. Ning Hsien Wang[6] published about +1420 the *Kêng Hsin Yü Tshê*,[7] a description of 541 natural products used in alchemy; he was regarded as a man of great learning, skilled not only in chemical art but also in medicine and agriculture. A third prince, living rather later, Chu Tsai-Yü,[8] published about +1610 the *Yo Lü Chhüan Shu*[9] (Collected Works on Music and Acoustics), having already shown in +1584 that the intervals for an equal-tempered scale are found by taking the twelfth root of two, and thus anticipating Stevin and Mersenne by several decades.[c]

Undoubtedly the greatest scientific achievement of the Ming was the culminating work of the *pên tshao* series, the *Pên Tshao Kang Mu*[10] of Li Shih-Chen,[11] which was finished in +1578 and appeared in +1596. Li Shih-Chen attained as high a rank *qua* scientist as it was possible for anyone to attain in isolation from the Galilean-Vesalian movement. About 1000 plants and 1000 animals are exhaustively described in sixty-two divisions, always, of course, with their pharmaceutical value in view, real or supposed (and it was real more often than hasty modern critics have usually been willing to admit). An appended work added more than 8000 prescriptions. Li Shih-Chen gives excellent discussions of distillation and its history, of smallpox inoculation, of the use of mercury, iodine, kaolin and other substances in therapeutics, etc. (see on, Sects. 25, 38, 39). Two other very important books of this time were the techno-

[a] Ming prefectures in Herrmann (1), Maps 54, 55, 56.
[b] Bretschneider (1), vol. 1, p. 49; Swingle (1), p. 193.
[c] See below, Sect. 26*h*.

1 縣志 2 徐霞客 3 周定王 4 周憲王 5 救荒本草
6 寧獻王 7 庚辛玉册 8 朱載堉 9 樂律全書 10 本草綱目
11 李時珍

logical treatise *Thien Kung Khai Wu*[1] of Sung Ying-Hsing[2] in which every kind of manufacturing process is described; and the *Wu Pei Chih*[3] of Mao Yuan-I[4] which deals fully with military technology. It has been thought that the former was under Jesuit influence, but after careful examination I am not inclined to believe this. Sung Ying-Hsing, who produced his book in +1637, failed five times in the imperial examinations, which may be considered a criticism of them rather than him; and Mao Yuan-I whose book had appeared nine years before, had been an unsuccessful commander. The most interesting of the smaller illustrated encyclopaedias which appeared was the *San Tshai Thu Hui*,[5] prepared by Wang Chhi[6] and his son in +1609; it has 106 chapters and a large number of drawings of all kinds of objects of scientific interest.

By the beginning of the +17th century, however, the position of the government was greatly deteriorating. Taxation had become extremely heavy, and many abuses had grown up, such as the creation of great domains for courtiers and princes, in the establishment of which large numbers of peasants were driven off their farms. Brigandage therefore greatly increased, and as the bands of dissidents joined forces the country drifted into a state of civil war. In +1636, Manchuria had been lost to a new State which the Manchu leader Nurhachi had succeeded in forming out of the tribes in +1618. Then in +1644 the most successful of the rebel leaders, Li Tzu-Chhêng,[7] appeared suddenly before the gates of the capital and was admitted through the treachery of a eunuch, whereupon the last Ming emperor committed suicide. Repeating former patterns in Chinese history, the Chinese general at Shanhaikuan, Wu San-Kuei,[8] who had strong personal reasons for hating the usurper, called in the help of Nurhachi. But once the Manchus had entered China in force, they could not be expelled again, and thus it was from +1644, the year of Li's short-lived triumph, and not from +1659, when the last Ming prince was destroyed in Yunnan, that the Chhing dated their dominion.[a]

It is at this point that we shall bring to an end this brief survey of Chinese history, intended only as a background to the numerous separate narratives of the history of science upon which we shall soon enter. There is a compelling reason for doing so.

The Society of Jesus had been founded by Ignatius Loyola in +1540. By a combination of circumstances in which factors connected with the foreign policy of France were not lacking (cf. Pinot, 1) the Jesuits were entrusted with the missions to East Asia. In the later years of the +16th century Francis Xavier went to Goa and to Japan, and though he saw the coast of China he never set foot on it.[b] But in +1582 Matteo Ricci, an Italian Jesuit, one of the most remarkable and brilliant men in history, reached Macao,[c] and in +1601 was able to proceed to Peking where he died in

[a] Maps of Chhing prefectures in Herrmann (1), 58, 59.
[b] Cordier (1), vol. 3, pp. 134 ff.
[c] *TH*, vol. 2, pp. 1765, 1768, 1775, 1784 ff.

1 天工開物 2 宋應星 3 武備志 4 茅元儀 5 三才圖會
6 王圻 7 李自成 8 吳三桂

+1610. Li Ma-Tou[1] (to give him his Chinese name) was not only an extraordinary linguist, mastering the Chinese language to perfection, but also a scientist and mathematician of eminence. Assimilating himself and his Jesuit colleagues to the manners and customs of Confucian scholarly society, he succeeded in obtaining a warm welcome at the imperial court, where he helped in calendar reform and aroused wide interest in all aspects of science and technology. Together with a few learned converts, especially Hsü Kuang-Chhi[2] (who later rose to cabinet rank), Ricci made translations of books on mathematics (e.g. Euclid), hydraulics and astronomy; and encouraged the compilation of books by others. Hsü Kuang-Chhi,[2] for example, wrote a large and excellent treatise on agriculture, the *Nung Chêng Chhüan Shu*.[3] In due course numerous scientific works by Jesuits and their Chinese collaborators appeared; some of these will be noticed below.[a] Still today one may visit the Library of the old Pei-Thang[4] at Peking,[b] and see the copies of the books of Euclid, Ptolemy's *Almagest*, Gemma Frisius, and Clavius on the astrolabe, which formed the nucleus of the collection and were well used in the early +17th century.[c] Manuscripts of the early Jesuits still exist, such as the careful work of le Cheron d'Incarville on the *Pên Tshao* recently described by Bernard-Maître (3).[d]

This is the point at which we meet the boundary of the plan of the present book. After the coming of the Jesuits, Chinese science fused with universal world science, and though its rise may have been slow during the 18th and 19th centuries, inhibited by the same factors in Chinese society which had hindered it through all earlier history, it is no longer easy to distinguish any particular style in the contributions made by Chinese thinkers and observers. The work of liaison begun by the Jesuits in the 17th century was continued by the Protestants of the 19th, until in our own time China took her place among all other nations as participant in the world community of science.

[a] Bibliography by Cordier (8). Detailed information on the whole movement will be found in the books and articles of Bernard-Maître (1, 2, 4, 5) and Hughes (3, ch. 5), and memoirs by Chang Yin-Lin (*1*), Peake (1), S. C. H. Liu (1) and, for medical science, the book of Fan Shih (*1*). An excellent introduction to the literature on the Jesuit period is given by Lach (3, 4).

[b] Map in Herrmann (1), 57.

[c] Verhaeren (1).

[d] Ricci's own papers have been edited by Venturi (1).

[1] 利瑪竇 [2] 徐光啓 [3] 農政全書 [4] 北堂

7. CONDITIONS OF TRAVEL OF SCIENTIFIC IDEAS AND TECHNIQUES BETWEEN CHINA AND EUROPE

(*a*) INTRODUCTION

ONE DAY about fifteen years ago I walked into the library of my friend Professor Gustav Haloun for one of the sessions in which I was reading with him that ancient, largely Taoist book,[a] the *Kuan Tzu.*[1] He suggested that we might rapidly pass over certain material connected with the chapter before us[b] as it seemed to embody nonsensical fables about animals, but I was for a closer examination of it. It turned out to contain a statement that certain marine animals were subject to a lunar cycle, increasing and decreasing in size as the moon waxed and waned. Great was my astonishment as I remembered that Aristotle said exactly the same thing.[c] While the details of this matter will be given in a later volume (Sect. 39, Zoology), the point to be emphasised here is that, although the many fragments which went into the composition of the *Lü Shih Chhun Chhiu* cannot be exactly dated, most of them are to be attributed to just after the time of Aristotle himself, i.e. the late −4th and early −3rd centuries.[d]

Aristotle's statements refer to the sea-urchin, and have been confirmed in our own time,[e] but the question immediately presents itself, what could be the relation between two such observations at the opposite ends of Asia appearing simultaneously? Could the observations have been independently made by Greek and Chinese fishermen? Or is it conceivable that a Greek-speaking Scyth might have conversed with a Chinese-speaking Hun about such matters, so that a rapid transfer of the idea took place over thousands of miles among peoples who had never even seen the sea? The latter possibility is difficult to believe.

Again, a few weeks before writing these paragraphs, I attended a meeting at the Royal Society of Medicine at which another old friend, Dr Charles Singer, spoke of Galen in relation to ancient anatomy. He emphasised that Galen himself never dissected the human body, nor did any of his teachers do so, yet it is certain that the Alexandrians had carried out dissections, beginning with Herophilus and Erasistratus

[a] Attributed to Kuan Chung of the −7th century; but put together in the −3rd or later and not connected with him.

[b] Ch. 37; it deals partly with 'sympathetic' effects and action at a distance. The parallel was ch. 45 of the *Lü Shih Chhun Chhiu* (cf. R. Wilhelm (3), p. 114).

[c] *De Part. Anim.* IV, 5 (ed. Didot, vol. 3, p. 280, 14ff.), 680*a* 31; *Hist. Anim.* 544*a* 16.

[d] The idea of a lunar influence on animal life may also be traced further back in China, for it is found in the *Hsia Hsiao Chêng* (cf. R. Wilhelm (6), p. 239) which may well be earlier than Aristotle.

[e] H. M. Fox (1). The lunar periodicity in reproduction affects gonad size.

[1] 管子

in the first half of the −3rd century. The early Christian fathers mostly speak of dissection as of something shocking which they themselves had not been personally in contact with.[a] I reflected on the strangeness of the fact that a similar rise and fall of ancient anatomy can be observed in China, beginning rather earlier with Pien Chhio, well attested in the time of Wang Mang (+9) and continuing a little later into the San Kuo period (*c.* +240), after which, as in Europe, it vanishes until the late middle ages. Then the Sung anatomists precede Mondino de Luzzi[b] by about a century, but fail to go further.

The object of these examples—and many other equally strange parallelisms could be cited—is to show the necessity of saying something, before discussing the history of thought and the individual sciences in China, about what one might call the credibility of the travel of ideas at different historical times between East Asia and West Europe. In the material later to be described innumerable points will arise which will tempt the question, how independent was this development? Could it have been affected by, or alternatively was it itself responsible for, some other event in the history of science and technology in India, in the Islamic lands, or in Western Europe? As a necessary preliminary, therefore, a brief discussion of the whole question of the conditions of possible mutual influences, as by trade-routes and translators, seems to be in place.[c]

(*b*) THE ORIGINALITY OF CHINESE CULTURE

In spite of the vast literature on the contacts between China and Europe, some guides to which will shortly be given, we still do not have enough facts on which to base a generalisation, and of course the further one goes back in history the worse the situation becomes. Inevitably much of what has been written has been little more than speculation. In the section on Language[d] we referred to the 18th-century 'Systems', based on early realisation of the ideographic nature of the Chinese language, that the Chinese 'were a colony of the ancient Egyptians'. In the following century this idea was canvassed with some persistence, but the now long discredited theories of a Terrien de Lacouperie or a Ball no longer need to be combated by upholders of the essentially autochthonous character of Chinese civilisation.[e]

Nevertheless the general attitude that a large proportion of characteristic developments in Chinese thought and practice were derived from western sources has died very hard. Let us take some examples. Henri Maspero[f] at first thought that the ancient geography of the Chinese, expressed in the *Shan Hai Ching*, had been

[a] For example, Tertullian, quoted in Needham (1), vol. 1, p. 25.
[b] Singer (1), p. 77; Sarton (1), vol. 3, p. 842.
[c] Many books on particular aspects will be mentioned as we go along, but this is a suitable place to refer to a Chinese work which covers the whole of the field, though not, of course, from our point of view—the six volumes of Chang Hsing-Lang (*1*). The smaller book of Fêng Chhêng-Chün (*1*) is complementary to this, since it deals with the only area not covered by Chang, namely the South Seas.
[d] P. 38 above.
[e] E.g. J. Ross (1); Chatley (4).
[f] (2), pp. 607ff. For his final opinions see (14), pp. 15ff., 37ff.

stimulated by a wave of foreign (Indian and Iranian) influence in the −5th century; and that the greater part of Chinese astronomy was of Western inspiration. The twenty-eight *hsiu* (divisions of the equator), the Jupiter cycle, the gnomon and the clepsydra, had, he thought, been introduced in the time of Darius, while a second wave, in the time of Alexander, had imported geometry, the duodenary cycle, the system of star cataloguing, etc. All this is now quite untenable, as we shall see, and in his later brilliant monographs on Chinese astronomy (3, 4) he abandoned such views, passing over in silence the questions of contacts and stimuli. The problem of the relations of the Chinese *hsiu* and the Indian *nakshatra* is indeed still with us (Sect. 20*e* below), and it has not been possible to grant historical priority for either of them. Their origin is probably Babylonian.

Other authors blandly assumed that nearly everything of value in Chinese science came from the West, e.g. Vacca (1), who did not hesitate to derive the Han hodometers from Heron of Alexandria (*fl.* +60),[a] though the first references to them in Chinese literature are associated with the names of Prince Tan of Yen[1] (−240 to −226) and Han Yen-Shou[2] (between −140 and −70). These early Han references do not describe the hodometer as something quite new, and even if some doubt remains as to exactly what kind of vehicle was meant, precisely described hodometers appear later in the Han[b] too soon to be accounted for by any transmission from Alexandria. Vacca also believed that the *Chou Pei* geometry[c] was Western in origin, that the early Chinese calculations of the value of π were inspired from the West, that Chu Shih-Chieh's[3] table of binomial coefficients in +1300 was derived from 'Umar al-Khaiyām, and so on. He ended with the really extraordinary statement that the first Chinese astronomer of any importance was the Buddhist I-Hsing[4] of the Thang—a whole millennium after the men who correspond to Hipparchus and indeed preceded him.[d]

Another interesting example, the work of the Austrian engineer Horwitz (1), will arise in detail in the Section on Engineering.[e] With other historians of technology, he noticed that some of the illustrations of machines in the Great Encyclopaedia of +1726 (*Thu Shu Chi Chhêng*) were evidently copies, rather badly redrawn, of pictures which he could identify in +16th-century European books. Since then, the line of descent of these pictures through the first works on engineering produced by the Jesuits in the early years of the +17th century has been traced, but the flaws in the argument were (*a*) the assumption that the mistakes were due to later Chinese copyists, and (*b*) that because the European source of an illustration had been traced, the machine must have been new to the Chinese when the Jesuits introduced it.

In the realm of philosophical theory and practice, determined efforts have been made to show that early Taoism owed much both to the Indian Upanishad literature for

[a] The date of Heron has long been a puzzle, but we accept the new arguments of Neugebauer (6).
[b] Cf. Sect. 27*c* below.
[c] Cf. Sect. 19 below.
[d] Cf. Sect. 20*f* below.
[e] Cf. Sect. 27*c* below.

[1] 燕太子丹 [2] 韓延壽 [3] 朱世傑 [4] 一行

its theory,[a] and to Indian yogism for some of its practices;[b] further, that Chinese Chhan Buddhism was an importation from India.[c] These views, however, as Creel says,[d] have never been really convincing. The Upanishads[e] are metaphysical commentaries on the Vedas, and date from the −8th to the −4th centuries,[f] so that they are little earlier than the first period of elaboration of Taoist doctrine. Their strongly marked metaphysical idealism, with its conception of the unity of the *brahman* and the *ātman*, the absolute and the self, is not at all characteristic of the Taoists; though the latter, as we shall see, greatly emphasised the unity of nature, and the incorporation of the individual within it. For the influence of Yoga practices,[g] especially the breathing exercises, which are certainly very ancient in India, upon early Taoism, a better case can be made out (Filliozat, 3). Some Taoist schools, at any rate, practised self-hypnosis by concentration on the inhaling and exhaling processes (Waley[h]), but it was not universal as Chuang Tzu has a passage condemning it. In any case the aims of this *samādhi* or *dhyāna* among the Taoists were entirely different from those of the Indian *ṛishis*. Both wished to master organic life and to attain 'supernatural' powers, but while the Indians sought for an ascetic virtue which would enable them to dominate the gods themselves (cf. Wilkins[i]), the Taoists sought a material immortality in a universe in which there were no gods to overcome, and asceticism was only one of the methods which they were prepared to use to attain their end. Besides contemplation of the universe during quietist self-hypnosis, they were also interested in the techniques of sexual intercourse by which they thought that life might be prolonged, and above all in the preparation of drugs which would render the taker immortal. Admittedly, both these other elements have parallels in ancient Indian and Iranian thought. Sex has been outstandingly significant in all Hindu myth and religion, particularly in the theme of the *Śaktis*,[j] and as Dubs (5) points out, the ancient intoxicating sacred drink (Indian *soma*, Iranian *hraoma*) was regarded as in some sense an elixir of immortality. But in general it must be said that the parallels which can be adduced remain always vague and uncertain, while the whole atmosphere and intellectual climate of Chinese thought is radically different from either the Indian or the Persian. Lastly, the dependence of Chhan Buddhist quietism on Indian vedantism and the story that it was brought by the monk Bodhidharma, has been shown by Pelliot (3) to be a late legend, designed only to give authority to the movement.

Another very common suggestion has been that the Yin-Yang dualism of Chinese thought (see Sect. 13*e*) was an importation of Iranian origin. Originally meaning the 'shady side' and the 'sunny side' of hills or houses, the words suddenly appear as

[a] Bagchi (1), p. 189; Wieger (2), p. 262; (3), p. 45; (4), p. 144.
[b] Conrady (1).
[c] Wieger (3), p. 159; (4), p. 524.
[d] (3), p. 92.
[e] Cf. Masson-Oursel *et al.* (1), p. 70; Renou (1), p. 63; V. A. Smith (1), pp. 17, 26; F. Edgerton (1).
[f] Cf. the discussion on dating between Needham (26) and Hora (1).
[g] Cf. Masson-Oursel *et al.* (1), p. 166; Renou (1), p. 266. Filliozat (2) combats the view that Yoga practices derived from Shamanism; a background of ancient pneumatic medical theory is much more likely.
[h] (4), p. 116.
[i] (1), p. 349.
[j] Woodroffe (1); Wilkins (1), p. 330.

philosophical terms about the −4th century, Yin standing for dark, weak, female, night, moon and so on; and Yang standing for bright, strong, male, day, sun, etc.—from these categories an elaborate theory of Nature grew up. A superficial similarity with Zoroastrianism is obvious, but I entirely agree with Waley[a] in his rejection of any direct influence. 'In Zoroastrianism', he says, 'darkness is essentially evil; the principle of light, essentially good.[b] But the fundamental conception of Yin and Yang was quite different; they were two independent and complementary facets of existence, and the aim of the Yin-Yang philosophers was not the triumph of light, but the attainment in human life of perfect balance between the two principles.' Waley is assuredly right in seeking the origins of the theory in indigenous divination practices.[c] He has still less difficulty in disposing of the suggestion that the theory of the five elements (*wu hsing*[1]) could have arisen from the Greek element theory, since in fact the basic Chinese conception was of *operations* in process, and static elementariness was implicit or derivative only. Finally, he discusses the relation between the paradoxes of Hui Shih[2] and the Eleatic paradoxes,[d] without attaining any definite conclusion—the correspondence is, indeed, another example of that extraordinary simultaneity between phenomena which we sometimes find at the two ends of the Old World. For the date of Hui Shih[2] is mid −4th century, and Eleatic Zeno's *floruit* is placed about −460.[e] The parallelism here is at least as remarkable as the case of the lunar periodicity of marine invertebrate reproduction, with which this section began, or the earliest appearance of the water-mill, which we shall study later[f] together with other mechanical examples.

Although we tend to think that everything must have had only one origin, we cannot rule out the possibility of completely independent and parallel lines of thought, especially where scientific theories, discoveries and observations are concerned.[g]

One might take the theories of atomism as an example. Its story in our own classical civilisation, beginning with such men as Leucippus and Democritus of Abdera, of the −5th century, and culminating in Epicurus and Lucretius of the late −3rd and early −1st, is well known to us.[h] Indian atomism seems to be later in date, the Jaina system of Umāsvāti showing its greatest strength about +50, and the Vaiśeshika *darśana* (theory) of Kaṇāda flourishing in the second half of the +2nd century.[i] But there are reasons, as Rey[j] urges, for believing that the roots of the theory of *paramāṇu* (atoms) go much further back in the history of Indian thought. Thirdly, in Chinese

[a] (4), p. 112.
[b] This dualism was even more strongly emphasised in Manichaeism, a religion which may be said to derive from Zoroastrianism, and which once spread from Albi to Sian. It was as much foreign to the Chinese as it was to Christendom, and in both culture-areas well recognised to be so, yet opposed for different reasons, naturally.
[c] See Sects. 13*g*, 14*a* below.
[d] See Sect. 11 below.
[e] B & M, p. 128.
[f] See Sect. 27*f* below.
[g] Cf. p. 227 below.
[h] B & M, pp. 138 ff.
[i] Grousset (1), p. 126; Sarkar (1), p. 33; Masson-Oursel *et al.* (1), p. 177; Sinha (1).
[j] (1), vol. 1, p. 428.

1 五行 2 惠施

physics atomism never arose, as we shall see,[a] but the geometry of the *Mo Ching*[1] (the Mohist Canon, which must have been put together somewhere in the neighbourhood of −370) seems to define a point as a line which has been cut so short that it cannot be cut any further.[b] Thus there seems really no reason why the concept of insecability should not have arisen in each of these three great civilisations independently. In all and each of them men were cutting and sawing lengths of wood and other material, and it would not be a very difficult thing for some reflective person to wonder what would happen if one went on cutting in half until the remaining object was so small that it could no longer be cut in half. Philosophical speculation could then work out all that the permutations and movements of such small objects could explain.

Another case which seems to me comparable is the Aristotelian doctrine of the 'ladder of souls' in which plants were regarded as possessing a vegetative soul, animals a vegetative and a sensitive soul, and man a vegetative, a sensitive and a rational soul.[c] I shall later show (Sect. 9*e*) that a very similar doctrine was taught by Hsün Tzu (Hsün Chhing).[2] Aristotle lived from −384 to −322, Hsün Chhing from −298 to −238. If transmission of the conception occurred, therefore, it must have been very rapid, though the conditions of travel at this period would make such movement most unlikely. We may prefer to believe that the idea originated independently, since it constitutes after all an obvious reflection on the *scala naturae*, a more or less intuitive recognition of the fact that some living organisms are much less complex than others, a foreshadowing, in fact, of the evolution-concept which ensues as soon as the ladder is realised to exist within time.

These facts add interest to a couple of parallelisms which L. Giles (3) has noted between Greek writers and the *Lieh Tzu* book, parts of which are very probably older than −400, though the actual existence of a philosopher of the name of Lieh Yü-Khou[3] is highly doubtful. In the first passage, a man gives thanks for feeling happy, enumerating reasons, (*a*) because born a human being, not an animal; (*b*) because born male and not female; (*c*) because born Greek (or Chinese) and not a barbarian, etc. This occurs in *Lieh Tzu*,[d] told of Confucius and Jung Chhi-Chhi,[4] the old man of Thai Shan, but a very similar story is also attributed by Plutarch in his *Life of Marius*[e] to Plato (therefore between +46 and +122); and by Diogenes Laertius[f] to Thales or Socrates (therefore about +220).[g] One wonders whether this type of story would not naturally arise from the conception of a ladder of souls, in which case the obvious similarity would not be evidence of transmission. Giles's other story, which concerns mourning for a child, and which occurs in *Lieh Tzu*[h] about Tungmên Wu of Wei,[5]

a See Sect. 26*b* below.
b See Sect. 19*h* below.
c Cf. Singer (1), p. 40, Fig. 18.
d Ch. 1, p. 11*a*; L. Giles (4), p. 26; R. Wilhelm (4), p. 5.
e § 46.
f I, vii, 33.
g There is a similar thanksgiving embodied in the Jewish daily liturgy.
h Ch. 6, p. 16*a*.

1 墨經 2 荀卿 3 列禦寇 4 榮啓奇 5 魏東門吳

as well as in Plutarch's *Moralia*,[a] is less elaborate and may not be more than a coincidence. Arendt (2) long ago pointed out similar parallels of a not very convincing nature, the most interesting being a criticism of men who are willing to accept political office without special training, though not undertaking the exercise of trades and professions when untrained—this occurs in *Mo Tzu*[b] and in Xenophon's *Memorabilia*,[c] where it is told of Socrates and Euthydemus. Mo Ti and Xenophon were contemporaries.

These are by no means the only instances of curious similarities. Thus Phelps has discussed parallels between the Confucian and Platonic treatment of the place of music in education. Martin,[d] too, has drawn attention to the argument about whether it is the duty of a son to inform against his father if the latter has been guilty of some crime, or to give evidence against him.[e] The fact that this occurs implicitly in the *Hsiao Ching*[1] (Filial Piety Classic)[f] is not very significant, since this book is generally regarded as having been forged in the name of the disciples of Confucius as late as the −1st (or even the +1st) century; but the argument itself is as old as the −6th, as may be seen from the *Lun Yü*[g] and *Mêng Tzu*.[h] Since Confucius died half a century before Plato was born, and since the teaching activity of Mencius was at its height only some twenty years after his death, there is certainly interest in the parallel which Martin draws between the Chinese formulations of the argument, and the discussion of the same question in the Platonic dialogue *Euthyphro*. The treatment is of course very different, but Martin adds an observation that the meaning of the name used as title is very similar to the Chinese expression *chih jen*[2] (straight man), which occurs in the same connection.

We may return for a moment to the general question raised in these paragraphs. Although this is not yet the place for final conclusions, there can be no doubt that China was, among the ancient civilisations of the Old World, the one which was most isolated from the others. The originality of its characteristic cultural patterns was therefore greater. But this does not mean that it did not participate in the spread of that civilising technological influence which radiated from the Fertile Crescent; furthermore, through all the successive centuries for three millennia, China was giving forth, and receiving, with ever-varying intensity, cultural and technical elements. Those who, with Price (1) and P. F. Cressey, go about to prove that Chinese civilisation has always been 'in the circuit' of the diffusion process, are struggling with an open door—for of this fact there can be no denial. It is, of course, equally necessary to make allowance for parallel development of ideas and techniques, independence of invention, and the like. In all these matters the repetition of generalisations is useless; what we need to know are more facts such as those to the elucidation of which the

[a] 610 D. [b] Ch. 47; Mei (1), tr., p. 224. [c] IV, ii.
[d] (3), vol. 2, pp. 199 ff.; (4).
[e] There is a strangely contemporary ring about this problem.
[f] Legge (1), p. 476. [g] XIII, 18. [h] VII (1), XXXV.

[1] 孝經 [2] 直人

present work is devoted. It is probable that our final conclusion will be that there was far more intercourse and reaction between the Chinese and their western and southern neighbours than has often been supposed, but nevertheless that the essential style of Chinese thought and culture patterns maintained a remarkable and perennial autonomy. This is the real meaning of the 'isolation' of China; contacts there were, but never abundant enough to affect the characteristic style of the civilisation, and hence of its science.

(c) RUMOURS OF CHINESE CULTURE IN THE CLASSICAL WEST

What knowledge the ancient peoples of the Mediterranean basin had of China was assembled long ago in the form of citations by Reinaud (2)[a] and Coedès (1). Naturally it concerned mostly silk and the trade in silk, and the geographical knowledge which arose directly out of this. But it is worth while to notice that some rumours of Confucianism seem to have reached Europe as early as the +2nd century.[b]

Bardesanes, a Syrian writer[c] who flourished at the end of that century and the beginning of the +3rd (a contemporary therefore of such men as Chêng Hsüan and Tshai Yung, scholars of the Late Han and early San Kuo period), is quoted by Eusebius[d] as having said:

> In every country men have instituted different laws, some being committed to writing and some not; I shall tell what I know of them, and what I can remember, beginning where the world begins (i.e. in the Far East). Among the Seres (the Chinese) the laws prohibit murder, prostitution, robbery, and the adoration of images. In all that immense country there is to be seen no temple, no prostitute, no adulterous woman, no robber brought before the magistrates, no murderer neither the victim of any murderer. The passage of glittering Mars across the meridian has no power there to bind the will of any man to put another to the sword. Nor can Venus when in conjunction with Mars force any of them to have commerce with each other's wives. Yet Mars is seen there in the heavens every night, and there is no hour either of the day or the night at which a child is not born among the Seres.[e]

This interesting passage was the basis for many repetitions and enlargements. The contemporary *Recognitiones Pseudo-Clementinae*[f] says, more frankly, 'Inflamed Mars has no power, as with us, over the free-will of men...'. Respect for the law is more potent among the Chinese than the constellation under which the individual is born.

[a] Reinaud was grinding a peculiar axe of his own, namely, that the Romans had a conscious plan of world conquest and therefore took lively interest in what the Indians and Chinese were doing.

[b] Of Taoism too, if certain assertions by classical authors about the longevity of the Chinese (e.g. Strabo, Lucian) are to be taken specifically (cf. below, Sect. 10).

[c] His dates were +154 to +222 (Sarton (1), vol. 1, p. 298). He was an astronomer who turned against astrology in his later years, and whose writings were the main source of the thought of Mani, the founder of the Manichaean religion.

[d] Patristic writer, +265 to +340 (Sarton (1), vol. 1, p. 357); *Praeparatio Evangelica*, VI, 10.

[e] Tr. Coedès (1), p. 77; eng. auct.

[f] Coedès (1), p. 78.

The *Book of the Laws of the Countries*, written apparently by a pupil of Bardesanes, adds that, although the Chinese are not subject to starry influences, there are yet rich and poor among them, sick and healthy, governors and subjects, 'because these things are given into the power of the rulers'.

There must surely have been some real transmission here, for a century and a half later, Caesarius,[a] the brother of Gregory of Nyssa, correctly stated that in China custom was followed rather than codified law.[b]

> In every country, among other peoples as with us, laws of princes exist, whether written or not; in some places they are written, while elsewhere there are customs (which have the force of law). For those who do not have (codes of) laws, ancestral customs predominate. Among such peoples, the first to be mentioned are the Seres (the Chinese) who inhabit the (eastern) extremity of the earth—they have for laws the customs of their fathers which prohibit prostitution, robbery, adultery, the adoration of images and the invocation of gods.... Among the Seres the laws of their ancestors are more powerful than the fates determined by the stars.[c]

This was afterwards copied several times, as by George Hamartolos, the Byzantine monkish chronicler,[d] about +840; and two other Byzantine writers, Cedrenos[e] and Phrantzes[f] in the +11th century. Still more remarkable is the rumour of Confucian atheism, which was apparently appealed to by Celsus,[g] the great pagan writer attacked by Origen.[h] The atheism of the Seres is mentioned several times in Origen's refutation.[i] It is of the greatest interest that a hint of Confucian rationalism and scepticism thus reached a world ridden by individual astrology and Gnostic superstition, though the picture of Seric virtue was distinctly idealised.[j] The transmission of all these ideas may be understood without difficulty, for they occur a couple of centuries after the opening of the Old Silk Road, and we shall shortly examine the records of contemporary Chinese historians which describe the visits of Roman-Syrian merchants to East Asian ports throughout this period.

Before going on, however, to speak of the trade-routes of historical times, and some of the travellers who are known to have gone along them, something may be said concerning exchanges that were more impalpable, either because they took place almost before history begins, or because of their literary or artistic character.

[a] Died +368. *Dialogues*, II, 109.

[b] Cf. below, Sect. 18.

[c] Tr. Coedès (1), p. 89; eng. auct.

[d] Sarton (1), vol. 1, p. 578; *Chronikon*, I, 19.

[e] Sarton (1), vol. 1, p. 776; *Hist. Comp.* 154A.

[f] *Chronikon*, III, ii, 49; Coedès (1), p. 91.

[g] *Fl.* +178 to +188; Sarton (1), vol. 1, p. 294.

[h] *Fl.* +185 to +254; Sarton (1), vol. 1, p. 317. See Thorndike (1), vol. 1, pp. 436 ff.

[i] *Contra Cels.*, VII, 62, 63, 64; Coedès (1), p. 82.

[j] Conversely the Chinese sometimes idealised occidental societies. A late +7th-century Taoist text (*TT* 873) on alchemy and geography, translated by Maspero (22), depicts the people of Roman Syria (Ta-Chhin, see p. 173 below) as both more virtuous and more wealthy than the Chinese. Similarly, as we shall see in Sect. 29*e*, in the +12th century both Chinese and Westerners had very complimentary, if not exaggerated, ideas of the greatness of each other's sea-going ships.

(*d*) THE CONTINUITY OF CHINESE WITH WESTERN CIVILISATION

In the first place, as already suggested, there was a certain continuity between China and Europe in the bronze age, not simply because of the presumed transmission of the technique, but because of an actual correspondence of forms of objects.

In a series of important papers Janse has demonstrated the essential unity of Europe and China in this respect from before the Shang dynasty (−1500) well into the Chou.[a] The correspondences described are impressive. One paper (Janse, 1) deals particularly with bronze and iron swords; it shows that each of the following forms occur widespread in both Europe and China, and sometimes in intermediate zones of the steppes: (*a*) two-edged swords with antenna-pommels[b] (see Fig. 26); (*b*) two-edged swords with ring (annulus) pommels (see Fig. 26); (*c*) scabbard belt attachments of bronze or jade of an identical and complex form;[c] and (*d*) scabbard strap attachments or cleats.[d] Certain other types, however, are peculiar either to China or to Europe. The annulus pommel continued in use in China in the Han time, as witness the reliefs in the Wu Liang tomb-shrine, one of which is reproduced here from the *Chin Shih So* (Fig. 25).[e]

Fig. 25. Wu Liang tomb-shrine relief (+2nd century) of a soldier with a ring-pommelled sword (*Chin Shih So*, Shih section, ch. 3).

Another paper (Janse, 2) was devoted to the curious cruciform tubes or buttons, which were probably harness pieces or ornaments, characteristic both of Chinese bronze age and of the Hallstatt civilisation in Europe (cf. p. 99 above). Many more similarities were set forth in a review (Janse, 3), in which the following further identities of form were described: (*a*) stone hammer-hatchets (i.e. like some modern hammers, having a hammer-head on one side and a hatchet or chisel head on the other); (*b*) the 'reversed spiral' motif of the Yangshao painted pottery (already referred to, p. 81); (*c*) triple-finned arrow-heads, found only in the Hallstatt region, south Russia and China; (*d*) zoomorphic pot-handles; (*e*) zoomorphic gold ornaments; (*f*) scabbard furnishings of special type; and (*g*) chalices of precious metals with

[a] See also de Takacs (1).

[b] Rostovtzev (1) figures later Sarmatian examples of the −3rd century, e.g. Pl. 24, Fig. 5, and p. 90.

[c] Janse (1) thinks that this design was transmitted from east to west.

[d] Rostovtzev (4), however, believes that these are really Han, and were introduced, with the long cavalry sword, from Persia.

[e] Janse (5), p. xxxiv, figures some of these swords from Han tombs in Indo-China, so that the distribution of the type was a continuous one from the Atlantic coast of Brittany to the Pacific coast of Tongking. They are very common in the Ordos desert (J. G. Andersson, 7).

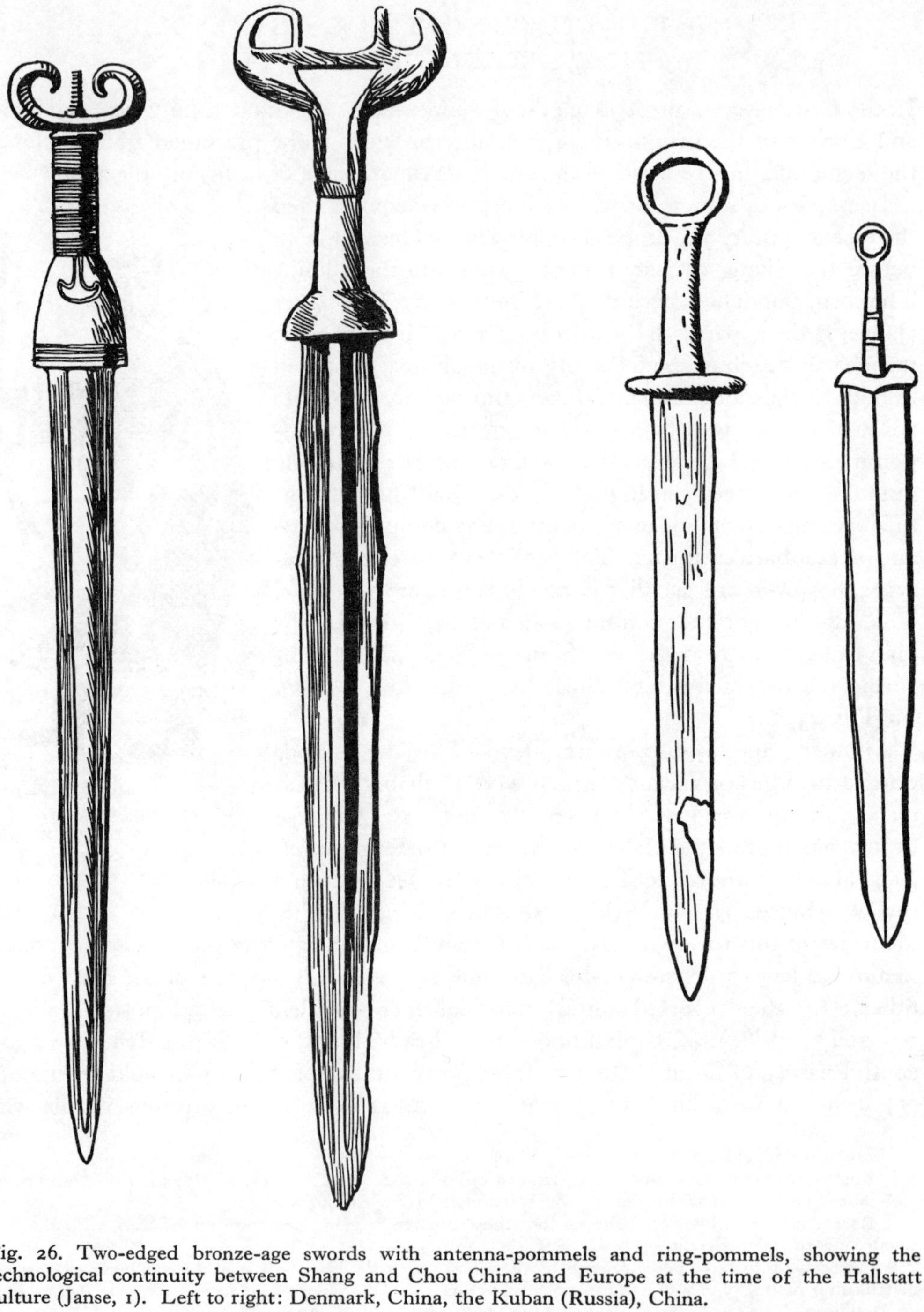

Fig. 26. Two-edged bronze-age swords with antenna-pommels and ring-pommels, showing the technological continuity between Shang and Chou China and Europe at the time of the Hallstatt culture (Janse, 1). Left to right: Denmark, China, the Kuban (Russia), China.

a sphere or ring on the stem. Lindblom (1) has added vessels with star-shaped lids, the wide distribution of which includes Africa. Besides all these, Janse (4) lists (*a*) a certain type of bronze lance-point; (*b*) torques or neck-rings;[a] (*c*) certain ceremonial axes (Fig. 27);[b] (*d*) harness buttons; (*e*) flat porphyry oblong ceremonial axes with rudimentary arms; and (*f*) bronze socketed celts.

The last-named object is a hollow bronze axe, so made that a wooden handle can be fitted into it. It was an epoch-making invention which helped to transform the forested landscape of the Old World. The first metal celts or axes had been solid and fixed like stone ones; then in the Middle East a hole was contrived in them for the shaft to pass through. But these axes took a lot of metal, available only to big States or in metal-rich areas, and the socketed axe saved no less than half the metal previously necessary. Probably it was invented on the edge of the forest country but not too far from the old metal centres, and spread both east and west. The socketed celt has given rise to controversy; Seligman (1, 2) first thought that it offered definite evidence of an eastward transmission reaching China as late as the −6th century, but later (3) had to admit that such hollow axes were datable in China as far back as the −12th, upon which Shiah (1) pointed out that since the earliest European types cannot go back much beyond −1300, an unlikely speed of diffusion would have to be supposed if it were held that the Chinese types were derived from the west European. The view now generally accepted (Childe, 12) is that there was a continuity of type from the Severn to the Huang-Ho, but that the point of origin was somewhere in between.

Another, even stranger, story concerns the ubiquity of the 'bird-chariot', i.e. a bronze or pottery image of a bird mounted on three wheels. Many collections of Chinese antiquities include specimens of these,[c] but they are known from Egypt and from many sites in Europe. Sometimes the latter have several birds perching on a kind of frame like a gun-carriage. Was the bird-chariot a toy or a religious cult-object? How far back does it go at each end of the Old World? These questions remain unanswered, but Laufer (25) has summarised all the information available.[d]

In these fields we find, therefore, the same situation that we have already encountered, namely, the almost simultaneous appearance of ideas and techniques at both ends of the Old World. Janse (3)[e] emphasises that in those early times there was no barrier of independent national State frontiers to hinder diffusion and exchange

[a] These torques, or neck-rings very like them, were reported by J. Anderson (1) in 1871 as still being worn by the girls of the tribes in the Shan hills, between China and Burma.

[b] Here the similarity is extraordinary. The shaft, socketed to hold the pole, has a non-functional blade on one side, and an animal in full relief on the other. There are even several variations of the same type, each of which can be matched from China to Europe.

[c] One is illustrated in the encyclopaedia of +1609, *San Tshai Thu Hui*, Chhi yung section, ch. 5, p. 9*b*. And I myself saw one, of unknown date, in the Lung-Fu Ssu market in Peking in 1952.

[d] Cf. Seligman (1, 4).

[e] The title of this valuable review was not perhaps very well chosen, the expression 'Empire of the Steppes' being better reserved for the Mongol dominion of two thousand years later. Here the point precisely is that there were as yet no national empires in existence which could prevent diffusion, the only limiting factors being presumably linguistic.

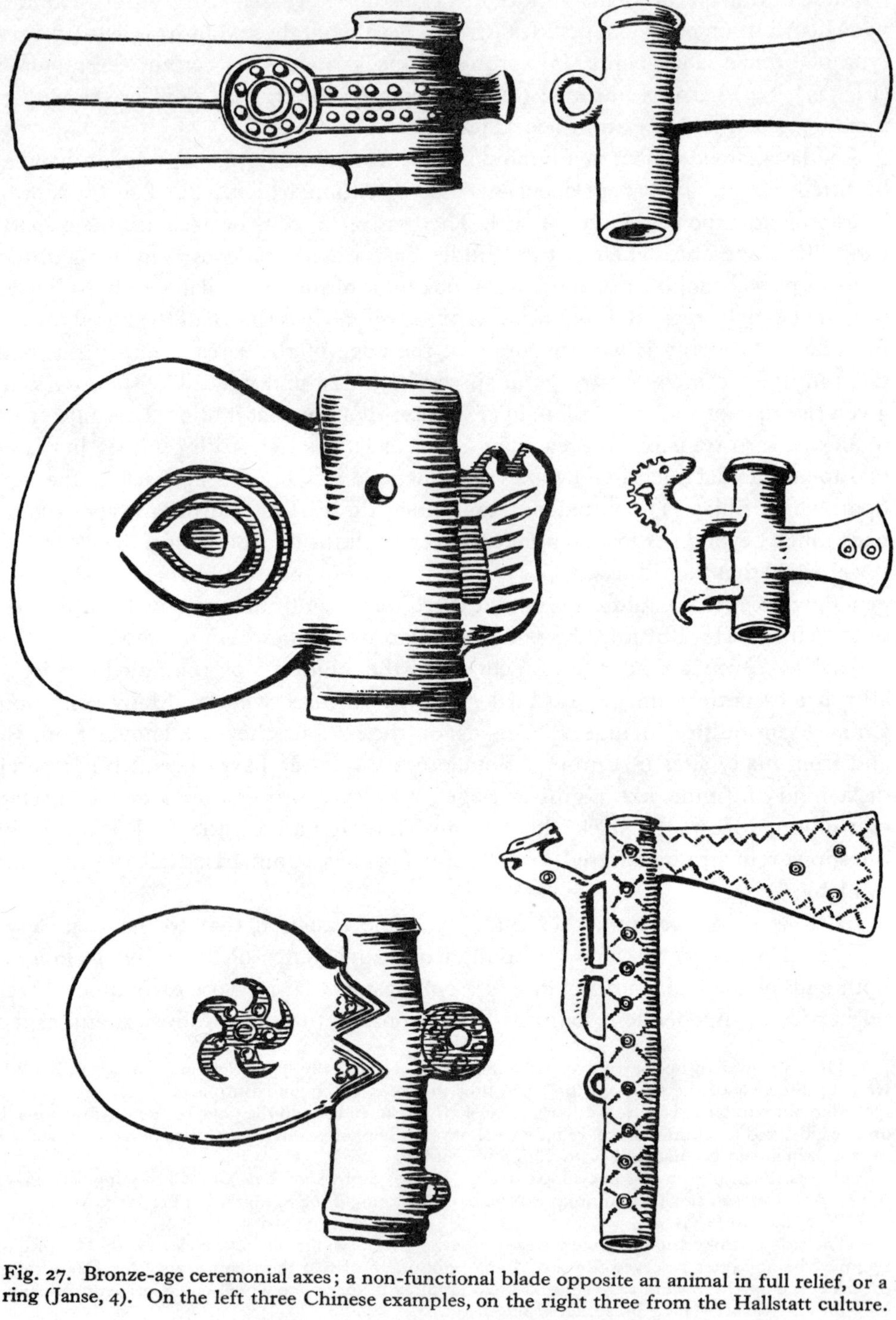

Fig. 27. Bronze-age ceremonial axes; a non-functional blade opposite an animal in full relief, or a ring (Janse, 4). On the left three Chinese examples, on the right three from the Hallstatt culture.

across the vast length of the steppes which extend from the Baltic and the Carpathians to the Ordos, and he illustrates this with an excellent map. He also makes (4) the acute observation that from the beginning of both the Hallstatt and the corresponding Chinese cultures there was an association of salt and iron (see on, Sect. 48). He notes that both were also interested in yellow amber.

(1) Literary, Folkloristic, and Art Parallels

From the figures or scenes depicted on these ancient works of art in bronze or precious metals or jade it is not a far cry to questions of folklore and ancient mythology. The interesting work of Andersson (7) on hunting magic as seen in the eurasiatic 'zoomorphic' style suggests connections with the earliest Neolithic or Palaeolithic cave-paintings, as in the Dordogne. Przyłuski (1) has devoted a fascinating paper to the 'unipeds'—divine beings with mermaid tails (see Fig. 28 for one from the Wu Liang tomb of Late Han time), representations of which are found in Gaul as much as in China. Fu-Hsi,[1] one of the mythical ancestor-gods, was often shown in this form;[a] Bishop (5) draws attention to the fact that the centre of his worship has always been in the Wei valley in the north-west of China, and I myself had occasion to note the mermaid tail still iconographically prominent when visiting the chief temple of Fu-Hsi at Thienshui in Kansu in 1945.

Armstrong, following the lead of Granet, has investigated, in an interesting series of memoirs, some common themes in Chinese and Western folklore. In the crane dance (1) he sees counterparts of the minotaur in the bull-headed Shen Nung[2] and Chhih-Yu,[3] and of the labyrinth in the underground passages of the tomb of Chhin Shih Huang Ti; finding that the dance itself was part of a cultural pattern of sacrificial and funerary ritual which extended from the Aegean through the Fertile Crescent to China and south-east Asia. The later association of the crane with Taoist mysticism is well known. Granet (1) had already identified the crane ritual with magic concerning thunder, rain, fertility and reincarnation. In two other papers (2, 3) Armstrong concludes that there was a eurasiatic community of ploughing ritual practices, ranging from the Chinese imperial ceremonial ploughing at one end to the 'triple-furrowed field' depicted on the shield of Achilles in Homer's description, or to our own Plough Monday, at the other. Then there is the magic connected with the herb *Artemisia vulgaris* or *alba*[b] (Fig. 29), prominent in the *Shih Ching* (Book of Odes), widely used not only in China but also in ancient and medieval Europe, and even in Mexico, as a constituent of incense and a powerful demonifuge, in spite of the unimpressive

[a] And so was his legendary sister (or consort), Nü-Kua[4] (Werner (1), p. 82). There is a valuable article by Wên I-To on the subject; he thinks that these beings were totem animals of primitive tribes who became incorporated into Chinese civilisation. The earliest literary references seem to be of Warring States time (*c.* −4th century); apparently the attachment of this shape to Fu-Hsi and Nü-Kua became fixed in the Former Han (−2nd).

[b] Armstrong (4); *ai*;[5] mugwort, southernwood; Bretschneider (1), vol. 2, nos. 429–36; vol. 3, nos. 72–7.

[1] 伏羲 [2] 神農 [3] 蚩尤 [4] 女媧 [5] 艾

character of the plant. Another paper (5) elucidates the affinities of Chinese bull ritual and spring ceremonies with parallel rites in ancient Crete and Egypt; and a sixth compares bird symbolism (especially that of the goose and swan) with that of the West. Students of comparative folk-tale structure debate whether the Mongol con-

Fig. 28. Wu Liang tomb-shrine relief (+2nd century) showing the deified culture-heroes Fu-Hsi and his sister-consort Nü-Kua as unipeds, with the carpenter's square and *quipu* (itself personified) as symbols of construction and order (*Chin Shih So*, Shih section, ch. 3). The inscription says: 'Dragon-bodied Fu-Hsi first established kingly rule, drew the (eight) trigrams, (devised) the knotted cords (*quipu*, for reckoning), in order to govern (all) within the (four) seas.'

quests were important in transmission to Europe (Cosquin, 1).[a] This field should be a fertile source for thought-provoking comparisons, as a good deal of work[b] has been

[a] Cf. the remarkable affinity of some of Chaucer's tales with ancient Indian stories; refs. in Tarn (1), p. 154. Cosquin dealt with the technologically interesting theme of the 'sorcerer's apprentice', showing that it did reach Europe from India, though probably not through Mongol intermediation.

[b] I mention the books of Hodous (1) and Werner (1), and the fifteen-volume *Superstitions de la Chine* of Doré (irritatingly so entitled, as it is an exhaustive description of Chinese folklore and mythology); but unfortunately little of the literature on this subject reaches satisfactory standards of sinological scholarship. Chinese scholars, however, have for some years produced the excellent journal *Folklore Studies* at Peking.

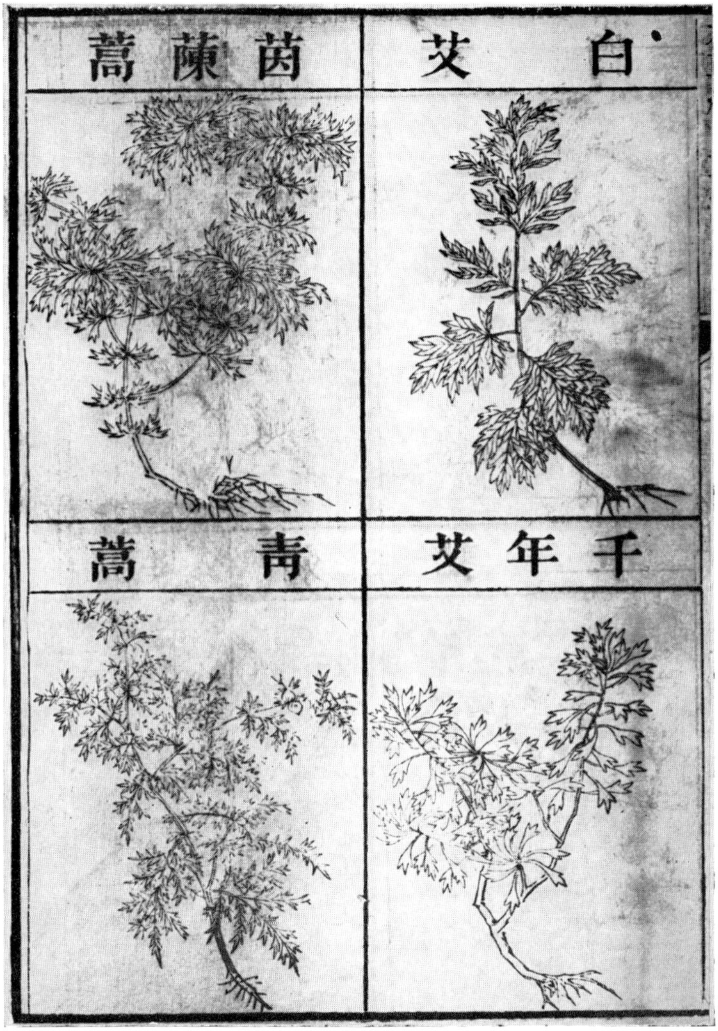

Fig. 29. Drawing of *Artemisia alba* (*pai ai*), from *Pên Tshao Kang Mu* (+1596), top right-hand picture. Li Shih-Chen's great pharmacopoeia recounts the many uses of this plant and its relatives in Chinese materia medica. They were well known as anthelminthics, and indeed this genus is the source of santonin, an alkaloid still most valuable against nematode parasites. The dried and powdered leaves of *Artemisia* formed the tindery basis of the moxa cautery, so much used in old Chinese medicine.

done on Chinese folklore and mythology; though the conclusions are bound to remain somewhat speculative.

Another attempt along the same line has been that of the Indian scholar Coyaji, who has been able to draw striking comparisons between the Persian epic of Firdausī, the *Shāh-nāmah*, and the Chinese epic or legend-cycle, the *Fêng Shen Yen I*.[1] The latter was put together by Hsü Chung-Lin[2] in the Ming (i.e. at some time in the +15th or +16th century). It deals with the overthrow of Chou Hsin, the last Shang emperor, by Wu Wang of the Chou, giving full rein to a rather Buddhist imagination in describing the magic wherewith the spirit-supporters of the Chou destroy the monsters which come to the aid of the Shang. Chou Hsin finally perishes in the flames, and Wu Wang rewards the spiritual beings and loyal ministers who have helped him by endowing them with titles, ranks and fiefs.[a] But the material of this legend-cycle is known to have been of far earlier date. Some of its ideas can be found in the *I Chou Shu*[3] (Lost Books of Chou), a work of which only ten chapters now remain, but which was found during the Chin dynasty when in +281 the tomb of An Li Wang[4] (one of the princes of the feudal State of Wei, who had died in −245)[b] was opened. Other books which were found at the same time were the *Chu Shu Chi Nien* (Records of the Bamboo Books, already referred to, p. 74), and the *Mu Thien Tzu Chuan* (see on, Sect. 27*d*). The magical epic, therefore, in its final state, is derived from pre-Han Taoist and post-Han Buddhist stories.

Coyaji now points out parallels of a rather striking nature with the *Shāh-nāmah*—King Kāus corresponds to Chou Hsin,[5] Sūdābah the wicked empress corresponds to Su Ta-Chi,[6] Syāwash the virtuous prince corresponds to Yin Chiao,[7] Rustum to Li Ching,[8] Suhrab to No Chha,[9] Keresāsp to the archer Yi,[10] while Palādwand's magicians have their due counterparts. There are many parallelisms also in cults, as of mountains, rivers, and the mushrooms of immortality for which Barzoe the physician searches. It would take us too far to follow them further, but the question of the date of the Iranian material remains. Firdausī of Tūs, one of the greatest of Persian poets, wrote the *Shāh-nāmah* between +990 and +1001 under the first of the Ghaznavid Sultans,[c] but the saga material he used was much older, and had first been written down in the Middle Persian prose works which Ibn al-Muqaffa' (d. +759)[d] had translated into Arabic. Nevertheless, it does not appear to be quite so old as the Chinese. But the subject requires further study and is only mentioned here as an example of possible cultural exchange.

Another Iranian-Chinese parallel, pointed out by Darmesteter, is that of the blasphemy of a king shooting arrows at Heaven; this seems to have come from Persia and was ascribed to several Chinese rulers, including the last prince of the State of Sung in −286.

[a] Nagasawa (1), p. 296; translation by Grube (1).
[b] Creel (1), p. 25; Wylie (1), p. 23.
[c] Brockelmann (1), p. 170.
[d] al-Jalīl (1), p. 107

1 封神演義
2 許仲琳
3 逸周書
4 安釐王
5 紂辛
6 蘇妲己
7 殷郊
8 李靖
9 哪吒
10 羿

When one passes from folklore themes to purely artistic motifs the relations between different civilisations seem even harder to trace, at any rate for those who are not art-historians. When Yetts (3) says that the winged animals seen in the Chinese patterned silks found at Loulan and by the Kozlov expedition in Mongolia 'proclaim their Mesopotamian ancestry' a certain hesitation may be excused. There is, however, one interesting case of diffusion which has been rather thoroughly worked out, and which, to a historian of science, is somewhat convincing. I mean the investigation of Reinach on the 'flying gallop'. It has recently been extended by Rostovtzev (2).

The ancients had four ways of representing horses in motion: (*a*) canter (only one hind foot on the ground); (*b*) rearing flexed (both hind feet on the ground, and the hind legs flexed); (*c*) rearing extended (both hind feet on the ground, and the hind

Fig. 30. The positions in the movement of a horse as seen by cinematography (above), and as depicted by ancient artists (below). The 'flying gallop' position is shown on the lower level on the extreme right (Reinach, 1 *a*).

legs extended); and (*d*) flying (all four legs extended, and all off the ground). Slow-motion cinematography in our own time proved that not one of the ancient positions except the first actually occurs in nature (Fig. 30), but the interest of the matter in the present connection lies in the historico-geographical distribution of the last form, the flying gallop.[a] Before the +18th century, at which time it was brought from China to Europe, this posture had been universally unknown in European art—with one exception, the Mycenaean, Minoan and Phoenician (*c.* −1000).[b] For some reason still obscure, this representation travelled east, and is found in Sassanid Persia, Bactria and China from the −3rd century onwards. Probably the impression of speed which it gives appealed to the Asian horse archers (see on, Sect. 30), such as the Parthians and Huns. In any case, the diffusion of this posture in art is a remarkable example of the contiguity of Chinese and Mediterranean or West Asian cultures.[c]

[a] This statement needs qualification in that horses, when jumping, and dogs may show the 'flying gallop' (Edgerton, 2).

[b] Edgerton adds, not very convincingly, ancient Egypt.

[c] Cf. also the elaborate work of Schuster on themes such as the 'triumphant equestrian' in Chinese and Persian embroidery motifs. Rostovtzev (2) believes that while the Chinese and Persian representations of the flying gallop are certainly connected, the Mycenaean had died out so long before the Persian began that a connection is unlikely.

It would take us much too far to go over all that the art historians have discovered about the mutual relations of Chinese art with that of other peoples. There are reviews dealing with the spread of Scythian and Sarmatian art,[a] with the passage of Greek art embodying Indian content eastwards through Bactria,[b] with the Iranian motif of the grapes and wine,[c] and with the coming of post-Renaissance painting and perspective to China at the time of the Jesuits.[d] Conversely, there had been a distinct influence of Chinese culture on the European art of the late Middle Ages and the Renaissance, especially in Italy, whither Marco Polo had returned.[e] Stylistic traits, and the actual painting of Asian types of personages, appear in Tuscan pictures of this period. Asian art-objects reached Europe, and Popes were buried in Chinese silks as other men of mark had been in Palmyra more than a thousand years before. In the eighteenth century the influence of Chinese art-forms on European design and decoration was extremely strong, as may be seen in the monographs of Reichwein (1) and Appleton (1). Relations between Chinese and Persian art were always close, especially so between the +12th and +14th centuries, when the influence was westwards.[f]

Fig. 31. The 'flying gallop' shown on a relief in the Wu Liang tomb-shrine, +147 (*Chin Shih So*, Shih section, ch. 3).

A subject not unrelated to the foregoing is the exchange of cultural influences in matters of clothing. The 'hennin' or tall conical sugarloaf hat with a veil of muslin hanging down from the top, worn by medieval European ladies (+15th century) derived from a fashion which the Buddhist pilgrim Hui-Sêng noted when he visited Bactria in +518, and which afterwards had a long run in China from the +11th to the +14th centuries.[g] It has also been suggested that the crinoline,[h] and the sticking

[a] Münsterberg (1); Rostovtsev (1), pp. 203ff., (6), (7), pp. 80ff.

[b] Cf. Wheeler (3); Salmony (3); Tarn (1). On the famous Graeco-Indian Gandhāra school of art, see the classical work of Foucher (2), the orientating bibliography of Deydier (1), and the latest exposition by Grousset (7). Combaz (1) has analysed the mutual relations of Mesopotamian, Persian, Indian and Gandhāra art and architecture.

[c] Hirth (9).

[d] Pelliot (27, 28); Laufer (28); Burling & Burling (1); Ferguson (5).

[e] Cf. Soulier (1), Münsterberg (2), Pouzyna (1), and others, with the critique by Olschki (6).

[f] Cf. Hornblower (1); Auboyer (1). When I was in Persia in 1945 I was struck by the phrase still in common use, 'as beautiful as a Chinese'. Cf. the beauties sent to the Persian court in the reign of Nushirwan (+567) from the Chinese emperor (Yule (2), vol. 1, p. 95).

[g] Schlegel (2); de Mély (7). To judge from a note in the *Sung Shu* (tr. Pfizmaier (58), p. 370), it seems to go back to the early +4th century in China.

[h] Schlegel (4).

of 'beauty-patches' on the face,[a] were originally Asian. Kondakov (1), too, considers that the 'scaramangion' or long stiff brocade robe worn by Byzantine nobles, was Chinese in origin. When at the cave-temples of Tunhuang, I was greatly impressed by the caparisoned horses of the Northern Wei dynasty (*c.* +450) and their armoured riders, astonishingly reminiscent of the knights of medieval Europe a thousand years later.[b] Sketches made at the time will be reproduced in the discussion of Military Technology (Sect. 30) below.

It is also likely that philological studies will throw some light on ancient cultural exchanges. Mutual borrowing of words between the Indo-European and the Sino-Thai groups of languages must have occurred, though the number of well-authenticated cases, as Forrest says in his new review,[c] seems exceedingly small. However, there are papers by Hirth (4), Conrady (2) and H. Jensen (1), which are worth looking into. For example, it is now generally accepted that the word *lo*[1] (anciently *lak*), which meant various kinds of fermented milk products, was a loan from the Indo-European root **glakt-* (Lat. *lac*; lactic, galactic). Milk was as foreign to autochthonous Chinese diet as it was basic to that of all the nomadic peoples. Our own impression of such studies is that they are most valuable when they have some direct relation with a concrete piece of technology, the movements of which can be traced by means of epigraphic illustrations and literary evidence.[d] Otherwise they are liable to remain rather speculative.

(2) The Names of China

Before going further, a short digression on the names by which China was known in the West seems indispensable. There is no doubt that the origin of the word Seres is *ssu*,[2] silk, transmitted to Europe as Gr. *ser* (σήρ), and that it therefore dates from approximately the beginning of the silk trade. Tarn[e] shows that its first occurrence is when Apollodorus of Artemita (−130 to −87) says that Euthydemus, a king of Greek Bactria, had made conquests (about −220) extending to the Seres.[f] We shall see presently what event this refers to; here the word did not mean the Chinese, but only Siberian tribes who acted as middlemen, and in its proper application to China is not found before the time of Julius Caesar and Augustus. This seems to show that some silk was finding its way to the West, through many intermediate hands, before Chang Chhien's journey and the opening of the Old Silk Road.

The glory of the First Unification took the name of the −3rd-century Chhin[3] dynasty to India, giving rise to the Sanskrit forms Cīnasthāna and Mahācīnasthāna,[g] whence derived Sina in Latin, and Thinae or Θίν (as in the *Periplus*),[h] Byzantine

[a] Schlegel (3). [b] Cf. Ecke (1). [c] (1), pp. 118 ff.

[d] Examples of this will be found in the Section on the Technology of Animal Power (27*d*) and in that on Metallurgy (36). [e] (1), pp. 44, 84, 108, 110. [f] Now only in Strabo XI, 516.

[g] Hence Pers. Chīn and Māchīn, like Gog and Magog, names which, indeed, appear on medieval maps purporting to show East Asia. Connections of Magog with Mongol and Mogul have been proposed but this cannot be sustained because Magog occurs in ancient Hebrew writings (e.g. Ezekiel). In the Alexander-Romance the words meant all kinds of 'outer barbarians'. [h] Yule (2), vol. 1, p. 183.

1 酪 2 絲 3 秦

Tzinista (as in Cosmas), al-Ṣīn in Arabic, and our own common word China.[a] Chinistan became Jenasdan in Armenian writings (Moses of Chorene).[b] Two other dynasties, much later, also gave rise to Western names; the Tho-pa[1] Wei[2] (+368 to +556) and the Chhi-tan[3] Liao[4] (+907 to 1119). The family or tribe name of the former ruling house got into Turkish, Persian and Arabic (through Thak-Bat) in forms such as Tabghaj and Tamghaj, after appearing in Byzantine times (Theophylact Simocatta, +7th century) as Taugas.[c] The tribal name of the latter ruling house, passing westwards overland, gave rise to Russian Khitai, still used, and the familiar European name Cathay. Marco Polo knew the southern part of China as Manzi; this was certainly because the northerners, who had lived under the Chin[5] dynasty, considered the southerners as rather barbarous, and applied to them the name of a certain kind of south-western primitive people, the Man tzu.[6]

There was much uncertainty in the European Middle Ages as to whether China and Cathay were one and the same, just as there had been earlier regarding Sina and the Seres, and for the same reasons, because the former in each case was associated with the sea route and the latter with the overland route. Yule wrote his famous book (2) around the problem of this identification, which, as he says, was fully solved by +1600.[d] It was in connection with this that there occurs the romantic story of the Jesuit lay brother, Benedict Goes (O Pên-Tu[7]). The Jesuits in India, at the end of the +16th century, were not sure whether Cathay and Khānbaliq (Cambaluc) were the same as China and Peking, though Ricci and his mission affirmed them so to be.[e] Goes set out therefore over the difficult and dangerous Himalayan passes to Kashgar, whence he proceeded slowly east along the Old Silk Road. It was at Qarāshahr that he first realised his geographical goal had been attained, when he met a party of Muslim merchants returning from the Chinese capital, where they had been quartered in the same hostelry as Ricci. Goes then entered China at Chiayükuan, but having no entry permit was immobilised at Suchow (Chiu-chhüan) and died there only a few days after the arrival of a Chinese Jesuit lay brother Chung Ming-Li,[8] who had been sent by Ricci to succour him.[f]

Western writers used many terms regarding China which it would take us too long to enumerate or explain, but the titles of the Chinese emperor at least may be mentioned; thus, for example, Theophylact's 'Taissan' is probably Thai Tsung[9] or Thien Shun,[10] and the expression 'Son of Heaven' (Thien Tzu[11]) was translated into Persian as Baghpūr, giving rise to Arabic Faghfūr and similar words. The capital Chhang-An appears from time to time in Western and Arabic writings as Khubdan or Khumdan, perhaps from *kung tien*,[12] the imperial palace.

[a] Isaiah's 'Sinim' cannot be allied with these, and must refer to somewhere else.
[b] Yule (2), vol. 1, p. 93.
[c] So Yule (2), vol. 1, p. 29.
[d] But in +1670 A. Müller Greifenhagius (2), apparently still doubtful, wrote an elaborate survey of the subject.
[e] See above, p. 149.
[f] Whole story in Yule (2), vol. 4

1 拓跋 2 魏 3 契丹 4 遼 5 金 6 蠻子
7 鄂本篤 8 鍾鳴禮 9 太宗 10 天順 11 天子 12 宮殿

(*e*) THE DEVELOPMENT OF OVERLAND TRADE-ROUTES

Hitherto we have been speaking partly of ages before any well-marked trade-routes existed. Now we must give attention to what is known of these routes and when they developed. Only in this way can one be equipped to approach the problems of diffusion and convergence in the history of science in the Old World.

The most handy book for the historian of science here is that of Hudson (1),[a] which is based on the more exhaustive treatments of various aspects of the subject by Yule (2), Herrmann (2–6), Schoff (1, 2), v. Richthofen (4) and others. It should be read in conjunction with that of Poujade (1), which deals with the transportation technology of the sea route. The easiest way for us to gain an overall picture of the subject is to look at the map (Fig. 32), which is based on that of Hudson.

For more than two thousand years it was the western half of this area which saw the greatest amount of coming and going. The people of the Indus Valley civilisation (Harappa and Mohenjo-daro, before −2500) were, as Childe (7) has explained in a short review, in touch with the Sumerian cities of Iraq. The Indus River is shown on the map between the two later cities of Taxila and Barbaricon. But nothing is known of the routes which connected Harappa and Sumeria—presumably they followed either the coast of Baluchistan by land or the Persian Gulf by sea.

When we come to the −5th century Herodotus gives us[b] an elaborate account of the Scythians, mentioning some ten separate peoples or tribes and describing their customs. It was obviously written on the basis of fairly detailed information, and it ends with the people called Hyperboreans. Moreover, it is not the only ancient record of the central Asian tribes which has come down to us, for fragments have been preserved of the *Arimaspea* of Aristeas of Proconnesus, which is at least one, and may be two, centuries older than Herodotus; these fragments describe the migrations of the peoples, one being said to have been pushed on by another.

Modern historical geographers such as Hudson have attempted to fix the locations of these tribes on the map. Thus the Thyssagetae (the fourth group in Herodotus' itinerary) would seem to have lived just south of the Ural mountains, and the Issedones (the eighth) somewhere just east of the Thien Shan. If the Arimaspi (the ninth) are rightly placed between the Tarim basin and the Gobi desert, the at first surprising conclusion is reached that the Hyperboreans were none other than the Chinese, dwelling in the Kuanchung domain and the lower Yellow River lands. This suggestion, first made by Tomaschek,[c] is not so strange as it sounds, for the ancient explorers had little means of ascertaining exact latitude and none whatever for finding the longitude. Moreover, as Hudson says, the Chinese did live 'beyond the north wind' in the sense

[a] I do not know of any book which packs more solid information into a small compass, and yet makes it readable with more pleasure, than that of Hudson; unfortunately, it is marred by a very inadequate index. The book of Soothill (2), in spite of its title, is mostly restricted to modern times.

[b] Bk. IV.

[c] And, indeed, as long ago as 1770 by de Guignes (2).

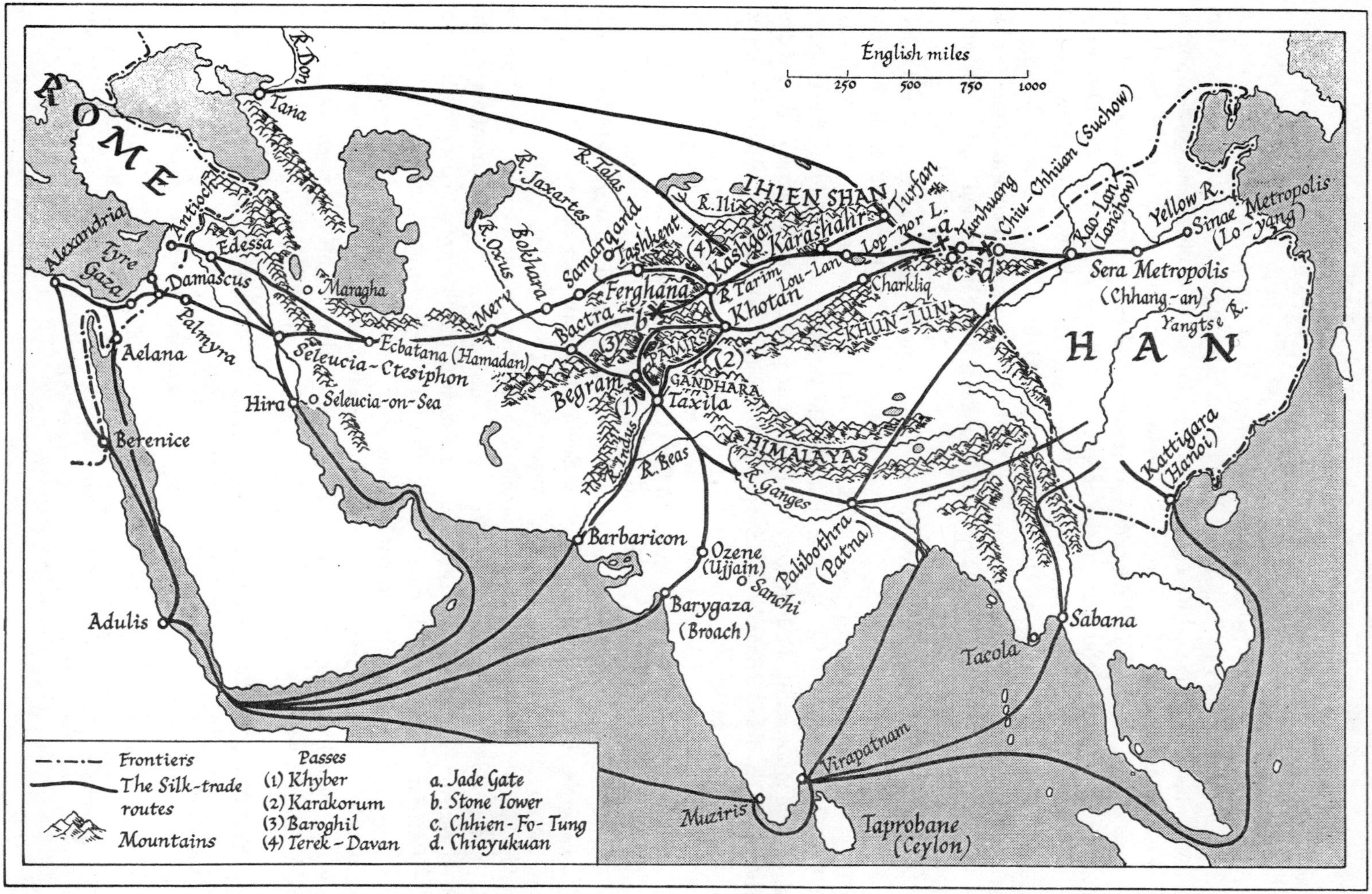

Fig. 32. Routes of trade between China and the West from the +1st and +2nd centuries onwards. Based upon Hudson (1).

that they enjoyed a comparatively mild climate on the other side of the terrible winter cold of Central Asia.[a]

Aristeas claimed that he himself journeyed as far as the country of the Issedones, and there is no reason to doubt that he did, but nothing is known of any travellers in the immediately succeeding centuries. Herodotus could obtain no contemporary information about the Hyperboreans. Soon after his time most of the Scythian peoples along the overland line vanished from the Greek horizon. Strabo (−63 to +20) knew nothing of the route or the peoples, and it was not until the +2nd century that Ptolemy, building on Marinus of Tyre[b] and the information which he had obtained from the agents of the silk merchant Maës Titianus, was again able to give an intelligible description of the country between the Caspian and China.[c] This was because the 'Old Silk Road' had in the meantime been opened and abundantly used.

The eastern route of Herodotus had been through Scythian lands, that is to say, from the Crimea and the Sea of Azov, past the south end of the Urals to the east end of the Thien Shan (i.e. to the north of the Caspian); he knew nothing of the lands east of Persia and south of the Caspian. The Greeks down to and including Alexander's time, were ignorant of the Pamirs, and the greater part of the Himalayas. Alexander, who in −329 crossed Bactria and Sogdiana in his great march of conquest,[d] stopped at the boundaries of Persia, founded a Macedonian city called 'Alexandria the Furthest' at modern Khojend on the banks of the Iaxartes (Syr Daryā), and then turned south for his attack on western India.[e] Chinese Turkestan, Siberia and Gangetic India all remained unknown to him and to the Western men of his age.

After the death of Alexander in −323 Bactria and Sogdiana were held for nearly two centuries by Greek rulers, first by Seleucid governors responsible to Antioch, and then by local kings whose capital was at Balkh (Bactra).[f] In −302 the Seleucids ceded Alexander's Indian conquests, including the Kabul valley, to the new Chandragupta Maurya empire, but the Greeks continued to man the old frontiers of the Persians, such as the oases of Bukhara and Samarqand. The great artistic tradition of the Bactrian Greeks which gave rise to the Graeco-Buddhist Gandhāra school by fusion with Indian motifs and content has already been mentioned. Always on the farthest horizon eastwards, for the Bactrians, were the Seres, but actual contact was not established until after the Greek kingdoms of that region had been overthrown by a nomadic people.

[a] Many scholars, however, such as Sinor (1), reject entirely the identification of the Hyperboreans with the Chinese, though it was seemingly favoured by Minns (1), p. 113, etc.

[b] *c.* +100; B. & M. pp. 634, 788.

[c] See McCrindle (6); Gerini (1); A. Berthelot (1) for the attempted identification of Ptolemy's Asian place-names.

[d] Documents in McCrindle (5); best modern account by Tarn (2).

[e] In India also he stopped at the boundaries of the ancient Persian empire (the Beas River), or rather his army stopped him, refusing to go any further (Tarn (2), vol. 1, p. 99). The Mongol army of Ogotai in Hungary a millennium and a half later affords a curious parallel.

[f] See the books by Tarn (1) and Rawlinson (2).

Soon after −175 a great commotion was caused in Central Asia because the Huns attacked and drove out from their traditional home in West Kansu, along the slopes of the Nan Shan, an ancient people, the Yüeh-chih[1] of Indo-European (Tokharian) speech.[a] These in turn, moving west, invaded the lands of the Wu-sun[2] people in the Ili basin, but did not remain there; they continued on their way, and about −130 overran the Greek kingdom of Bactria, where they settled. Later on they spilled over into the Indus Valley, as Persians and Greeks had done before them, and established the 'Indo-Scyth' or 'Śaka' dynasty. They had not long taken over Bactria when Chang Chhien reached them as ambassador from the Chinese emperor Wu Ti.

Chang Chhien's story is told by Ssuma Chhien in chapter 123 of the *Shih Chi*,[b] and is of such interest that it warrants something more than the brief reference already given it in the historical section (p. 107). Translations of this chapter by Hirth (2) and de Groot (1) are available.[c] From the outset it is clear that the mission of Chang Chhien had the object of gaining allies against the Huns.

> Chang Chhien[3] was a native of Hanchung; during the Chhien-Yuan reign period (−140 to −134) he was a '*lang*' (titular officer of the imperial household; a yeoman). At that time the Son of Heaven made enquiries among those Hsiung-nu[4] (Huns) who had surrendered (as prisoners) and they all reported that the Hsiung-nu had overcome the king of the Yüeh-chih and made a drinking-vessel out of his skull. The Yüeh-chih had decamped and were hiding somewhere, all the time scheming how to take revenge on the Hsiung-nu, but had no ally to join them in striking a blow. The Chinese, wishing to declare war on, and wipe out, the Huns, upon hearing this report, desired to communicate with the Yüeh-chih, but the road having to pass through the territory of the Hsiung-nu, the Emperor sought out men whom he could send. Chang Chhien, being a *lang*, responded to the call, and enlisted in a mission to the Yüeh-chih...[d]

Almost at once he was made a prisoner and brought before the Shanyu or Khān, who held him for ten years, but eventually he escaped and continued his journey to Bactria and Sogdiana. By the time he reached his destination,[e] the Yüeh-chih had given up the idea of a retributive war against the Huns, but he remained there a year before starting on the return journey. This involved another year's imprisonment among the Huns though he tried to avoid them by passing through the Chhiang[5] or Tangut country, probably on the Tsaidam plateau. Upon his return to China in −126 with a Tartar wife and his faithful Tartar servant Kan Fu,[6] he was made

[a] Their race and language constitute most difficult problems, on which see Tarn (1), pp. 288, 533. It is certain, however, that they were the Tokhari of Strabo and Ptolemy.

[b] Also *Chhien Han Shu*, chs. 61, 96A, 96B.

[c] Cf. Herrmann (2); Hudson (1), pp. 53 ff.; McGovern (1), pp. 131 ff.

[d] Tr. Hirth (2).

[e] The dates are not quite settled; Haloun (4) believed that Chang Chhien could not have set out before −133.

1 月氏 2 烏孫 3 張騫 4 匈奴 5 羌 6 甘父

an Imperial Chamberlain, and later a Marquis, but only to be degraded on account of real or alleged military incapacity when participating in an expedition against the Huns. Finally, he went out once more as ambassador to the Wu-sun, and though his mission again did not succeed,[a] he was appointed Ta-Hsing[1] (Director of a Department in the Office of Foreign Affairs) and died in −114.[b]

One of the most interesting points in Chang Chhien's travels was the following. In making his report to the emperor, he said:

> When I was in Ta-Hsia[2] (Bactria), I saw there a stick of bamboo from Chiung[3] (Chiung-chow in Szechuan), and some cloth from Shu (Szechuan). When I asked the inhabitants of Ta-Hsia how they had obtained possession of these, they replied, 'The people of our country buy them in Shen-Tu'[4] (India).[c] Shen-Tu may be several thousand *li* to the south-east of Ta-Hsia. The people there have fixed abodes, and their customs are very much like those of the Ta-Hsia, but the country is low, damp, and hot. The people ride on elephants to fight in battle. The country is close to a great river. According to my calculation, Ta-Hsia must be 12,000 *li* south-west of China. Now the country of Shen-Tu being several thousand *li* to the south-east of Ta-Hsia, and the produce of Shu being found there, that country cannot be far from Shu.[d]

Chang Chhien had in fact clearly realised the existence of a trade-route between Szechuan and India via Yunnan and Burma or Assam. But while small quantities of goods probably came over it from very ancient times, its difficulties were such that an important expedition sent out at his suggestion by the emperor Wu Ti, to find a way round to Bactria without having to cross either Hunnish or Tibetan territory, 'had to be abandoned, in spite of heavy expense incurred'.

Chang Chhien brought back information about Parthia (An-Hsi[5]),[e] Seleucid Media and Syria, possibly also Egypt (Li-Kan[6]),[f] and Babylonia (Thiao-Chih[7]), as well as

[a] McGovern (1), pp. 145 ff.
[b] His biography is in *Shih Chi*, ch. 111, and especially *Chhien Han Shu*, ch. 61.
[c] A strange name; the later one used was Thien-Chu.[8]
[d] Tr. Hirth (2).
[e] This name is undoubtedly derived from Antioch in Margiana (Merv), see Tarn (1), p. 281.
[f] The derivation of this name, and the region to which it was intended to apply, have given rise to much argument. Its orthography fluctuated considerably (Hirth (1), p. 170), as can be seen; characters which should be read 'Li-Hsien' and 'Li-Chien' appearing as well as 'Li-Kan'. Hirth thought that it referred to Rekem (another name for Petra). Dubs (30) states a case for the correct pronunciation of its name as Li-Chien. If this is accepted, the view that it was a corrupt transliteration of Alexandria becomes phonetically plausible, for the terminal *k* in the first component (as pronounced in the Thang and earlier) could well have stood for the *x*, as the Chinese heard it. The question of exactly which of all the sixteen Alexandrias was meant, remains, but since the *Chin Shu* and the *Wei Lüeh* both say that Li-Kan was another name for Ta-Chhin,[9] it may well have been the greatest of all Alexandrias, that of Egypt. This was Pelliot's opinion (31). Ta-Chhin (see below, p. 186) was the term used by the Chinese from the −2nd century onwards for Syria, the Hellenistic Orient in general, and vaguely for the whole of the Roman empire. Some believe that the name Ta-Chhin may have derived from Ṭaiyi' (see Hitti, 1), one of the greatest and oldest of the Arab tribes. This word later became generally used for Arab merchants (personal communication from Dr Lieber, Jerusalem).

1 大行 2 大夏 3 卭 4 身毒 5 安息
6 黎軒; 犂軒; 犂鞬 7 條支 8 天竺 9 大秦

Graeco-Iranian Bactria (Ta-Hsia[1]),[a] Ferghana (Ta-Yuan[2])[b] and Sogdiana (Khang-Chü[3]).[c] His mission was followed by a number of others, the emperor sending out ambassadors in all directions, partly to persuade small States to become tributary vassals, and partly to purchase certain commodities which the Chinese particularly wanted (such as the 'blood-sweating horses of Ferghana', cf. below, Sect. 30). A third motive was probably the accumulation of intelligence about foreign countries. Most of the envoys followed Chang Chhien's example in making collections of strange plants and other natural objects. As Pan Ku said, in the time of Han Wu Ti, 'specimens of strange things began to arrive from every direction (*i wu ssu mien erh chih*[4])'.[d]

A careful investigation of these was made by Laufer (1). The grape-vine (*Vitis vinifera*)[e] and alfalfa (*Medicago sativa*)[f] are clearly attributed to Chang Chhien in the *Shih Chi*. But other introductions probably came a good deal later, such as the chive (*Allium scorodoprasum*), the coriander (*Coriandrum sativum*), the cucumber (*Cucumis sativus*), the fig (*Ficus carica*), the safflower (*Carthamus tinctorius*), the sesame (*Sesamum orientale*), the pomegranate (*Punica granatum*) and the walnut (*Juglans regia*);[g] at various times between the +3rd and the +7th centuries. Of these eight plants and trees, no less than five have names which include the character *hu*,[5] which meant Central Asian or Persian. Only later tradition, as embodied in such books as the *Chhi Min Yao Shu*[6] (probably +5th century) and the *Po Wu Chih*[7] (originally +3rd), ascribed all such introductions to Chang Chhien himself. It might be added here that there was of course a radiation, at least equal, of plants and trees from China outwards. The orange seems to have originated from the borderlands of China and Tibet north of the Himalayas, and it had started on its travels long before; the peach and the pear reached India in the time of Kanishka (+2nd century), hence they were known there as the *cīnanī* and the *cīnarājaputra*.[h] Many centuries later the western mountain regions of China were to provide an altogether surprising proportion of the flowers now cultivated in Western gardens, such as the rose, peony, azalea, camellia, chrysanthemum, and so on.[i]

Chang Chhien's report was certainly available to Ssuma Chhien and seems afterwards to have been made into a separate book, as the Sui dynasty bibliography had

[a] This name used to be derived from Tokhari and Tokharistan (e.g. in Yule (2), vol. 1, p. 37), but such an origin cannot be upheld. It is phonetically impossible having regard to the pronunciation of Chinese in the −2nd century, and historically impossible because the Tokhari were still nomadic Yüeh-chih, while Bactria was a land of fortified cities and villages. Haloun (7) showed that the name goes back many centuries before Chang Chhien's time, applied by the Chinese to a fabled northern people. Presumably the great envoy identified the two. See on the whole matter Tarn (1), pp. 295 ff.

[b] This name used to be derived from Skr. *yavana*, the Greeks (e.g. Hirth, 9; Yetts, 13), but such an origin is very improbable (Tarn (1), pp. 474 ff.). No alternative hypothesis exists.

[c] Description in *Chhien Han Shu*, ch. 96A.

[d] *Chhien Han Shu*, ch. 96B, p. 23*a* (Wylie, 10).

[e] Cf. Grumm-Grjmailo (1).

[f] R288 and 397 respectively.

[g] In succession, R672, 217, 60, 601, 21, 97, 250 and 619.

[h] Bagchi (1), p. 58.

[i] See the book of E. H. M. Cox (1).

1 大夏 2 大宛 3 康居 4 異物四面而至 5 胡
6 齊民要術 7 博物志

a *Chang Chhien Chhu Kuan Chih*[1] (Record of Chang Chhien's Expeditions Beyond the Passes), and this book is cited in preference to the *Shih Chi* by the author of the *Ku Chin Chu*,[2] written about +300. Another book, the *Hai Wai I Wu Chi*[3] (Record of the Strange Things of Countries Overseas),[a] was also in later centuries ascribed to Chang Chhien. Both have since been lost.

The notable mission of Chang Chhien was the beginning of the development of the silk trade, which may be said to have arisen partly as a by-product of Wu Ti's energetic diplomatic and military policy in Central Asia.[b] The first through caravans from China to Iran date from −106. But one of the most interesting things about the mission from the point of view of the history of science is that it was a Chinese success in geographical exploration where the Greeks had earlier failed.[c] A little before −200 Euthydemus, King of Bactria, feeling the lack of gold which had formerly come from Siberian sources, made an effort to reach the Seres along a road north of the Thien Shan, but he did not succeed, though the blue-eyed red-haired race which the expedition met with gave rise to a tradition that the Seres were people of that kind.[d] Fifty years later, the expedition of Euthydemus was no doubt remembered in Bactria, and may have accounted for the warmth with which Chang Chhien, a Seric Livingstone, was received.[e]

(*f*) THE DEVELOPMENT OF THE MARITIME TRADE-ROUTES

Thus far we have been discussing the 'tunnelling' from both sides along the land route, either by way of the Scythian north or the Bactrian centre. One must next speak of the sea route, and this involves reference to Greek and Roman relations, not with Central Asia, but with India.[f]

The earliest information which the Greeks had of India was brought by the physician Ctesias of Cnidus who was at the Persian court in −416 and back in Greece in −398.[g] Unfortunately, his description was much more fantastic and marvellous than that of a duly sceptical physician would be expected to be. His text[h] contains

[a] Bretschneider (1), vol. 1, p. 25.

[b] Hudson (1), p. 66; Sarton (1), vol. 1, p. 197. The Chinese would probably have described it as defensive, since the object was to gain the help of the Yüeh-chih against the perpetual menace of the Huns on China's north-western borders.

[c] Tarn (1), pp. 109 ff., 274.

[d] Pliny, *Hist. Nat.* VI, 88.

[e] Tarn (1), p. 87.

[f] On this subject there are two useful books, those of Warmington (1) and Rawlinson (1). The latter makes the more use of Indian sources, but both are now old. Cf. the interesting paper of Barnett (1). Most of the important texts have been assembled by McCrindle (1), who had an 18th-century precursor in Wm. Robertson (1).

[g] Actually the first Greek to speak of India was Hecataeus of Miletos (−549 to −486), who mentions seven Indian place-names. Herodotus also had a little to say about it (III, 98 ff.).

[h] Translated by McCrindle (2).

1 張騫出關志 2 古今註 3 海外異物記

references to 'weak water',[a] 'fountains of liquid gold' (gold smelting from ores), and 'tree-garments' (cotton).

The greatest Greek traveller in India, the veracity of whose statements was vindicated after long disbelief, was Megasthenes,[b] Seleucid ambassador to Sandracottus (Chandragupta Maurya) between −302 and −288. He gives much anthropological and zoological information, mentions cotton and the 'gold-digging ants',[c] describes the practices and opinions of the 'Brachmanes and Sarmanes' (*brāhmaṇas* and *śramaṇas*, i.e. Buddhist monks), and even refers to Malaya under the name of Chryse.

Megasthenes was by no means the only, or the last, ambassador from Greece to India. About −250 Dionysius, sent from Ptolemy Philadelphos of Egypt, presented his credentials to Bindusāra or his son the great Asoka.[d] The Persians may also have sent embassies, or at least philosophers and priests.[e] Buddhist missions seem also to have been despatched to Hellenistic countries (Wüst, 1; V. A. Smith;[f] Thomas[g]). The name of at least one such apostle of the Yavanas has come down to us: Mahārakshita. Then in the −2nd century there was further contact due to the invasion of India by the Graeco-Bactrian king Demetrius and his general and son-in-law Menander (the Milinda of Indian records) between −184 and −167.[h]

Abundant literary traces of Graeco-Indian contact between the −2nd and the +2nd centuries have been unearthed by such scholars as Festugière (1, 2), Filliozat (4, 5) and Bevan (1). The *Life of Apollonius of Tyana* contains much which must have been derived from a knowledge of the brahmins and ascetics of India (Meile, 1; Goosens, 1). The refutation of the Brahmins by St Hippolytus contains sound Upanishad doctrine (Filliozat, 6). The pneumatic system in Greek medicine is closely related to the theories about *prāṇa* in older Indian books (Filliozat, 1). Pepper appears, named, in the Hippocratic corpus, lac[i] somewhat later, and the twenty-eight *hsiu* (equatorial mansions of the moon) figure in Gnostic texts. Gnosticism itself, and Neo-Platonism too, are full of Indian traits (Kennedy, 1, 2; Bréhier, 1; de Lubac, 1; etc.). Conversely, the water-mill is supposed to have been introduced to the Indians by a Greek, Metrodoros, but the story is not found until a very late date.[j] More certain, but of less technological interest, is the spread of Greek plastic art in the realm of Buddhism effected by the school of Gandhāra already mentioned. There was much more in this, however, than merely artistic convention, for earlier Buddhism had not felt the need of statues or paintings of Gautama at all; now, however, the world-denying philosophy was being supplemented, even overlain, by the system of devotion (*bhakti*) to a personal saviour. This was one of the great differences between

[a] Cf. the Geological Section (23) below.

[b] He may be read Englished in McCrindle (3), from the fragments collected by Schwanbeck.

[c] Cf. the Metallurgical Section (36) below; this ancient fable is bound up with the Siberian sources of supply and therefore related to the expedition of Euthydemus.

[d] V. A. Smith (1), pp. 76, 97.

[e] V. A. Smith (1), p. 79.

[f] (1), pp. 98, 134.

[g] (3), p. 499.

[h] See Tarn (1), pp. 129, 414, and de la Vallée Poussin (1), pp. 239 ff.

[i] On lac see conveniently Burkill (1), p. 1290.

[j] In Cedrenos (+11th century), vol. 1, p. 516.

Hīnayāna and Mahāyāna Buddhism.[a] Much discussion[b] has taken place on the extent and efficacy of the Western influence, but it seems certain that first the Greek mystery religions and then Christianity, had some hand in it.[c]

A focal point in the intercourse of peoples on the western half of the map is the anonymous book entitled *The Periplus of the Erythraean Sea*[d] (i.e. the Indian Ocean, Persian Gulf and Red Sea), written about +70 (the time of Heron and Wang Chhung) by a Graeco-Egyptian seafaring merchant whose home port was Berenice on the western shore of the Red Sea. It is one of the most interesting books of antiquity, for its author, while clearly not an educated man in the academic sense, speaks of the routes, the ports and the commodities carried and bought and sold,[e] with the certainty of one who had himself taken part all his life in such activities. The development of the trade between India and Rome via Egypt had taken place by a succession of stages.[f] Before the middle of the −1st century, the Egyptian ships had simply coasted round the southern shores of Arabia and Baluchistan to arrive at Barbaricon and Barygaza (Broach). But at a date which Charlesworth (1) placed about −15, though it may well have been several decades earlier (Tarn (1) thinks as early as −85), it was discovered that use could be made of the monsoon winds to allow of rapid direct passages from the straits of Bāb al-Mandab (the opening of the Red Sea) to Barygaza and all points south, e.g. Muziris, Nelcynda and Comari, at the tip of the Indian subcontinent, whence a further passage to Ceylon (Taprobane) only involved coasting.[g]

We may now look at the historical development of the whole sea route to and from East Asia.[h] From the +1st to the middle of the +3rd century Western ships (Roman in name, but actually Graeco-Egyptian) were reaching ports all round India,[i] and towards the latter part of the period a few went as far as Kattigara, which may have been in Indo-China or even perhaps somewhere in Kuangtung.[j] Indian and Singhalese ships were also covering parts of the same sea-lanes.[k] There has been much talk of Roman-Syrian 'factories' in India, and some archaeologists may have strained the evidence a little too far, but there is solid ground for accepting certain settlements, such as Vīrapatnam (near modern Pondicherry), as having been closely connected

[a] See below, Sect. 15.

[b] Well summarised in Tarn (1), pp. 395 ff.

[c] V. A. Smith (1), p. 134; cf. Needham (6), p. 9.

[d] Translations by McCrindle (4) and Schoff (3).

[e] Warmington (1, pp. 145 ff.) is particularly good on the nature of the products, animal, vegetable and mineral. Some of them, such as the perfume from the leaves of some plant, which came into commerce as 'malabathrum', have caused great difficulty in identification (cf. Yule (2), vol. 1, p. 184). It was doubtless correctly identified with *tamalapattra* by da Horta in +1563, but whether the plant was *Laurus cassia*, or, as Laufer (27) decided, after ransacking the Chinese botanical literature, patchouli (*Pogostemon*), remains to be determined.

[f] Cf. Herrmann (6); Tarn (1), p. 368.

[g] The discovery was early attributed to a merchant captain, Hippalos, but he may have been a more or less legendary culture-hero, or even a wind-god (see Tarn, 1).

[h] Summarised by von Richthofen (3) and more recently by Mills (3) and Filliozat (4).

[i] Meile (2).

[j] Schoff (4). Hudson (1) accepts Hanoi, but there have been many arguments about it; cf. the monograph of Gerini (1).

[k] Lévi (5).

with the Roman trade (Wheeler, 1). Fragments of pottery found there had come from Arretium in Tuscany about +24; just as a mirror-handle still preserved had found its way from Gandhāra to Pompeii before +79 (Filliozat, 4). And coins of Antoninus Pius have been found at Go-Oc-Eo in Indo-China (Coedès, 3). The reasons for the decline of 'Roman' navigation in the South Seas and the Pacific are no doubt complex, but by the +2nd century the Indian colonial expansion to Java, Sumatra, and Cambodia, was reaching a climax,[a] and it may be that Indian shipping, which from this time on regularly visited Chinese ports, was hostile to the small number of ships from the Far West.

Only after the +3rd century does Chinese long-distance navigation appear, and not until the +13th reach full development. Chinese ships sailed to Penang in Malaya about +350, and Ceylon at the end of the +4th century;[b] by the +5th they were probably coming to the mouth of the Euphrates in Iraq[c] and calling at Aden.[d] Ammianus Marcellinus[e] refers to Chinese merchandise at the annual fair of Batanea on the Euphrates around +360.[f] Such contacts[g] seem to have continued until about +900, when a decline set in.[h] Prisoners from the Talas River returned home from the Gulf in +762 in Chinese junks.[i] About +850 Sulaimān the Merchant refers[j] to the port of Sīrāf on its north coast as a terminus of Chinese shipping.[k]

Now came the rise of the Islamic Arab shipmasters.[l] In +758 they were strong enough to burn and loot Canton, just a century after the first Arab embassy to China (+651). In the +9th century they habitually circumnavigated Malaya, frequenting Kuangtung in considerable numbers and establishing 'colonies' or 'factories' there, especially at Canton (Khanfu) and Hangchow (Khanzai), as their predecessors the

[a] Cf. Grousset (1), p. 155.

[b] See Lévi (2).

[c] The early arrival of Chinese ships at the head of the Persian Gulf rests mainly on a statement by al-Mas'ūdi (vol. 1, pp. 216ff.) who wrote about +947, but many, including Reinaud (1), Warmington (1) and Hudson (1, p. 113), have accepted its probability. Cf. Yule (1; 2, vol. 1, p. 83) and Hourani (2). The *Sung Shu* (ch. 97), written about +500, says that traffic goes on with the people of Ta-Chhin (Syria) and Thien-Chu (India), adding that the *chou*[1] (perhaps Chinese ships) and the *po*[2] (perhaps foreign ships) 'make connection along the routes, and the merchants engage in commercial intercourse' (cf. Hirth (1), pp. 46, 102). It seems that Hirth & Rockhill (p. 15) were too positive in asserting that no Chinese ships ever reached the Persian Gulf before the Thang.

[d] Schwarz (1) and Duyvendak (8) have written on the Chinese discovery of Africa.

[e] XIV, 3.

[f] Of course it need not have come all the way in Chinese bottoms.

[g] A description of the route between Iraq and India was given by the Christian monk Cosmas Indicopleustes in +525, just about this period, before the rise of Islam. On him see McCrindle (7); B & M, p. 1045; Yule (2), vol. 1; Coedès (1), pp. 132ff. He is more important for geographical theory than for questions of this kind.

[h] It is quite striking that while in the Thang time the Chinese pilgrims voyaged mostly in non-Chinese ships; in the late Sung and Yuan the European travellers always used Chinese junks and were very impressed by them.

[i] Pelliot (32).

[j] Sauvaget (2), pp. xxxix, 6, 7, 8, 41, 43. Cf. Ferrand (5).

[k] Hourani (1) points out, however, that the expression 'China ships' (*marākib al-Ṣīn, sufūn Ṣīniyah*) common in Arab geographers and travellers, *may* have had the same significance as 'East Indiamen', or 'China Clippers'. But the expression *al-sufūn al-Ṣīniyah* (Chinese ships) is also used.

[l] Their lingua franca was probably Persian (Schafer, 2), corresponding to Sogdian on the overland route. On the whole question of Arab and Persian seafaring see Hadi Hasan (1) and Hourani (1).

1 舟 2 舶

Syrians and Graeco-Egyptians had done before them,[a] on a much lesser scale, in the +3rd. The Arabs also knew Korea (Sila) and Japan (Wakwak). Huzzayin (1) and Hourani (1) have described this Arab trade in much detail, and Ferrand (1, 2) has collected the texts of Arabic merchants and travellers relative to eastern and south-east Asia in translated form.[b] This was the time (+10th to +13th centuries) for which we have the most important Chinese accounts of maritime commerce and navigation, especially the *Chu Fan Chih*[1] (Records of Foreign Peoples) by Chao Ju-Kua,[2] of which there is an English translation with abundant notes by Hirth & Rockhill.[c]

The Arab names for Chinese ports had such wide currency in medieval times that it is worth examining them for a moment. Canton (Kuangchow[3]) was Kuangfu, or by corruption Khanfu, or Sīn al-Ṣīn. Chiaochow,[4] also in Kuangtung, was known as Lukin, from Lungpien,[5] a place nearby, hence al-Wakin. Chhüanchow,[6] the greatest port of them all, in Fukien, was Djanfu, Zaitun, Zeytoun, Zayton, etc. a name now known[d] to have been derived from the avenues of *tzhu-thung*[7] trees[e] planted round it by Liu Tshung-Chiao,[8] governor of the city in +962. Hangchow[9] in Chekiang appeared as Quinsay, Khinzai or Khanzai, terms derived from the expression *hsing tsai*,[10] 'temporary residence', of the imperial family after the loss of the northern capital. The Arab quarter (Fan fang[11]) had a headman (*qadi*) selected from among the foreigners, who was responsible for public order. Some families settled in China and their descendants entered the civil service; of such origins was Phu[f] Shou-Kêng,[12] Commissioner of Merchant Shipping at Zayton between +1250 and +1275, just before the arrival of Marco Polo. Like the latter, Phu Shou-Kêng served the Mongol dynasty. He had to collect customs dues, issue licences to Chinese merchant captains, supervise the foreign colony, see to the prohibition on the export of currency, and exercise general control over the affairs of the port.

After the end of the +12th century Arab navigation in Pacific waters gave place to Chinese, and in the +15th there came the short-lived period of Chinese maritime supremacy under the Ming dynasty, which has already been referred to.[g] This again brought Chinese sea-going junks to Borneo, the Philippines, Ceylon, Malabar and even East Africa.[h] Finally, the Portuguese explorations at the beginning of the +16th century opened the modern era. Meanwhile, what was happening to the overland routes?

[a] One of the evidences that the late Roman-Syrian navigators to China had 'factories' near Canton is that they planted henna there (cf. below, Sect. 38, and Hirth (1), pp. 268 ff.). This we know from the account in the *Nan Fang Tshao Mu Chuang* of the 'finger-nail flower', *chih chia hua*[13] (R 248; Stuart (1), p. 232).

[b] It is necessary to use Ferrand's books with some caution.

[c] The parts concerning the Islamic countries had previously been translated into German by Hirth (11).

[d] Kuwabara (1) has much information on these subjects. The detailed investigation of the life and times of Phu Shou-Kêng is also due to him.

[e] The Indian coral tree (*Erythrina indica*); R 384.

[f] It is thought that this family name was derived from the Arabic prefix Abū.

[g] Cf. p. 143 above and Sects. 22, 29 below.

[h] Again see Duyvendak (8).

1 諸蕃志　2 趙汝适　3 廣州　4 交州　5 龍編　6 泉州　7 刺桐
8 留從効　9 杭州　10 行在　11 蕃坊　12 蒲壽庚　13 指甲花

(g) THE OLD SILK ROAD

We left the roads towards the end of the −2nd century, after the overthrow of Greek rule in Bactria by the Yüeh-chih, and about the time of Chang Chhien's embassy. It was just after this, i.e. about −106, that the trans-Asian silk trade became regularised (see Map in Fig. 32). The ancient westerners knew of two Chinese capitals, Sera Metropolis (probably Chhang-an, though some have placed it at Lanchow or even higher up the Silk Road, at Kanchow), and Sinae Metropolis (Loyang). Hudson points out[a] that this was because contacts with the Former Han dynasty, whose capital was at Chhang-an (Sian), were by the land route, while the chief contacts with the Later Han dynasty, whose capital had been moved to Loyang, were by the sea route at least as far as Tongking. This led to doubt (as we have seen) whether the Seres and the Sinae were the same nation.

In any case silk leaving China had to go first along the foot of the Nan Shan through the oasis cities of Liangchow, Kanchow, Suchow (Chiu-chhüan) and Anhsi to Tunhuang, at which point two roads diverged, one keeping just north of the Tibetan Tsaidam plateau, and the other (opened about +5) crossing the desert past Turfan and Qarāshahr to run just south of the Thien Shan. In either case, the Tarim basin (Taklamakan desert) was avoided. A third road left the northern route a short distance west of Tunhuang, at old Yumên, and struck across the desert to Loulan, a city located near the lake of Lop Nor; but after +400 increasing desiccation brought about the abandonment of the city and the closure of this road. All routes met again at Kashgar, after which there were further alternatives. One went over the Terek-Davan pass into Ferghana and then through Samarqand to Antioch in Margiana (Merv). Another led from Kashgar to Balkh (Bactra) in Bactria and thence to Merv, crossing the Pamirs by a way which is not now certainly identified, but which probably involved the Taun-Murun pass. It was along this road that one passed the fortification known as the 'Stone Tower', so prominent in Ptolemy and other ancient geographers; it stood on the Bactrian frontier (see Herrmann, 2). After Merv the road was much easier, passing through Iran by Hecatompylos (the old Parthian capital) and Ecbatana (modern Hamadan) to reach Seleuceia-Ctesiphon on the Tigris (just below modern Baghdad), whence there were abundant routes westwards in Syria.

Hudson[b] divides the whole route into four sections: (*a*) as far west as the Pamirs, i.e. to the western boundaries of modern Sinkiang; (*b*) from the Pamirs to the Merv oasis, i.e. Bactria or Sogdiana, according to whether the southern route by Balkh or the northern route by Samarqand was followed; (*c*) from Merv to Seleuceia in modern Iraq; (*d*) from Seleuceia to the Roman frontier. Of these four stages the first and the last were the most difficult ones. The second was Yüeh-chih or Hunnish under Kushan and Kidara dynasties from −140 to +560 when the Turks took it over; on the whole therefore conditions were fairly stable for trade.[c] Similarly, the third was

[a] (1), p. 86. [b] (1), p. 79. Cf. Schoff (5). [c] Codrington (1) describes its geography.

essentially Persian, first under the Arsacid Parthians from −129 to +224 and then under the Sassanids until the Arab conquest. The first and the fourth, however, were under constantly fluctuating allegiance. Between Kansu and Kashgar, trade was doubtless often only possible by silent barter between a series of nomadic tribes, especially Scyths, Huns and Mongols, with Tibetans and many little Central Asian city States also involved.[a] But sometimes during this period Chinese sovereignty was very real as in the time of the great governor Pan Chhao.[1] At the other end of the line the various buffer States between Rome and Parthia, such as Osrhoene and Palmyra, had the opportunity of enriching themselves as middlemen in this trade.[b]

It was precisely the existence of so many middlemen and potential middlemen along the overland route which encouraged the tapping of it by sea at intermediate stages. It is possible that some of the silk trade avoided stage (*d*) by circumnavigating Arabia down the Persian Gulf and up the Red Sea,[c] but probably not much of it went by so roundabout a route. It is certain that a connection was established which cut out not only stage (*d*), the Euphrates buffer-States, but also stage (*c*), Parthia itself. This was the transhipment at Barbaricon and Barygaza, described by the author of the *Periplus*. Silk was brought over the passes of the Hindu Kush and through the Khyber pass to Taxila[d] (near modern Rawalpindi) and thence down the Indus[e]—such a route was possible because the Kushan empire (Kuei-Shuang[2]) from early in the +1st century controlled north-west India as well as Bactria and Sogdiana.[f] Much further east of this region, and less well suited to trade with the Mediterranean, yet still on the coasting route, were two other outlets for silk from China. The first was the Ganges. By some route which is rather obscure, but which is definitely mentioned by Ptolemy,[g] trade came over the Tibetan plateau and down perhaps by way of Ladakh or Sikkim or other Himalayan passes to Palimbothra (Pataliputra, now Patna) on the Ganges, whence it met western ships at the mouth of the river, as the author of the *Periplus* says. The second was the connection between Yunnan and Burma, which is attested by Chinese sources, and which led to the ports of Sabana (near Moulmein) and Tacola (near Rangoon). Lastly one must list the circumnaviga-

[a] Cf. McGovern (1) and Drake (1). Silent barter on the Old Silk Road is often referred to by ancient Western writers, e.g. Pomponius Mela (*c.* +50), *Chorogr.* I, 60 (Coedès (1), p. 11; Yule (2), vol. I, p. 196); Pliny, *Hist. Nat.* VI, 20; and Ammianus Marcellinus (*c.* +330), *Res Gestae*, XXIII, vi, 64 (Coedès (1), p. 97).

[b] Cf. Rostovtzev (8).

[c] The story of Kan Ying (below, p. 196) would point to this.

[d] Chinese jade and Japanese scallop-shell have been found at Taxila (Tarn (1), p. 364).

[e] See the special study by Foucher (1).

[f] V. A. Smith (1), pp. 128 ff. and de la Vallée Poussin (1), p. 308. The greatest Kushan king was the famous Kanishka. They were descendants of the Yüeh-chih who had conquered Bactria, and whose Śaka language they still spoke. Cf. *Hou Han Shu*, ch. 118, p. 11*b*. In 1937–9 the French expedition to Begram, north of Kabul, unearthed there two walled-up stores of transit trade-goods (terminal date about +250). There were carved ivories from India, lacquer ware from China, bronzes, glass and plaster medallions from Syria and Alexandria. Publication of the full description of this material has long been eagerly awaited (Wheeler, 2).

[g] *Geogr.* I, 17.

[1] 班超 [2] 貴霜

tion of Malaya, and the arrival of Roman Syrians and Greek Egyptians at Kattigara itself, in or very close to Chinese territory. We are told that the pioneer of Roman commerce in the Pacific was one Alexander in the early +2nd century; he wrote an account of his travels which was used by Marinus of Tyre and Ptolemy.[a]

Apart from the silk itself, other commodities travelling over the great trade-route, whether by land or by sea, were relatively few. The most interesting of them was 'Seric iron' mentioned by Pliny,[b] but discussion of it must be reserved for the Metallurgical section.[c] There were also skins, cinnamon bark and rhubarb; the last-named seems to have had a special route of its own, through the tribes north of the Caspian. The Mediterranean region had little to export for which there was any Chinese demand. Glass, however, which in Roman times had reached a high degree of technical perfection in its Lebanese (Phoenician) home and in Egypt, had a market throughout Asia.[d] Wool and linen textiles seem also to have travelled east to some extent. These things, however, even reinforced by smaller amounts of miscellaneous products, in no way sufficed to balance the trade, and 'for Rome's oriental foreign trade as a whole', says Hudson,[e] 'we know that there was a serious adverse balance which was made up by bullion or specie payments'.[f] It may have amounted to as much as the equivalent of one million pounds sterling annually for the whole of the east-west traffic, and some consider that this drain was a major factor in the economic decline of the Roman world. It does not follow of course that the gold reached China, and there is more reason to think that it was absorbed by the middlemen States, which had products of their own, as well as Roman products, to barter with the Chinese.[g]

(1) The Road, the Nomads and the Wall

We have now, I hope, sketched the trans-Asian contacts in classical times sufficiently to enable the reader of this book to assess for himself the plausibility of travel of any given ideas and techniques to or from China. What of the situation at the break-up of the Roman empire?

The relations between Rome and China at the time of the barbarian migrations and invasions have been discussed in a special book by Teggart (1). By a consideration of the chronological correspondences in Roman and Chinese historians, he has concluded that there was indeed a transmission of shocks (analogous to those mentioned by Herodotus in which the Scythian tribes pushed each other from place to place, cf. p. 170 above), and that it took effect over the whole length of Asia. During the +1st century, wars in Sinkiang and Mongolia between Chinese and Huns were constantly

[a] Hudson (1), p. 89.

[b] *Hist. Nat.* XXXIV, 41, 145; cf. Edkins (9). [c] Sect. 36 below.

[d] An Alexandrian glass vase with a head of Athena, of the −2nd century, has been found in Honan (Rostovtzev (5), p. 513), though its genuineness is apparently still in doubt (Tarn (1), p. 536).

[e] (1), p. 98.

[f] This situation was precisely repeated in the days of the East India Company (see Sect. 27*k* below) and was one of the chief causes of the Opium Wars in the early nineteenth century.

[g] Warmington (1), pp. 272 ff., has a detailed discussion of the adverse balance. Roman coins (+14 to +275) found in Shansi have been described by Bushell (1).

followed by disturbances on the Danube and the Rhine. Such disturbances also followed with great regularity conflicts between Romans and Parthians in Syria or Armenia.

The barbarian invasions of Alans, Goths and other steppe peoples, which ended the Roman imperial age, are naturally important in any discussion of trans-Asian contact routes.[a] The causes of these Germanic invasions have long been contested. Leaving aside such pseudo-explanations as 'the desire for change' or 'longing for adventure' which have satisfied some writers,[b] the classical theory has been over-population,[c] although this solution of the problem finds no favour with the population experts.[d] There is not enough evidence that exhaustion of the soil brought about by rudimentary agricultural methods led to land-hunger (L. Schmidt, 1); in fact, the Germans were more advanced in their agricultural machinery than the Romans (cf. Sect. 41). It is hard to see how it can be attributed to the domestication of the horse (Peake & Fleure, 1) which must have occurred more than a millennium earlier.

The most widely favoured type of theory has been that which sought to attribute the barbarian migrations to changes in the physical environment of the nomads, particularly a progressive, or cyclical, desiccation of the Gobi and steppe regions. The idea seems to derive from the researches of the travellers v. Richthofen[e] and Pumpelly (1) in 1866 and 1877, and was furthered by Kropotkin's paper (1) of 1904. Stated in moderate and not too precise a form, it had some attractiveness; but it was over-elaborated by Ellsworth Huntington (1, 2), whose theory of the 'Pulse of Asia' (1907), reinforced by followers such as Brückner and Curry, was too precise to carry conviction. 'A regular succession of climatic cycles', says Curry's summary, 'approximately 640 years in duration, each including on the average something like 300 years of increasing aridity, has produced a series of alternating periods of migration and consolidation in Europe and Asia, where the effects can be traced between −2300 and +1600.' But most of those who have studied the subject have not been convinced. Aurel Stein (2, 4, 5) particularly emphasised the difficulty of obtaining any reliable estimates of changes of climate during historical times. That desiccation is still going on along the borders of Central Asia is fairly certain; when I was in farthest north-west Kansu in 1943 we had Stein's maps with us, and we repeatedly noted that areas which were marked 'steppe' or 'scrub' had turned to sandy or gravel desert during the intervening thirty years. In those parts, too, one hears many stories of the drifting of sand so as to cover city-walls, temples and *phailous*, as at Yulin on the edge of the Ordos. The lowest level of cave-temples at Tunhuang was largely buried in sand when I was there a dozen years ago, though some of these have now been uncovered. However, to argue from such facts regarding conditions eighteen centuries ago is rather hazardous.[f]

[a] Gibbon was not unconscious of the relevance of events in China and on the Chinese borders, e.g. the rise of the Tho-pa Wei dynasty (vol. 5, pp. 210ff.).

[b] E.g. Wm. McDougall (1), p. 223.

[c] Bury (2, p. 6); Barker (1).

[d] E.g. Carr-Saunders (1), pp. 297 ff.

[e] (2), vol. 1, p. 24.

[f] Cf. Schomberg (1), and the interesting paper of J. W. Gregory (1) on the world desiccation problem.

The safest course is probably to admit that climatic changes may have played some part, but that other factors were also at work. The effective barrier offered to south-eastwards Hun migration by the Great Wall from about −230 till +300 has been considered a factor in turning the train of nomadic buffer shocks westwards.[a] Only after the +3rd century did numerous Hunnish, Turkic and Mongol peoples swarm over north China and establish dynasties there. Though Teggart does not smile upon this opinion, it receives some new plausibility from his own correlations. At any rate, it is significant that after the Germanic and Gothic migrations Europe attained a rather stationary state during the early medieval period, that is to say, during the centuries when the Central Asian peoples had scope for their expansion in China—one thinks of the Northern Wei dynasty lasting till the middle of the +6th century; then after the middle of the +8th there was the rise of the Uighurs and the Chhi-tan, and in the +10th the Liao and later the Chin. Not until the +13th did Europe again feel the force of outward expansion from the heartland in the shape of the Mongols.

The great migrations into Europe naturally followed the broad Scythian way north of the Caspian, but though we have to take them into account in assessing the possibilities of travel of ideas, they never completely interfered with the silk route. When Roman power was waning, the sea route was intercepted by the growth of the little-known Abyssinian kingdom of Axum,[b] which, by means of its port Adulis, at the lower end of the Red Sea, was able to attain middleman status as important as that of the Persians. Early in the +4th century, with the foundation of Byzantium, Alexandria lost nearly all its remaining importance as an entrepôt, and new routes were developed leading straight from Merv through Armenia without touching Mesopotamia at all. Indeed, the silk industry became one of the foundations of Byzantine prosperity.

Teggart attributes much of the barbarian uneasiness which found expression in the Germanic migrations to interferences with the great trade-route by Chinese or Roman wars. Certainly in the new situation the Byzantines were in a difficult position, since other circumstances forced them into wars with the Persians, the very people on whom they relied for supplies of silk. In +552, therefore, ten years after the silk industry had been declared a State monopoly by Justinian, an event of great political as well as commercial significance occurred, namely, the introduction of the silkworm, *Bombyx mori*, into Europe. There are various accounts of it. Procopius[c] says that certain Indian monks offered to bring it from 'Serinda'; Theophanes[d] relates that the eggs of the moth were smuggled in a hollow cane by a Persian from the country of the Seres. Hudson[e] resolves the discrepancy by supposing the monks to

[a] Von Richthofen (2), vol. I, p. 445; Minns (1), p. 121; Foord (1); Rostovtzev (1), p. 114; Lattimore (4); E. D. Ross (1, 2). No Western writer in antiquity ever mentions the Great Wall with the possible exception of Ammianus Marcellinus (*c.* +360), *Res Gestae* XXIII, vi, 64 (Coedès (1), xxvii, 94). Yule (2), vol. I, p. 16, thought that only barriers of mountains might be meant, but the text does use the word 'agger'. Cf. Sect. 28*c* below.

[b] Hudson (1), p. 105.

[c] *De Bello Gothico*, IV, 17; Coedès (1), pp. xxviii, 127; Yule (2), vol. I, p. 203.

[d] In Müller's *Fragmenta Hist. Graec.* IV, p. 270 (Yule (2), vol. I, p. 204).

[e] (1), p. 121.

have been Persian Nestorians who had been to India before, and that Serinda was Cambodia or Champa; the other view is that it was Khotan or Kashgar. Most writers have overlooked what seems to me the most interesting part of the problem, namely, who were the people who knew enough of the technology of silk growing, reeling, twisting, etc., to transplant the industry successfully from Cambodia or Sinkiang to Syria and Lydia? Anyone could bring a hollow cane. In any case, the transplantation was successful.

By the end of the +6th century Turkish domination of Central Asia had begun; they exchanged embassies with Byzantium, and the return Byzantine one, headed by Zemarchus, left an account of what its members saw.[a] Although it did not lead immediately to the opening up of a new route, since the Byzantines now worked their own silk, Hudson sees much significance in it, since it was the first time for more than a thousand years that anyone from Europe had travelled as far east as Aristeas of Proconnessus, i.e. north of the Caspian and the Thien Shan.[b] One result of this mission seems to have been a remarkably accurate account of China, by Theophylact Simocatta, an Egyptian Greek, who wrote (as we have seen) about the Taugas (i.e. Tho-pa, the Chinese), in +628. He was indeed well-informed, since he knew that there had been a great battle in which a 'black-coated' army had passed over a great river to destroy a 'red-coated' army.[c] This had in fact happened when the Sui crossed the Yangtze and defeated the Chhen, capturing Nanking in +589.[d] The converse of this, which also illustrates the extent to which correct information could travel, is the account in the *Chiu Thang Shu*, in its chapter on the Arabs (Ta-Shih[1]),[e] of the sieges of Byzantium by the Caliph Mu'āwiyah between +671 and +678, and (when they failed) the subsequent arrangement by Joannes Petzigaudias of a tribute payment from the Arab side.[f]

Chinese records speak of an embassy from Byzantium (Fu-Lin[2])[g] in +643 (see on, p. 205). This new name now began to displace the old term Ta-Chhin, which had been used for Syria, and perhaps by implication the whole Roman empire, and later the eastern Greek successor States of Alexander, from the −2nd century onwards.[h] After the last of these embassies, in +742, there is a rather long gap when it is difficult to say how much intercourse was going on across Central Asia. Disturbances in Sinkiang

[a] Menander Protector, in Müller, *Fragmenta Hist. Graec.* IV, p. 235 (Yule (2), vol. 1, pp. 205 ff.).

[b] From the interesting evidence presented by Dubs (6), a direct military contact between Roman and Chinese soldiers is to be surmised for −36, but it will be more convenient to postpone discussion of this (p. 236 below; cf. Sects. 22, 30). The legionaries were serving a Central Asian prince.

[c] Yule (2), vol. 1, p. 30; Coedès (1), pp. xxix, 138.

[d] See *TH*, vol. 2, p. 1253.

[e] Perhaps from Tazi and Tadjik (Hirth (9); Yule (2), vol. 1, p. 88), men of the tribe of Ṭaiyi'.

[f] Yule (2), vol. 1, p. 48.

[g] Formerly derived from εἰς τὴν πόλιν, like Istanbul (according to al-Mas'ūdi), but now certainly from Rūm (From, Hrom, Eastern Rome).

[h] It is still interesting to read the polemic between Forke (1) and Herrmann (7) as to exactly what it meant. From the time of the *Wei Shu* at least, i.e. +5th century, the capital had been known as An-Tu[3] (Antioch).

[1] 大食 [2] 拂菻 [3] 安都

were caused by the growing power of the Tibetans, who from +670 had held Kashgar for twenty-two years. Their rise to political importance had taken place early in the +7th century, under their great king Srong-btsan Sgam-po who had taken a Chinese princess to wife and established Buddhism in his country. In spite of a memorable expedition of the Chinese general Kao Hsien-Chih[1] in +747 across the Pamirs to Gilgit, which achieved temporarily its object of preventing the junction of the Tibetans with their Arab allies, the country could not be held, and at the Battle of the Talas River (+751) the Thang Chinese were (as noted already, p. 125) decisively defeated by the new Muslim power. This closed the land route. Nevertheless, what was lost in ease of travel by land was gained by sea, since, as already mentioned, there now began the great age of Arab navigation to China. The embassies which reached China in +1081 and +1091, though said to be from Fu-Lin, were certainly not Byzantine, but Seljuq Turkish, from the ruler of Baghdad, who of course possessed Byzantine territory taken at the Battle of Manzikert (+1071).[a]

(2) Persians in China

Throughout the Thang period, however, Persian merchants were travelling in China, having presumably come by sea if the land route happened to be closed. They left their mark in the fiction and reminiscences of the dynasty, and we owe to Schafer (2) an examination of hints which could be of much importance for the transmission of scientific ideas and techniques. The first embassy had arrived in +455, with gems, carpets and aromatics; the last was in +984 from the Sāmānids of Khurāsān. In Thang China, if we may judge from the literary references, Persian merchants were popularly regarded as magicians, alchemists and wealthy dealers in precious stones of magical power. Their great centre was the western market at Chhang-an, with its gem-stores, wineshops, Zoroastrian temples and Nestorian churches.[b] In the tales preserved in such collections as the *Thai-Phing Kuang Chi*[2] (Miscellaneous Records collected in the Thai-Phing reign-period) the Persian (foreign, *hu*[3]) merchants often pay great sums for jewels or minerals which their Chinese possessors did not know were valuable. Lumps of gold from alchemists' furnaces command a high price. In one story the Persian, wearing Taoist robes, enlists the help of the young hero in the performance of an alchemical ritual destined to produce the elixir of immortality. Names of individual sinified descendants of Persians, who may well have been significant in Persian-Chinese scientific contact, have come down to us; there were several families called Li. Li Hsien,[4] younger brother of the poet Li Hsün,[5] was widely known as a Taoist alchemist and herbalist skilled in the use of arsenicals, and a notable chess player.[c] Their ancestors had come to China in the Sui, and their sister

[a] See Hirth (1), p. 298.

[b] The best monograph on Western foreigners at the Thang capital is that of Hsiang Ta (3).

[c] We know about him from the *Mao Thing Kho Hua*[6] (Discourses with Guests in the Thatched Pavilion) by Huang Hsiu-Fu,[7] ch. 2, p. 10*a*.

[1] 高仙芝 [2] 太平廣記 [3] 胡 [4] 李玹 [5] 李珣
[6] 茅亭客話 [7] 黃休復

Li Shun-Hsien,[1] also a poet, became one of the ladies of the Prince of Shu in the Wu Tai period.[a] Li Hsien wrote a *Hai Yao Pên Tshao*,[2] or Pharmaceutical Codex, which dealt with foreign drugs, and probably also with marine products.[b] A work which did definitely deal with the drugs employed in Persian medicine was the +8th-century *Hu Pên Tshao*,[3] by Chêng Chhien.[4] Unfortunately, neither of these has survived except in the form of quotations in later pharmaceutical codices. Then there was the physician Li Mi-I,[5] who, though of Persian origin, went on to Japan in +735 and participated in the Nara cultural renaissance.[c] A little before this (+714) we hear of Chi-Lieh[6] (perhaps Cyriacus), a Nestorian or Manichaean priest or monk, 'forging' or imitating foreign rarities for the court.[d] His technical skill (*chhiao*[7]) was great, but the Confucian officials disapproved of it, and the emperor was not amused.

(3) Pax Mongolica

The land route, if reduced in significance during the +11th and +12th centuries,[e] was destined to attain an undreamed-of importance under the Mongols. Their defeat of the Russian confederacy in +1221 threw Europe open to them, and sixty years later, as has been already described (p. 140), the whole of Asia and Eastern Europe, from Shanhaikuan to Budapest and from Canton to Baghdad, was united under one political authority. Not surprising was it, therefore, that travel should become relatively easy.[f] A merchant's handbook of the +14th century says,[g] 'The road which you travel from Tana (at the mouth of the Don) to Cathay is perfectly safe, whether by day or by night, according to what the merchants say who have used it.' What was more surprising was that this was what the Mongol conquests had led to. Ralph Fox, speaking of Chinghiz Khan, says graphically:[h]

> The Mongol horseman, who spoke always in images of purest epic poetry, taken from the life of his steppe, master of the four fierce hounds who drank dew and rode on the wind, for all his close friendship with the Uighur and Uzbek merchants, could never have imagined that his life and conquests would have ended in the ventures of the cautious Master Francis

[a] Chang Hsing-Lang (*1*), vol. 4, p. 99.

[b] Bretschneider (1) thought the former, and attributed it, following *PTKM*, ch. 1 A, p. 7 *a*, to his brother, whom Li Shih-Chen placed in the +8th century (+769 to +780). It seems from the other sources, however, that Li Hsien and his brother and sister were flourishing in the early +10th.

[c] See Takakusu (3), p. 7.

[d] *Tshê Fu Yuan Kuei*, ch. 101, p. 7*a*; ch. 546, p. 3*a*. This Chi-Lieh is not to be confused with the Nestorian bishop of the same name who came to China in +732 and is mentioned in the inscription of the famous Nestorian Stone (+781); see Saeki (1), pp. 75, 168.

[e] There was never a complete barrier, of course. The typically Chinese designs found in St Mark's at Venice date from just this time, and were probably copied from textiles obtained through Byzantium (Einstein, 1).

[f] And all the intellectual contact which this implied; see the interesting summary of Goodrich (1), pp. 174, 176.

[g] The *Libro di Divisamenti di Paesi* of Pegolotti, *c.* +1340.

[h] (1), p. 257.

1 李舜弦 2 海藥本草 3 胡本草 4 鄭虔 5 李密醫
6 及烈 7 巧

Balducci Pegolotti,[a] riding in his comfortable wagon, with his woman, his dragoman, and his little convoy of merchandise. Master Francis was the future....No nomad horseman, watching from his shaggy red pony that quiet caravan amble by, could have imagined that it was the distant forerunner of conquerors more merciless than the four hounds of Temujin.

At this point the present account may be concluded. The relations of China with the West now enter a modern phase, and have been referred to at some length already in the historical section (pp. 142ff.). Here only a few further specific references may be given. The two most important books for the study of this period—the open road under the Mongols, the sealing off under the Ming, and the moderate reopening under the Chhing—especially the earlier part of it—are those of Bretschneider (2) and Yule (2). The latter deals with the accounts of Odoric of Pordenone,[b] John of Monte Corvino,[c] John Marignolli,[d] Ibn Baṭṭūṭah,[e] Benedict Goes,[f] and other travellers.[g] Marco Polo has already been referred to.[h] From this time on, the transmission of ideas or techniques across or around the central regions of the Old World is relatively easy to understand.[i] But, unfortunately, the most interesting and important transmissions occurred in earlier times.

This is by no means to say that little is to be expected from intensive researches into this period. In the section on Textile Technology, for instance, we shall find strong indications of detailed technical transmissions from China to Italy, though so far as I know Marco Polo himself never mentions textile machinery. So also Olschki (6) has described a little-known trade in Tartar (= Mongol and Chinese) slaves, which provided domestic servants in Italian homes during the +14th and +15th centuries. Between +1366 and +1397, for instance, no less than 259 Tartars, mostly though not exclusively young women, were sold in the slave-markets of Florence (Livi (1) and Zanelli (1) have studied the sources). The influx seems to have begun about +1328, when Marco Polo's own servant Peter the Tartar was granted Venetian citizenship, and to have ended with the fall of Byzantium in +1453. There is evidence that much racial mixture occurred, and the Mongol or Chinese girls probably contributed some useful genes to the European population; whether they and their companions in servitude contributed also in the transmission of certain ideas or techniques, future investigation alone may show. In the previous century the position had been the other way round. The Mongol Khans had accumulated a considerable number of European technicians

a One of the best accounts of Pegolotti and his merchants' guide-book is that of Beazley (1), vol. 3, pp. 324 ff. Sarton, however, (1), vol. 3, p. 772, avers that he never went to Cathay.

b In vol. 2. There is also a good summary of Odoric's journey in Beazley (1), vol. 3, pp. 250 ff.

c In vol. 3. Cf. Beazley (1), vol. 3, pp. 161 ff.

d Beazley (1), vol. 3, pp. 288 ff.

e Yule (2), vol. 4, pp. 80 ff.

f Yule (2), vol. 4, pp. 198 ff.

g There is a popularisation of Yule's accounts by Roberts (1), who adds the Ming and Chhing periods. The book of Olschki (1) has not been available to us. Cf. Sarton (1), vol. 3, pp. 187, 203.

h P. 141 above.

i There is no space here to review the literature dealing with modern travellers over these routes. Besides the Old Silk Road along the Nan Shan, already described, there grew up a direct desert route from Peking to Kashgar; on this see Teichman (1) and Lattimore (5). Grenard (2) has a careful study of the roads in Sinkiang.

at their courts. Of these men the best known so far is William Boucher, a Parisian goldsmith and mechanic who served Kuyuk and Mangu between +1246 and +1259 at Karakoron, and to whom a special monograph has been devoted by Olschki (4). Such culture contacts may well account for the presence of fleurs-de-lys motives on the robes of Buddhist statues, as de Mély (7) has reported, but in spite of the fact that many details of Boucher's life and work have come to light,[a] there is little which indicates positively the transmission of any mechanical art or natural principle. However, all these are but examples of the close relations between East and West in this period of the Pax Tartarica or Mongolica. They call for closer examination.

Then again there is the graphic and charming account in the narrative of Odoric of Pordenone of his visit to a Buddhist abbey near the West Lake at Hangchow, where he watched an assembly of apes being carefully fed by a monk:

> He took two great buckets full of scraps from the table, and opening the door of a certain shrubbery which was there, we went therein. Now in this shrubbery there is a little hill covered with pleasant trees and all full of grottoes. And as we stood there he took a gong, and began to beat upon it, and at the sound a multitude of animals of divers kinds began to come down from the hill, such as apes, monkeys, and many other animals having faces like men, to the number of some three thousand, and took up their places round him in regular ranks. And when they were thus ranged round about him, he put down the vessels before them and fed them as fast as he was able. And when they had been fed he began again to beat the gong, and all returned to their retreats.
>
> So I, laughing heartily, began to say, 'Tell me, prithee, what this meaneth?' And he answered, 'These animals be the souls of gentlemen, which we feed in this fashion for the love of God.' 'But', quoth I, 'no souls be these, but brute beasts of sundry kinds.' And he said, 'No, forsooth, they be nought else but the souls of gentlemen. For if a man be noble his soul entereth the form of some one of these noble animals; but the souls of boors enter the forms of baser animals and dwell therein!' And say what I list against it, nought else would he believe.[b]

If thus Odoric could tell his brethren at Padua in +1330 about the doctrine of the transmigration of souls,[c] it is hard to believe that other matters, more important for our interests but not considered of sufficient interest to be written down then or later, did not also pass from east to west in these centuries.

[a] Mention will be made of them in the appropriate places (Sect. 27).

[b] Tr. Yule (2), vol. 2, p. 203. While not doubting the genuineness of Odoric's experience, it may be noted that an exactly similar story had been told sixteen centuries earlier (Megasthenes, frag. 11; Aelian, *Nat. Anim.* XVI, 10).

[c] It is interesting to notice the twist which the European friar gave to the doctrine—its ethical content was lost on him, and social class rather than virtue was what determined position in rebirth. It does not seem very likely that this was due to the ignorance of the monk who was his informant; the incongruity of the word 'gentleman', indeed, is what gives the whole passage its piquancy.

(*h*) CHINESE-WESTERN CULTURAL AND SCIENTIFIC CONTACTS AS RECORDED BY CHINESE HISTORIANS

The arteries and veins—or perhaps one should rather say the capillary vessels—of communication from east to west across the Old World have now been outlined. But they have not been sufficiently peopled with living individuals, and unless this is done it is difficult for the imagination to visualise whether, and if so, how, intellectual exchanges at given times could have occurred. I propose, therefore, taking the *China and the Roman Orient* of Hirth as a scaffolding,[a] to give some quotations from Chinese sources concerning contacts.

Hirth assembled and translated seventeen accounts of embassies and mercantile missions, mostly contained in the official histories, and the results are summarised in Table 7. This is by no means complete, but affords a sketch of the references to the coming and going of individual people, and of the products or characteristics of the western lands in which the Chinese took particular interest. It may be significant that considerably more embassies are recorded as having come to Chinese capitals than as having been sent out from them, but it is certain that many of the former were really travelling merchant groups.

The embassies from China to Parthia (An-Hsi[1]) described in the *Chhien Han Shu* and *Hou Han Shu* as well as the *Shih Chi* must have followed closely upon the mission of Chang Chhien already described (p. 172), and must have been, therefore, about −120. Such intercourse continued throughout the Han period, but there is a description of a mission from Chi-Pin[2] in −30, interesting for a number of reasons. There has been disagreement as to where Chi-Pin exactly was; it probably included modern Afghanistan, Gandhāra, Kashmir and the upper Indus basin.[b] The *Thung Chien Kang Mu* says:[c]

When, under the Emperor Wu, communication with the Western countries began, only the kingdom of Chi-Pin refused to submit, considering that the soldiers of Han could never reach there. Its king even put to death several Chinese envoys. In the time of the Emperor Yuan[d] the King of Chi-Pin sent an embassy to apologise, but the emperor lacked interest in foreign relations and refused to receive it. After the Emperor (Chhêng)[e] had ascended the throne, a further embassy was sent, with offerings and apologies, and it was planned (to accept them and) to send a mission to conduct the ambassadors home with all honour.

[a] It is a classical book. Critique by Pelliot (6); see also Edkins (2). The relevant chapters in the *Hou Han Shu* and the *Wei Lüeh* were translated and annotated more fully by Chavannes (6, 15).

[b] See McGovern (1), pp. 209, 484, and the paper by Lévi & Chavannes (1); the problem seems to be settled in Tarn (1), pp. 469 ff. The name meant the Kushan empire, which included Kapiśa and Kashmir, and its derivation comes from the Greek place-name Kophen, near Kabul in Afghanistan (cf. Herrmann (2), Map).

[c] Ch. 6, p. 115*a*; the original text is *Chhien Han Shu*, ch. 96A, p. 12*a* (tr. Wylie (10), p. 37).

[d] −48 to −33.

[e] −32 to −7.

[1] 安息 [2] 罽賓

Table 7. *Chinese-Western contacts mentioned in Chinese historical sources*

Source	*Shih Chi*, ch. 123 史記	*Chhien Han Shu*, ch. 96A 前漢書	*Hou Han Shu*, chs. 116, 118 後漢書	*Wei Lüeh* 〔魏略〕 in *San Kuo Chih*,[a] ch. 30 三國志	*Chin Shu*, ch. 97 晉書	*Sung Shu*, ch. 97 宋書	*Liang Shu*, ch. 54 梁書
Date written, *c.*	−90	+100	+450	*c.* +264	+635	+500	early +7th, approx. +629
Referring to events, *c.*	−100	−100	+97	+220 to +264	+265 to +419	+420 to +478	+502 to +556
Embassy Ch→West	(Parthia)	(Parthia)	.	.	.	*	.
Embassy West→Ch (often only merchants)	*	*	* (+120)	* (+134)	* (+284[b])	*	* (An-Tun, +166) (Chhin Lun, +226)
Matters mentioned:							
eggs (ostrich?)	*	*	.	.	.	.	.
jugglers	*	*	*	*	.	.	.
description of *Ta-Chhin* (Roman Syria)	.	.	*	*	*	.	*
exploration of Kan Ying	.	.	*	.	*	.	.
asbestos	.	.	*	*	*	*	.
night-shining jewel	.	.	*	*	*	.	.
coral	.	.	*	*	.	.	*
byssus	.	.	*	*	.	.	.
storax	.	.	*	*	.	.	*
artificial gems	.	.	*	*	.	*	.
amber	.	.	*	*	.	.	*
'weak water'	.	.	*	*	.	.	.
crystal columns	.	.	*	*	.	.	.
gold ball clepsydra	.	.	.	.	.	.	.
trepanning for blindness	.	.	.	.	.	.	.
unravelling of Chinese silk	.	.	.	.	.	.	.

Source	*Wei Shu*, ch. 102 魏書 & *Pei Shih* ch. 97 北史	*Thang Shu*, 唐書 *Chiu* ch. 198 舊	*Thang Shu*, 唐書 *Hsin* ch. 221 A, B 新	Nestorian Stone	*Sung Shih*[c] ch. 490 宋史	*Chu Fan Chih* 諸蕃志	*Ming Shih* ch. 326 明史
Date written, *c.*	+572	*c.* +950	+1061	+781	+1345	+1225	+1739
Referring to events, *c.*	+386 to +556	+618 to +906		+719	+960 to +1279	contemp.	+1368 to +1643
Embassy Ch→West	.			.	.	.	*
Embassy West→Ch (often only merchants)	.	* (+643, 667,[d] +701, +719)	* (+666,[e] +719)	.	* (+1081,[f] +1091)	.	* (+150, +1371)
Matters mentioned:							
eggs (ostrich?)	.	.	.	.	.	.	.
jugglers	.	.	*	.	*	.	.
description of *Ta-Chhin* (Roman Syria)	*	.	.	*	*	.	.
exploration of Kan Ying	.	.	.	.	.	.	.
asbestos	.	.	.	*	.	.	.
night-shining jewel	*	*	*	*	*	.	.
coral	.	*	*	.	*	*	.
byssus	.	*	*	.	*	.	.
storax	.	.	.	*	*	.	.
artificial gems	.	.	.	.	.	.	.
amber	.	*	*	.	*	.	.
'weak water'	.	.	.	*	*	*	.
crystal columns	.	*	.	.	*	*	.
gold ball clepsydra	.	*	*	.	*	.	.
trepanning for blindness	.	.	*	.	*	.	.
unravelling of Chinese silk	.	.	.	.	*	.	.

a Quoted in the commentary.

b The paper mission.

c As the account in the *Sung Shih* is very short, the data compiled in its column are taken also from the contemporary *Wên Hsien Thung Khao* of Ma Tuan-Lin (+1319).

d This mission offered theriaca.

e 'Repeatedly.'

f Really Seljuq.

But Tu Chhin[1] (an influential high official)[a] said to Wang Fêng[b] (prime minister) 'Friendly intercourse with barbarian nations is advisable only where communications are reasonably easy. The Hsientu[2] passes will always be an obstacle to relations with Chi-Pin. The friendship of that country could not benefit the western cities (dependent on us), and even if it should be hostile it could do them no harm. We were first friendly with the Chi-Pin people, then they did us harm, hence we broke off relations with them. Moreover, the so-called ambassadors who have just arrived are neither nobles nor men of mark, but really merchants who want to engage in commerce; their offerings and apologies are only a pretext. To despatch a return embassy would have no meaning; it would only show that we had been deceived. (In order to reach Chi-Pin,) after passing the Phi-Shan[3] mountains, our envoys would have to traverse four or five countries, each of which is full of robbers. Then one must cross the Greater and the Lesser Headache Mountains,[4] chains of naked and burning rocks, so named because they cause headache, dizziness and vomiting. Then comes the San-Chhih-Phan[5] gorge, thirty *li* long, where the path is only 16 or 17 inches wide, on the edge of a precipice, and where the travellers have to be tied together with ropes. From here through Hsientu it is 3000 *li* and more, on a road full of dangers. The sage kings of old devoted themselves to the nine provinces and the five concentric zones, caring nothing about what was outside. Yet now it is proposed to order the escort home of these barbarian merchants, thus exposing soldiers and imperial envoys to indescribable fatigues and dangers. Such useless enterprises should not be the policy of an enduring dynasty. If the officers of the escort have already been nominated, let them at any rate accompany the Chi-Pin people no further than the Phi-Shan mountains.' Wang Fêng supported this memorial of Tu Chhin. So they were allowed to trade and sent away with presents. And afterwards envoys from Chi-Pin arrived in China every few years.[c]

The passage shows first that not one only, but a number of semi-diplomatic semi-mercantile missions passed between North-west India or Afghanistan and China in the −1st century. In the second half of the previous century, as already mentioned (p. 172), the Yüeh-chih nomads had conquered the Greek Bactrian kingdoms initiated by Alexander the Great. But before that, in the first half of the previous century, the Yüeh-chih, in the course of their westward migration, pushed by the Huns, had occupied the Ili River basin,[d] dislodging from there the Śakas or Sacae, one of the ancient Scythian-Sarmatian tribes. These people, fleeing south and west, had passed through Greek Sogdia and Bactria to Iran on the one side (where they gave their name to Śakastan, modern Seistan), and to Afghanistan and North-west India (Gandhāra and the Indus Valley) on the other, where they founded the Chi-Pin State. These particulars are necessary in order to realise that the Chi-Pin missions may well have been one of the means of Greek-Chinese culture contact.

The background of these Chi-Pin embassies has been confused, but the elucidation of the Greek side has now been completed by Tarn and others.[e] According to Pan Ku,[f]

[a] G2055. [b] G2167.

[c] Tr. auct., adjuv. Wieger (*TH*, p. 556); McGovern. Wieger identifies the Phi-Shan mountains with the Karakach pass between Khotan in the Tarim basin and the high Tibetan plateau (the Headache Mountains); the San-Chhih-Phan would be the Karakorum pass. Hsientu is the equivalent of Hsüantu[6] and certainly means the Hindu Kush Mountains.

[d] McGovern (1), p. 127. [e] Tarn (1), pp. 339 ff. [f] *Chhien Han Shu*, ch. 96A, pp. 11*a* ff.

[1] 杜欽 [2] 縣度 [3] 皮山 [4] 大小頭痛山 [5] 三池盤石阪 [6] 懸度

a king of Chi-Pin called Wu-Thou-Lao[1] killed (about −55) some Chinese envoys, and his son sent by way of apology a plenipotentiary who was escorted back by a Chinese general Wên Chung.[2] The son then plotted to kill him too, so he allied himself with one Yin-Mo-Fu,[3] son of the king of Jung-Chhü,[4] attacked Chi-Pin, killed the faithless ruler, and established Yin-Mo-Fu as king of Chi-Pin and vassal of China. Subsequently Yin-Mo-Fu himself killed another Chinese envoy, and again an embassy was sent to make excuses; this was the one which Yuan Ti would not receive. The last embassy, which Tu Chhin advised Chhêng Ti to despatch home without official recognition, was that of about −30. It now appears that Yin-Mo-Fu was none other than the last Greek king in India, Hermaeus, who ruled a small part of the former Euthydemid Bactrian dominions in the Indus basin from about −50 to −30. Jung-Chhü is Yonaki, i.e. 'Greek-town', and Wu-Thou-Lao, probably a mistake for 'brother of the king' (ἀδελφοῦ) on the coins of Spalyris, stands for the contemporary Śaka house ruling Kabul and the Kophen valley. It is, indeed, remarkable that the last vestiges of Greek rule in India existed solely because of Chinese diplomatic support. After −30 all the domains in north-west India reverted to the Śakas, and in +19 to Parthian overlordship exerted from Seistan, which Pan Ku[a] knew as Wu-I-Shan-Li,[5] i.e. Alexandria Prophthasia, so called in the Chinese traditions from the name of its capital. It may, of course, be said that the envoys from Chi-Pin must have been preoccupied only with political matters such as the saving of the remnants of Greek culture in India, and certainly they came from what must have been a very provincial outpost, but their private interests may have been much wider. What ideas and techniques could they have introduced to the Chinese one may well wonder, and what did they bring back with them as they moved in single file through the Himalayan passes?

The whole series of incidents has a marked bearing on the problems of transmission. By the time of the later Chi-Pin missions the Silk Road had been open for nearly a hundred years. And far away in the West it was the end of what Singer (2) calls the Middle Alexandrian period. Rather characteristic of this time is the almost simultaneous appearance of certain inventions (the hodometer, the water-mill, etc.) in China and in the Mediterranean area. But in which direction such ideas and techniques travelled, or whether they were genuinely independent developments, still remains wrapped in deepest mystery.

The Chi-Pin contacts also illustrate the marked disinclination of the Chinese to travel far outside what they felt to be their natural geographical boundaries. Tu Chhin is here clearly describing the symptoms of mountain-sickness, a mysterious malady which may well have been felt as a warning not to transgress the boundaries of the *oikoumene*. Later, when speaking of the 'isolation' of Chinese civilisation (Sect. 47), we shall have to remember this passage. But the name of one Chinese has come down

[a] *Chhien Han Shu*, ch. 96A, p. 13*a*.

[1] 烏頭勞 [2] 文忠 [3] 陰末赴 [4] 容屈 [5] 烏弋山離

to us, who in these early times penetrated as far west as the adventurous Aristeas of Proconnessus had penetrated to the east—Kan Ying.[1]

His story is in the *Hou Han Shu*:[a]

In the 9th year of the Yung-Yuan reign-period of the Emperor Ho (+97) Pan Chhao[2] (governor-general of Central Asia) sent Kan Ying as an ambassador to Ta-Chhin (Roman Syria) and he reached Thiao-Chih[3] (Babylonia, Mesopotamia). When about to take his passage across the sea, the sailors of the western frontier of An-Hsi (Parthia) said to him 'The sea is vast and great; with favourable winds it is possible to cross it within three months, but if you meet with slow winds it may also take you two years. For this reason those who go to sea take on board a supply of provisions for three years. Moreover, there is something in this sea which is apt to make a man homesick, and several have thus lost their lives.' When Kan Ying heard this, he went no further.[b]

The passage has been much discussed. Wieger (1) observes characteristically that the remora of filial piety deprived Nerva or Trajan of the pleasure of Kan Ying's visit.[c] More to the point is Hudson's comment[d] that the Parthians were extremely anxious to avoid the occurrence of any direct contact between the Chinese and the Romans, as it would have revealed the profiteering of the Parthian merchants and might have led to measures to cut the middlemen out. They therefore disguised from Kan Ying the fact that the quickest way to Syria and Rome was across the desert or by upper Mesopotamia overland, and instead frightened him by exaggerated tales of the perils and slowness of the maritime route down the Persian Gulf and round Arabia into the Red Sea. But for us the important thing is again the possibility of intellectual contact and the passage of ideas. And for one Kan Ying whose story got into the official histories, there may have been a dozen other envoys or merchants who perhaps did not penetrate so far to the west, yet far enough to give or to receive ideas and descriptions of techniques.

The disinclination of the Parthians to allow the establishment of contact was no imagination of Hudson's, for the author of the *Wei Lüeh* (quoted in the commentary to chapter 30 of the *San Kuo Chih*, and written about +400) distinctly says so.

The people (of Ta-Chhin, Roman Syria) are tall, upright in their dealings, like the Chinese,[e] but wear foreign dress; they call their country another China. They always wished

[a] Ch. 118, pp. 2*b*, 9*a*; cf. *TH*, p. 720.

[b] Tr. Hirth (1), p. 39; cf. Chavannes (6), pp. 159, 177. The reference to the chapter number in Hirth (1), pp. 34, 36, 37, 40, is a misprint.

[c] An alternative version, in *Chin Shu*, ch. 97, says: '...there was something about the sea which caused one to long for home; those who went out could not help being seized with melancholy feelings; if the Chinese envoy did not care for his parents, his wife and his children, he might go' (tr. Hirth (1), p. 45).

[d] (1), p. 83.

[e] This admirable recognition of the equivalence of the two civilisations reminds us of the expression used by Andrew Corsalis writing to Lorenzo de' Medici in +1515, when he said of the Chinese that they were of great skill and 'di nostra qualità' (cf. Yule (2), vol. 1, p. 180).

[1] 甘英 [2] 班超 [3] 條支

to send embassies to (us in) China, but the An-Hsi people (Parthians) wanted to make profit out of their trade with us, and would not allow them to pass their country.[a]

It may be asked whether any Chinese envoys ever reached Rome itself, even if only disguised merchants, like so many Roman Syrians who went to China. The answer seems to be definitely no, in spite of a passing reference in the historian Florus[b] to an embassy from the Seres; this is not now accepted as sufficient evidence.

(1) Jugglers and Acrobats

The embassies of +120 and +134 were not from the Far West but from Shan (on the Burmese border) and from Kashgar respectively.[c] The former, however, offered Syrian acrobats, who performed before the Emperor An on New Year's Day, +121. Perhaps the passage of jugglers and acrobats to and fro merits more attention in the history of science than it has yet received, when we remember how much of the early mechanics of such men as Ctesibios (−2nd century) and Heron of Alexandria (+1st), as also of their Chinese counterparts[d] such as Ting Huan[1] and Ma Chün,[2] was occupied with mechanical toys for palace entertainments, devices of illusion, stage play machinery, and the like. The jugglers appear right at the beginning of Chinese-Western relations, for they are mentioned in the Parthian missions of Chang Chhien's time (*c.* −120). Those of +120 could, says the *Hou Han Shu*, 'conjure, spit fire, bind and release their limbs without assistance, interchange the heads of cows and horses, and dance cleverly with up to a thousand balls'.[e] Similarly, we read in the later *Sou Shen Chi* (Reports on Spiritual Manifestations) of +350 (Bodde, 9, 10) about wandering magicians and conjurors from India. This is a point which seems well worth further investigation. For ancient and medieval people there was not much difference between jugglers, alchemists, mechanicians, leeches, star-clerks and all dealers in magic and gramarye. We have indicated in the table all the sources which speak of jugglers and acrobats.

In +166 occurred the famous 'An-Tun[3] embassy'. The *Hou Han Shu* says:

The kings of Ta-Chhin (Roman Syria) always desired to send embassies to China, but the An-Hsi (Parthians) wished to carry on trade with them in Chinese silks, and for this reason they were cut off from communication. This lasted till the 9th year of the Yen-Hsi reign-period of Emperor Huan (+166), when the king of Ta-Chhin, An-Tun, sent a mission which, from the frontier of Jih-Nan[4] (Annam), offered ivory, rhinoceros horns, and tortoise-shell. From that time dates the (direct) intercourse with that country. The list of their tribute contained no jewels whatever, a fact which throws doubt on the tradition.[f]

[a] Tr. Hirth (1), p. 70; cf. Reinaud (2), pp. 203, 237.
[b] IV, 2. Yule (2), vol. 1, p. 18; Reinaud (2), p. 106; Coedès (1), p. xv.
[c] Hirth (1), pp. 36, 72.
[d] Cf. below, Sect. 27.
[e] Tr. Hirth (1), p. 37.
[f] Ch. 118, p. 10*b*, tr. Hirth (1), p. 42; cf. p. 173; cf Chavannes (6), p. 185; cf. *TH*, p. 755.

1 丁緩 2 馬鈞 3 安敦 4 日南

While there seems good reason for identifying An-Tun with Marcus Aurelius Antoninus, everything else points to the conclusion that this was a group of merchants and no accredited embassy. Parthian power was then low; Seleuceia and Ctesiphon had been captured by the Romans in the previous year, and what probably happened was that a trade mission, sent on from Antioch or Alexandria by the sea route, exchanged its Western goods on the way, and offered to the Chinese court products of south-east Asia which it had acquired on the spot. For some time after this, the direct sea route brought merchants of the Roman Empire to the maritime borders of China. What was transmitted besides wares?

We even know the name of one of these merchants. The *Liang Shu* says:

During the 5th year of the Huang-Wu reign-period (of the Wu State of the San Kuo Period) (+226), a merchant of Ta-Chhin, whose name was Chhin Lun,[1] came to Chiao-Chih[2] (Tongking in Indo-China). The prefect of Chiao-Chih, Wu Mo, sent him on to the Wu Emperor,[a] who asked him for a report on his native country and its people. Chhin Lun prepared a statement and replied. At about this time Chuko Kho[3] nephew of Chuko Liang) chastised Tan-yang[4] and captured some dusky dwarfs. When Chhin Lun saw them he said that in Ta-Chhin such men were rarely seen. The Emperor thereupon sent male and female dwarfs, ten of each, in charge of an officer, Liu Hsien of Hui-chi, to accompany Chhin Lun, but Liu Hsien died upon the road, and Chhin Lun returned direct to his own country.[b]

The interest of this story is the definite evidence that at least one Syrian trader spent a period of time in China in +226 and returned safely home. What ideas or techniques did he take with him in either direction? It was the very end of the Alexandrian period, the time of Diogenes Laertius and Achilles Tatius; Ptolemy's work had been finished for some time. Is it possible that Chhin Lun could have been interested in cartography?

The other recorded western embassy of the +3rd century was the one which we have referred to in the Table as the 'paper' mission. The *Chin Shu*[c] says that it came between +280 and +290, and fuller details are in the almost contemporary *Nan Fang Tshao Mu Chuang*, which says, in its entry under 'honey-fragrance paper' (*mi hsiang chih*[5]):

Honey-fragrance paper is made of the bark of the *mi-hsiang* tree; its colour is greyish, and it has spots giving it the appearance of fish-spawn. It is very fragrant, but strong and pliable; it may be soaked in water without spoiling. In the fifth year of the Thai-Khang reign-period of the Chin dynasty (+284), Ta-Chhin (Roman Syria) presented 30,000 rolls.[d]

[a] This was Sun Chhüan, who had marked scientific interests, especially in geography and cartography (cf. Sect. 22). As we shall later see, the Chinese tradition of quantitative cartography arose just as the Greek tradition temporarily died. Chhin Lun's visit occurred just at the moment when we might visualise a transmission of the idea of meridians and parallels; it could have been nothing more.

[b] Ch. 54, tr. Hirth (1), p. 48; cf. pp. 47, 306.

[c] Ch. 97, p. 9*a*.

[d] Ch. 2, end, tr. Hirth (1), p. 272.

[1] 秦論 [2] 交趾 [3] 諸葛恪 [4] 丹陽 [5] 蜜香紙

Since paper was certainly unknown in Europe or the Middle East at that time (see Sect. 32), and since papyrus would not have been appreciated (as this offering was) by the Chinese who had paper, Hirth (1) supposes convincingly that just as in former cases, the Syrian or Alexandrian merchants in question had disposed of all their western goods before they reached Annam, and then, in order to pay their way into Canton, purchased local goods, perhaps at cheap rates, to send as gifts to the imperial court. We know that the *mi-hsiang* tree (*Aquilaria agallocha*),[a] which, though mentioned in this book, first entered the pharmacopoeia in the *Pên Tshao Shih I* of *c.* +725, was a native of Annam. This mission could have been another means of propagation of cartographic or other ideas.

After this there is a very long gap. For the next 360 years there are no Chinese records of direct contact with the West. This is not to say, of course, that no channels were open, for we read, for example, of the Alans (Sarmatians) established near the Caspian Sea, sending tribute to the Northern Wei in +435.[b]

(2) The 'Night-Shining Jewel' and the False Gems

This is therefore a convenient point to look again at Table 7 and examine a few of the things which interested the Chinese so much about the Roman empire that they found a place in the official histories. The significance of jugglers and acrobats has already been pointed out. Asbestos cloth was by no means new to the Chinese;[c] known as the 'cloth which can be cleansed by fire' (*huo huan pu*[1]), it had been obtained as early as the Chou period from India or Central Asia. It was undoubtedly the origin of many 'salamander' legends, but by the Liang time its true origin was known, for it was called 'stone wool' (*shih jung*[2]).[d] Another Syrian product, the 'night-shining jewel' (*yeh kuang pi*[3]), which so fascinated the Chinese that it was mentioned in all the principal sources, is more difficult to explain. Hirth[e] cites a number of ancient Western authors who speak of certain precious stones which are luminous in the dark. The most likely mineral appears to be chlorophane,[f] which in spite of some of its synonyms such as pyrosmaragd, is not an emerald or a beryl (beryllium aluminium silicate) but a fluorspar (calcium fluoride), many varieties of which show strong phosphorescence and fluorescence on being heated or scratched in dim light.[g] Riddell (1) has traced the possible connection of this stone with the Indian cobra-stone or *naga-kallu*, alleged to be used by cobras to attract fireflies, and to be the origin of the sacred jewel famous in Japan under the name of *hoshi-no-tama*.

[a] R 252*a*.
[b] *TH*, p. 1094.
[c] Cf. Sect. 25*f* below.
[d] *KCCY*, ch. 27, p. 23*a*.
[e] (1), p. 243.
[f] So named by T. von Grotthus in 1794.
[g] This thermo-luminescence, which may be due to traces of arsenic sulphide incorporated in the fluorspar, gives a violet light sometimes enabling print to be read at a distance of six inches (Mellor (2), vol. 3, p. 693). It was first described by Elsholtz in +1677, about the same time as Robert Boyle's 'icy noctiluca'.

[1] 火浣布 [2] 石絨 [3] 夜光璧

Coral and pearls from the Red Sea (cf. Klunzinger (1) and Kallenberg, 1), and amber from the Baltic or Sicily (cf. Schneider, 1), it is not surprising to find among the wares of Syrian merchants trading to China. But the recognition by the Chinese of the artificial nature of some of the gems from Syria is of interest, in view of the prominent place taken by 'imitative' processes in the beginnings of European chemistry. Hirth[a] is surely right in surmising that much of the wealth of gold and jewels described in ancient and early medieval accounts of East Roman palaces and temples was really gilt copper and coloured glass. In any case, Syria was centrally situated among gem-producing districts, so that models for imitation were readily forthcoming. The *Hou Han Shu* already voices the opinion that most of the gems were artificially made,[b] and since glass[c] was undoubtedly an old Phoenician invention, its use would explain also the 'crystal columns', descriptions of which, as existing in Ta-Chhin, are found in several of the sources of Table 7, and in others as well. In the light of these facts it is interesting that the finds at Vīrapatnam, the Roman-Indian emporium near Pondicherry, have included large quantities of gems and false gems, some made of glass, dating from the first half of the +1st century (Filliozat, 4; Wheeler, 1). Vīrapatnam may well have been an entrepôt on the China route. The Chinese knew that the Indians had access to Hellenistic jewellery.[d]

(3) Byssus and Storax

Byssus was certainly the name of a fabric of some kind known to the peoples of the ancient Mediterranean, especially the Greeks and Egyptians. Most authorities agree[e] that it was either cotton or a mixture of cotton and linen, the cotton having perhaps been brought from India. But the term 'byssus' was also applied to the threads spun by certain marine lamellibranchiate molluscs, e.g. *Pinna nobilis* and *squamosa*, which they use to attach themselves to the rocks at the bottom of the shallow littoral waters in which they live.[f] These threads are not chitin (a polysaccharide), as was once thought, but a tanned protein.[g] In the Hellenistic age it was found that they could be dried and woven into fabrics, and in later times there grew up a whole corpus of fables about them, stemming both from this odd industry and from Aristotle's (very scientific) discussion of zoophytes, or creatures on the borderline between plants and animals. Among modern scholars these fables caused great confusion, but the story has been restored to rationality by Laufer (6).

[a] (1), p. 237.

[b] Hirth (1), p. 43. Chavannes (6), translating ch. 118, prefers to make the text say that the juggling tricks, not the gems, were not genuine, but this seems to involve some forcing of the words. Besides, it would have been stating the obvious.

[c] See below, Sect. 26*g*.

[d] As witness *Hou Han Shu*, ch. 118, p. 12*a* (Chavannes (6), p. 193): 'The Indians are in communication with Ta-Chhin and get gems from there.'

[e] Pauly-Wissowa, vol. 3, (1), pp. 1, 1108; Daremberg & Saglio (1), art. 'Byssus'; G. A. Faber (1); G. Schaefer (1).

[f] Cf. Cooke (1); Yonge (1) and C. H. Brown (1).

[g] Winterstein (1), vol. II, 2, pp. 43 ff. in art. by L. Fredericq.

The *Hou Han Shu*, in its account of Roman Syria, says: 'Further, they (the Ta-Chhin people) have a fine cloth said to originate from the down of a "water-sheep" (*shui yang tshui*[1]), and also they have a stuff made from the cocoons of wild silkworms.'[a] And the story is carefully preserved through the centuries (though very few people east of Persia can ever have seen any of these fabrics) until the *Hsin Thang Shu* says, as late as +1060: 'They weave the hair of the "water-sheep" into cloth which is called "cloth from the west of the sea" (*hai hsi pu*[2]).'[b] The only variant on this occurs in the early +14th century *Wên Hsien Thung Khao*, where Ma Tuan-Lin has 'cloth from the sea' (*hai chung pu*[3]), more correctly still.[c]

Now the *Pinna* byssus textile was not an old art, for even Pliny and Aelian say nothing about it.[d] But there are five mentions of it in the *Periplus of the Erythraean Sea* of about +70, under the name of *pinikon* (πινικόν),[e] and Laufer suggests that it was really a by-product of the Persian Gulf pearl fishery. Tarentum in Italy, which now has long been the main centre of the industry (Yates), was apparently not involved in late Hellenistic times. Other mentions of it then occur in Alciphron[f] (+2nd century), Tertullian[g] (*c.* +200), and later in Basil the Great[h] (*c.* +350) and Procopius (+6th century). Evidently the Syrian merchants lost no time in exporting this curious textile to China.

So far the descriptions had kept close to the facts, but by the +8th century a magic crop of fables was in full development. The *Chiu Thang Shu*[i] has another tradition, speaking of lambs engendered in the soil, sprouting on stalks like umbilical cords. This occurs also in the *Hsin Thang Shu*,[j] and in a commentary on the *Shih Chi* written in +737 which was found by Chavannes.[k] A thousand years later the editors of the *Thu Shu Chi Chhêng* encyclopaedia made a section for the 'earthborn sheep' (*ti sêng yang*[4])[l] in which they quoted from Tuan Kung-Lu's[5] *Pei Hu Lu*[6] (Northern Family Records) of +875, and from Wu Lai's[7] *Yuan Ying Chi*[8] (The Vast and the Minute) of the early Liao dynasty, about +1000. The earliest Arabic version of the legend seems to be in the geographical works of Abū Isḥāq al-Iṣṭakhrī (*fl. c.* +950),[m]

[a] Ch. 118, p. 9*b*, tr. Laufer (6). Owing to the laconic nature of the Chinese it is hard to be sure what exactly was meant, but I prefer Laufer's rendering, from which those of Hirth (1), Chavannes (6) and Schlegel (6) all diverge.

The wild silkworms can be quickly disposed of. This silk was an ancient product of Assyria and Syria; it was made not by reeling (since the cocoons were collected after the moths had emerged) but by combing and spinning. Pliny (XI, 77) and Isidorus (XIX, 22, 13) refer to it as 'bombycinae'. Its only connection with Cos is that Aristotle says a woman of Cos invented the process (Yates (1), p. 163).

[b] Ch. 221B, p. 11a (tr. Hirth (1), p. 59).

[c] Here Laufer (6) suspects Arabic influence, in which tongue the fabric was called *ṣūf al-baḥr*, sea-wool.

[d] It is not quite sure whether Aristotle even mentions the byssus fibres of *Pinna*, for at *Hist. Anim.* 547 b 15 there is a textual difficulty.

[e] The translation of Schoff (3) leads astray here.

[f] *Epistolae*, I, 2, 3.

[g] *De Pallio*, in Migne, *Patrologia Latina*, vol. 2, col. 1093.

[h] In Migne, *Pat. Lat.* vol. 29, col. 161.

[i] Ch. 198, p. 16*b*, tr. Hirth (1), p. 54.

[j] Ch. 221B, p. 11.

[k] (6), p. 183.

[l] *Chhin chhung tien*, ch. 112; *hui khao*, ch. 2, p. 16*b*.

[m] Mieli (1), p. 115.

1 水羊毳 2 海西布 3 海中布 4 地生羊 5 段公路
6 北戸錄 7 吳萊 8 淵穎集

which gives us the transitional form. Byssus is here an animal which comes out of the sea, leaves its wool on the shore, and is chased and eaten by crabs.[a] Later forms of the legend are really extraordinary; the Talmud makes the animal a man earth-bound by an umbilical stalk, and other Arabic authors confuse the byssus textile with a textile made from the down of certain birds.[b]

By the +14th century European travellers in Asia (such as Odoric of Pordenone (+1317 to +1330)[c] and John Mandeville (+1332 to +1356)) are picking up from Persian folklore the tale of the 'Scythian Lamb', which acquires the dignity of a Latin zoological name *Agnus scythicus*, all ready for Linnaean systematists. Its true identity remained a mystery, however, until modern times. In a brilliant book *The Vegetable Lamb of Tartary*, Lee sought to show that the Scythian Lamb was really the cotton plant, but though the texts which he collected were valuable, this identification cannot be accepted.[d] Cotton was well known practically everywhere by the +14th century, and though the Chinese probably knew little about it until the Sung (Goodrich, 3) there is nothing reminiscent of it in their presentations of the earth-sheep fables.[e]

Summing up this digression, then, if such it is, one may say that both Mediterranean and Chinese writers had accurate knowledge of the *Pinna* byssus textile industry up to about the +6th century, but that afterwards there was a legendary development which brought the *Pinna* ashore, turned it into a lamb or sheep, if nothing more extraordinary, and completely mystified scholars of the eighteenth and nineteenth centuries. We may turn now to storax.

Storax (*su ho*[1]) is a kind of gum which has always been produced in the Middle East, but it is a little hard to understand why it was so much prized by the Chinese. The learned pharmacist Daniel Hanbury devoted particular attention to its history,[f] concluding that the original and classical storax was a fragrant resin produced by the tree *Styrax officinalis*, a native of the Levant; but that the liquid storax which replaced it in later commerce was an ointment-like product prepared by subjecting the bark of the tree *Liquidambar orientale* to heat and compression. This tree is a native of

[a] These crabs are the precursors, Laufer suggests, of the cuirassed horsemen who, in the Chinese legends, ride around the vegetable lambs, frightening them so that they spontaneously break their umbilical stalks, and wander away to form flocks.

[b] Actually, a textile including combed bird-down was made in Persia, and also among the aboriginal tribes in South China. The *Chiu Thang Shu* (ch. 37, p. 18*b*) says that a princess An-Lo (d. +710) had some; cf. also the *Hsin Thang Shu* (tr. Pfizmaier (67), p. 27). *The Lang Huan Chi* (+10th or +11th century) speaks of some kind of gold brocades made of 'phoenix feather gold'. Perhaps the Arabs confused *Pinna* with *penna* (feather). The making of garments from feathers constitutes yet another link with Amerindian technique. Cf. the association of feathered dress with Taoist genii (Sect. 10 below).

[c] Yule (2), vol. 2, p. 241. Odoric likened it to the barnacle-geese (cf. Sect. 39 below) in which, of course, he believed.

[d] Lee was at any rate right in concluding that the fern rhizomes carved in China which had intrigued 18th-century scholars had nothing to do with it.

[e] Schlegel (6) made confusion worse confounded by introducing the camel into the story as well as cotton. Yule (2), vol. 1, p. 202, and Chavannes (6) were on the right track.

[f] (1), pp. 129 ff.

[1] 蘇合

Asia Minor.[a] The Chinese sources show that much interest was taken in other incense materials from the Middle East. Perhaps their main use was as drugs; even modern books on materia medica class storax with the balsams of tolu and Peru—their volatile constituents such as benzoyl cinnamate and other cinnamic derivatives render them intestinal disinfectants and externally valuable against *Sarcoptes scabiei* and other parasites.

There remain three points of scientific or technical interest in the Chinese accounts of Syrian affairs.

(4) Striking Water-Clocks

From the Thang onwards the accounts note the existence of important water-clocks in the West, with a mechanism so arranged as to allow golden balls to drop one by one into a receptacle and thus mark the passing hours.[b] It was only the striking mechanism which aroused interest—with the clepsydra itself the Chinese had been familiar since rather remote antiquity. Now it so happens that we are well informed about these striking water-clocks in the Near East. Cresswell (1) has described some Persian and Arabic manuscript illustrations of clock automata dating from the close neighbourhood of +1350; these could be identified as parts of the well-known treatise on automata of Abū al-'Izz Ismā'īl ibn al-Razzāz al-Jazarī[c] finished in +1206, the *Kitāb fī ma'rifat al-ḥiyal al-handasīya*[d] (Treatise on the Knowledge of Geometrical (Mechanical) Contrivances). The descriptions refer to the great striking clepsydra with moving figures which was set up on the east gate of the Great Mosque at Damascus, where it was seen by the traveller Ibn Jubair al-Kinānī[e] in +1186. His account, translated by le Strange,[f] tells us that at each hour of the day and night two weights of brass fell from the mouths of two brazen falcons into brazen cups (which, being perforated, permitted the return of the balls to their reserve). These must have been the 'golden balls' of which the Chinese records speak. Above the falcons was a row of doors as many in number as the hours of the day; at each hour a bell struck and the doorway of the completed hour closed. At dusk all the doors opened again. Above the row of doors was a row of lamps, as many in number as the hours of the night; at each hour during the night one of the lamps was lit, giving a red glow, until all were illuminated. At dawn all the lamps went out. Eleven workmen were employed to keep the machine (*al-mikannīyah*) in order. Other manuscript illustrations show an automaton orchestra of five figures sitting and standing in front and below (Fig. 33). This '*ṭabl-khana*' was a royal privilege, and by its drums connected perhaps with old Persian traditions still living in the drum-towers so universal in China.

[a] Later, the Chinese (to judge from R462) must have obtained their storax from the related species *Liquidambar altingiana*, which grows in south-east Asia, and the Malay name of which, *rasa-māla*, was corrupted into the 'rose malloes' of late European commerce in Asia. The real meaning of *rasa-māla* is excrement, and the name arose because of an old fable that storax was the dung of lions; this reached the Chinese, but they disbelieved it. The whole story has been well summarised by Burkill (1), pp. 117 ff., who gives the name of the tree as *Altingia* spp.

[b] Hirth (1), pp. 53, 213.

[c] Mieli (1), p. 155; Sarton (1), vol. 2, p. 632.

[d] Translated by Wiedemann in very scattered locations; bibliography in Sarton, *loc. cit.*

[e] Sarton (1), vol. 2, p. 412.

[f] (2), p. 249.

Obviously these +12th-century water-clocks could not be the source of the description in the *Chiu Thang Shu*, written in the middle of the +10th. But we know that in these machines the Arabs were continuing a Byzantine tradition, for Diels (2) has shown that the striking clepsydra, described by Procopius of Gaza some time between +473 and +535, closely resembled the +12th-century one at Damascus. Whether the Chinese descriptions refer to the clock at Gaza or, as Hirth (1) suggests, to one at Antioch, is not clear, but there were probably several of them in Byzantine Syria.

The remains of at least one of these striking water-clocks still exist today, namely at the Bū'anāniya *madrasah* (college) at Fez in Morocco. Michel (9) has given a photograph of the building with its row of twelve windows and twelve bronze gongs, formerly illuminated and sounded successively by automatic action through the watches of the night.

(5) Trepanation and Theriaca

Then there was the question of trepanning for blindness. The *Hsin Thang Shu* says (ch. 221B): 'They (the people of Ta-Chhin) have clever physicians, who by opening the brain and extracting worms, can cure *mu-shêng* (a sort of blindness) (*Yu shan i, nêng khai nao chhu chhung i yü mu shêng*[1])',[a] and is repeated by the *Wên Hsien Thung Khao*. This is of considerable interest as I believe it to be the solitary instance of any attention consciously paid in Chinese writings to early Western medical science. The same operation was understood about the same time (+3rd century) in China, if the story based on the *San Kuo Chih*[b] about Hua Tho's offer to operate on the head of the Wei emperor Tshao Tshao is to be trusted. As is generally known, the practice of trepanation goes back as far as the late Palaeolithic,[c] and Hirth (1) was able to find in Hippocrates a definite description of its use for certain forms of blindness,[d] thus in the book on vision: 'When sight is lost without any apparent disease of the eyes, one should make an incision in the parietal region, dissect the soft parts, trepan the bone, and evacuate the liquid which will come out; this is the treatment, and thus these patients are cured.'[e] Presumably the loss of sight had been caused by a cyst or benign tumour pressing upon the brain tissues. The 'worms' were an addition of the Chinese writer. This information about the medicine and surgery of the West may have reached China through Indian channels since, as Przyłuski (5) has shown, the Asoka legend has interesting parallels. A Buddhist sūtra (N 1367), of the Kushan period, subsequently translated into Chinese, has the son of Asoka being cured of blindness by Bactrian physicians at Gandhāra. Another tale (*Sūtrālaṃkāra*, no. 45) makes the prince the son of a Chinese emperor.

[a] Tr. Hirth (1), p. 59.
[b] *San Kuo Chih, Wei Shu*, ch. 29, pp. 1*b* ff.
[c] Cf. art. by Forgue in Laignel-Lavastine (1), vol. 2, pp. 350ff.; also Parry (1); Sigerist (1), vol. 1 pp. 110ff.
[d] Littré, vol. 7, p. 26 and vol. 9, p. 159.
[e] Tr. Littré, eng. auct.

[1] 有善醫能開腦出蟲以愈目眚

Fig. 33. Illustration of a striking water-clock from a MS of al-Jazarī's treatise on mechanical contrivances (+1206). At the top, the signs of the Zodiac exhibited, then figures successively appearing and lamps successively illuminated, below that, the golden balls dropped into brazen cups from the beaks of brazen falcons to strike the chime; lastly an automaton orchestra of five musicians (Cresswell, 1). Such Arab striking water-clocks (derived from Byzantine origins) were described by Chinese historians n the +10th century, and the Chinese had themselves been constructing them since the +7th.

This transmission may well have been effected by Nestorian missionaries, whose activities may also explain the Chinese reference to theriaca, which, says the *Chiu Thang Shu*,[a] was offered by an embassy from Byzantium in +667. *Ti-yeh-chia*[1] (old sound *tê-ya-ka*), or theriaca,[b] was one of the great preoccupations of ancient and early medieval Western pharmaceutics. Originally, as used by Nicander of Colophon (*fl.* −275), it meant, naturally enough, the healing by antidotes of all kinds of animal poisons,[c] but Mithridates, King of Pontus (−132 to −63), conceived the idea of a universal antidote,[d] and thus started the fashion of those preparations with innumerable ingredients all of uncertain value which are found in Andromachus of Crete (Nero's physician, *fl.* +60),[e] Pliny (600 ingredients),[f] and even Galen.[g] In later times, all kinds of things were incorporated into it, such as gall, myrrh, opium and hemp. According to the *Pên Tshao Kang Mu*,[h] it was first mentioned in the *Thang Pên Tshao* of the +7th century; it came from the Western countries and was known in Canton during the Sung. The gall of the pig was one of its constituents, its taste was bitter and cold, and while not poisonous itself it cured 'the evil effects of all diseases'. But in spite of this statement, Li Shih-Chen rightly gave it a very insignificant place. Hirth had in his possession a manuscript *Pên Tshao* of +1506 with a coloured illustration showing the presentation of the red and black theriaca pills to the emperor. On the whole we may say that the relatively poor reception of both Byzantine theriaca and Syrian 'gems' showed commendable scepticism on the part of the Chinese.[i]

We have now come to the Byzantine and Thang periods, when the expression Ta-Chhin drops out and is replaced by Fu-Lin,[2] Byzantium. The theriaca mission was of +667, but there was at least one earlier one, that of +643,[j] and at least two later ones (+719, +720). In the same year (+719) there was a mission from the Sogdian and Bactrian kingdoms, then about to be overrun in the Arab conquests, and, according to the *Tshê Fu Yuan Kuei*:

The king of Chih-Han-Na[3] (Jaghānyān[k]) Ti-Shê,[4] sent with (the ambassadors) a certain astronomer Ta-Mu-Shê,[5] suggesting that his abilities should be tested, that he should be interrogated upon his religion[l] and that he should be allowed to exercise it in a special temple where he would live upon the offerings of his congregation.[m]

[a] Ch. 198, p. 16; Hirth (1), pp. 56, 277.

[b] The identification, which is quite convincing, we owe to Hirth.

[c] Sarton (1), vol. 1, p. 158.

[d] Sarton (1), vol. 1, p. 214. Much in the same way as vitamins are combined for use today in capsule form.

[e] Sarton (1), vol. 1, p. 261.

[f] *Hist. Nat.* XXIX, 1 (8), 24; XX, 24 (100), 264; XXIX, 4 (21), 70.

[g] *De Theriaca.*

[h] Ch. 50B, p. 45*b*.

[i] One question, that of 'weak water', indicated in Table 7, is postponed for discussion in Section 23 on Geology.

[j] It raises questions about the Nestorian patriarchs, into which I shall not go here.

[k] Tokharestan.

[l] He seems to have been a Manichaean (cf. p. 128). See Grousset (1), p. 352.

[m] Ch. 971, p. 3*b*; tr. auct., adjuv. Wieger, *TH*, p. 1398.

[1] 底也伽 [2] 拂菻 [3] 支汗那 [4] 帝賒 [5] 大慕闍

It remains to be seen what possible ideas Ta-Mu-Shê could have brought or carried away in the field of astronomy (or astrology).

Since, as we have seen (p. 186 above), the embassies of +1081 and 1091 were from the Seljuq Turkish Sultans of Baghdad, and since it would not now be long before the Mongol unification of the Old World with all that that implied for ease of transmission, the period of direct Chinese-Roman relations, such as they were, closes. Only one strange and belated diplomatic contact demands notice. In +1371, not long after the expulsion of the Mongols and the establishment of the Ming, a Byzantine ambassador, Nieh-Ku-Lun (probably Nicholas Cumanos) was in China.[a] He was sent home honourably with a letter from the Chinese to the Roman emperor, and a Chinese official, Phu La, bore him company at least part of the way. It is tempting to speculate on the reason for his journey to China; perhaps he was really only a merchant; but conceivably the Byzantine government, apprehensive of the growing danger of Turkish conquest, sent out a mission of enquiry to the home of the newly developed technique of explosive weapons in warfare. In the event, the Turks used it more effectively. Before speaking of the relations between China and Islam, what of the possible passage of science and technique between China and India down to the end of the Thang?

(*i*) CHINESE-INDIAN CULTURAL AND SCIENTIFIC CONTACTS

Already we have seen early evidences of connections;[b] for example, the Szechuanese products which Chang Chhien recognised about −130 as having reached Bactria by the Yunnan-India route, and the appearance of Buddhism in China between +70 and +160 (pp. 173, 112 above). Hints exist that other things besides cloth and bamboo staves passed through the forested mountains of Yunnan and Assam from the beginning of the Han time onward. After all, the word Cīnasthāna came from Chhin. The *Huai Nan Tzu* book (*c.* −120) contains[c] a remark that Yü the Great 'when he went to the Country of the Naked People, left his clothes before entering it and put them on when he came out, thus showing that wisdom adapts itself to circumstances'. It is claimed[d] that this story occurs also in ancient Buddhist writings told of Gautama or a Bodhisattva. If so, this parallel would be analogous with those Graeco-Chinese parallels mentioned at the beginning of this section. Here again it could conceivably be independent, since primitive peoples in hot climates existed south of China as well as east of India.

In +90 there were diplomatic and other contacts between the Chinese and the Kushans (Kuei-Shuang,[1] descendants of the Yüeh-chih and the Śakas) who under the

[a] *Ming Shih*, ch. 326, p. 17*b*.

[b] Cf. the summary of Bagchi (1). The book of Radhakrishnan (1), though well intentioned, seems to entertain several basic misconceptions concerning Chinese thought, as well as historical mistakes of various kinds. In general, Indian authors seem prone to overestimate the importance of Buddhism in China.

[c] Ch. 1, p. 6*b*.

[d] Bagchi (1), p. 8.

[1] 貴霜

great Kanishka were about to conquer the whole of northern India.[a] Throughout this time the Khotan region was a meeting-place of Indian, Iranian, Greek and Chinese civilisations. It is therefore quite natural to find mention of China in the contemporary *Laws of Manu*.

The great period of Buddhist interchange, however, begins after the middle of the +4th century.[b] In +386 the Indian monk Kumārajīva (Chiu-Mo-Lo-Shih[1]) came to China from Central Asia and powerfully propagated mahāyanist doctrine,[c] but though famous he was only one of a long and numerous succession.[d] There were also many embassies from Ceylon during the +5th century, and from Kapila and the Gupta Kingdom.[e] More interesting for our purpose is the great series of Chinese Buddhist scholars and monks who made the pilgrimage to India from this time onwards, suffering untold dangers and hardships en route, but bringing back large numbers of texts for translation.

(1) The Buddhist Pilgrims

The first great name among these is that of Fa-Hsien,[2] who left for India in +399 and returned in +414. His book, the *Fo Kuo Chi*[3] (Records of Buddhist Countries), has often been translated.[f]

The second mission of interest was that of the two monks Hui-Sêng[4] and Sung Yün[5] of the Northern Wei dynasty, who were away in Udyāna and Gandhāra between +518 and +522. They each wrote an account of their travels, the *Hui-Sêng Hsing Chuan*[6] (Travels of Hui-Sêng) and the *Wei Kuo i Hsi Shih-I Kuo*[7] (Eleven Countries West of Wei), but both have perished—fortunately what is perhaps a conflated abridgement does still exist, the *Sêng Hui-Sêng shih Hsi Yü Chi*[8] (Record of Western Countries by the monk Hui-Sêng) preserved as chapter 5 of Yang Hsüan-Chih's[9] *Loyang Chhieh Lan Chi*[10] (Description of the Buddhist Temples of Loyang) written about +530.[g] These monks had some kind of diplomatic status.

The greatest of all the Buddhist pilgrim-scholar-diplomats was Hsüan-Chuang[11],[h] born in +603, who was in India between +629 and +645, and died in +664. His *Ta Thang Hsi Yü Chi*[12] (Records of the Western Countries in the time of the Thang)

[a] *TH*, p. 716; McGovern (1), pp. 249, 251; V. A. Smith (1), pp. 129–43.

[b] For a picture of India during the Buddhist pilgrimage period consult de la Vallée Poussin (2) and Prasad (1).

[c] *TH*, pp. 1011, 1041.

[d] Biographies, but no characters for Chinese names, in Bagchi (1).

[e] Summarised in Yule (2), vol. 1, p. 67.

[f] Rémusat (1); Beal (1); Legge (4); H. A. Giles (3).

[g] There is a translation by Beal (1) but that of Chavannes (3) is more reliable.

[h] Much argument has centred round the proper transcription and romanisation of his name, for the character *hsüan* afterwards became taboo and was replaced by *yuan*.[13] Hence his name may be met with in all kinds of variants, e.g. Yuan-Thsiang, etc. We retain Chuang for the second character, though there is reason to think that the contemporary pronunciation was Tsang.

1 鳩摩羅什 2 法顯 3 佛國記 4 惠生 5 宋雲
6 惠生行傳 7 魏國以西十一國 8 僧惠生使西域記
9 楊衒之 10 洛陽伽藍記 11 玄奘 12 大唐西域記
13 元

is available in translation.[a] The information contained in his own account of his travels is much supplemented in the biography written by his disciple Hui-Li:[1] *Ta Thang Tzhu-Ên-Ssu San Tsang Fa-Shih Chuan*[2] (Life of the Master of the Law and the Tripitaka, of the Great Loving-Kindness Temple of the Thang).[b] Hsüan Chuang's character has endeared him to subsequent generations.[c]

The fourth important name is that of his younger contemporary, I-Ching[3] (born +635, in India between +671 and +695, died +712). This monk left a record not so much of his personal travels as of the Buddhist rites and practices as he found them in India and south-east Asia in his time, the *Nan Hai Chi Kuei Nei Fa Chuan*[4] (Record of Buddhist Practices sent home from the South Seas)—there is a translation by Takakusu (1) and an abstract by Beal.[d] Even more valuable, however, is his work *Ta Thang Chhiu Fa Kao Sêng Chuan*[5] (Records of the High Monks who went out to seek for the Books of the Law in the Thang time) of about +705, translated by Chavannes (4).[e] Another +8th-century pilgrim to India was the Korean monk[f] Hui-Chhao[6]. The last of the series of pilgrimages was brought up by Chi-Yeh[7] who was in India from +964 to +976 with many other monks, and whose record is preserved in Fan Chhêng-Ta's[8] *Wu Chhuan Lu*[9] of +1177.[g] This was the expedition which led to the printing of the Tripitaka in China[h] of which there is an account in the *Fêng Chhuang Hsiao Tu*.[i] The last dated monument erected by a Chinese Buddhist pilgrim in India is of +1033.[j]

I doubt whether this material has ever before been examined from the point of view of the history of science. The whole story has often been told,[k] on account of its interest for the history of religions, and by no one more charmingly than by Grousset, whose book *Sur les Traces du Bouddha* is to be recommended. Let us see whether any light may be obtained from this whole field for Chinese-Indian scientific relations. The monks were of course interested in Buddhist religion and theology almost to the exclusion of anything else, and we cannot be surprised, therefore, to find little to our purpose in Fa-Hsien.[l] In the story of Hui-Sêng and Sung Yün,[m] however, there is a remarkable passage.

[a] Julien (1) and Beal (2), on which the commentaries of Watters (1) and Pelliot (4) should be consulted.

[b] This has been translated by Julien (1) and Beal (3).

[c] Modern biography by Su Yuan-Lei (1). The most recent life of him in English is by Waley (16). The accounts of his travels formed the basis for the famous allegorical novel *Hsi Yu Chi* (translated by Waley (17) under the title of *Monkey*), which was written, or at least brought into its final form, by Wu Chhêng-Ên early in the Ming. See Nagasawa (1), p. 246. The character who gives his name to the translation is of course the Indian god Hanuman in a Chinese guise.

[d] (3), p. xlii.

[e] Abstracted by Beal (3), p. xxvii.

[f] See Fuchs (4).

[g] Account of a Journey by Boat to Wu. Tr. Huber (1).

[h] See Carter (1), p. 62.

[i] Maple-Tree Window Memories. Ch. 2, p. 19*a*.

[j] Bagchi (1), p. 79.

[k] E.g. by Nehru (1), pp. 154ff.

[l] Summary in *TH*, p. 1047.

[m] Summary in *TH*, p. 1190. Sung Yün was a native of Tunhuang.

1 惠立 2 大唐慈恩寺三藏法師傳 3 義淨 4 南海寄歸內法傳
5 大唐求法高僧傳 6 慧超 7 繼業 8 范成大 9 吳船錄

When the two monks had audience of the King of Udyāna (Wu-Chhang;[1] a kingdom of the Upper Indus valley, probably modern Swat):

he questioned Sung Yün and said, 'Are my honourable visitors from the Land of the Sunrise?' Sung Yün replied, 'Our country is bounded on the east by the great sea; from this the sun rises according to the will of Tathāgata.' Then the King asked again, 'What sages has your country produced?' Sung Yün explained to him in detail the virtues of the Duke of Chou, of Confucius, of Chuang Tzu and Lao Tzu; and then he described the palaces of silver and halls of gold standing upon Mt. Phêng-Lai,[2] and the holy immortals and saintly men who assemble in them. Finally he spoke of the skill of Kuan Lo[3] at predicting the future, of the medical science of Hua Tho,[4] and of the magical procedures of Tso Tzhu;[5] and all these things he explained in due order. The King said, 'If all this is as you say, it must indeed be the realm of Buddha, and I would like to be reborn there.'[a]

Thus Sung Yün, although a Buddhist, chose, in describing the merits of his country, Confucian, but especially Taoist, and hence 'proto-scientific', claims. The immortals of Phêng-Lai were (as we shall see, Sects. 10, 33) particularly associated with the earliest alchemy, Kuan Lo (cf. Sect. 26*i*) was a divination-expert (+209 to +256) who comes into the history of the discovery of the magnetic compass, Hua Tho (+130 to +220) was one of the greatest of Chinese physicians (see Sect. 44 below), and Tso Tzhu (+155 to +220) was a famous Taoist magician. Thus both the Taoism and the science of China were highly regarded by Chinese representatives abroad in the +6th century.[b]

A century later the same note is struck again. The story of Hsüan-Chuang[c] contains several interesting points. Grousset (3) rightly emphasises the Confucian rationality and humanitarianism of this Buddhist monk in face of certain of the cruel extravagances and sinister aspects of Hindu cults.[d] Hui-Li informs us that in the great monasteries where Hsüan-Chuang stayed, such as Nālanda, medicine, astronomy, mathematics and magic were all studied.[e] But the most interesting occurrence was what happened when Hsüan-Chuang decided to return home. Hui-Li says:

The monks of Nālanda, when they heard of it, begged him to remain, saying: 'India is the land of Buddha's birth, and though he has left the world, there are still many traces of him. What greater happiness could there be than to visit them in turn, to adore him and chant his praises? Why then do you wish to leave, having come so far. Moreover, China is a country of *mlecchas*, of unimportant barbarians, who despise the religious and the Faith.

[a] Tr. Chavannes (3), p. 408, eng. auct.; adjuv. Beal (1).

[b] It is therefore all the more interesting that a Sanskrit translation of the *Tao Tê Ching* was prepared in +647 for Bhāskara Kumāra, King of Kāmarūpa (Assam); see Pelliot (8), and below, Sect. 10.

[c] Summary in *TH*, p. 1344.

[d] E.g. the comments of Hsüan-Chuang on the suicide temple (Grousset (3), p. 129; Julien (1), vol. 2, p. 278) and the sermon against the disagreeable practices of Sivaist ascetics (Grousset (3), p. 187; Julien (1), vol. 1, p. 225).

[e] Grousset (3), p. 156; Beal (3), pp. 112, 153.

[1] 烏萇 [2] 蓬萊 [3] 管輅 [4] 華佗 [5] 左慈

That is why Buddha was not born there. The mind of the people is narrow, and their coarseness profound, hence neither saints nor sages go there. The climate is cold and the country rugged—you must think again.'

The Master of the Law replied, 'Buddha established his doctrine so that it might be diffused to all lands. Who would wish to enjoy it alone, and to forget those who are not yet enlightened? Besides, in my country the magistrates are clothed with dignity, and the laws are everywhere respected. The emperor is virtuous and the subjects loyal, parents are loving and sons obedient, humanity and justice are highly esteemed, and old men and sages are held in honour. Moreover, how deep and mysterious is their knowledge; their wisdom equals that of spirits. They have taken the Heavens as their model, and they know how to calculate the movements of the Seven Luminaries; they have invented all kinds of instruments, fixed the seasons of the year, and discovered the hidden properties of the six tones and of music. This is why they have been able to tame or to drive away all wild animals, to subdue the demons and spirits to their will, and to calm the contrary influences of the Yin and the Yang, thus procuring peace and happiness for all beings....How then can you say that Buddha did not go to my country because of its insignificance?'[a]

Again, therefore, a Chinese abroad commends the scientific achievements of his country, though in this case with a Confucian rather than a Taoist flavour; at any rate even astronomical instruments are mentioned. Admittedly the speech was reconstructed somewhat later by Hsüan-Chuang's biographer, but Hui-Li was a direct disciple and must have known the mind of his master. 'With what verve, in this speech,' says Grousset (3),[b] 'Chinese precision, scientific spirit,[c] and organisation are opposed to the political incapacity and pragmatic indifference of India!'

When we find a similar attitude in I-Ching we begin to realise something of what was perhaps a psychological disinclination on the part of the Chinese to believe that other countries could have anything valuable to add to the sciences as they had so far developed in their own country. I-Ching was more interested in medicine than the other pilgrims, and three chapters[d] of his book deal with medical matters, as has been discussed by Barth (1). I-Ching mentions the 'eight sections of medical science', which are the eight parts of the Ayurvedic medical system,[e] but while speaking well of Indian physicians, says that the best herbs are nearly all to be found in China rather than India.[f] He says:

In China there are more than 400 kinds of herbs, minerals, stalks and roots, most of which are excellent and rare in colour and taste, and very fragrant in their smell; thereby we can

[a] Tr. Julien (1), eng. auct.; adjuv. Beal (2). Cf. Waley (16), p. 57.

[b] P. 189.

[c] Grousset's actual word is 'scientisme', which is much better, though I have not been able to find an English word for it. Science it cannot quite be, since in the +7th century neither China nor India had natural science in the full modern sense; he has in mind the Chinese proto-sciences and correlative naturalism, and what Creel (3) called 'Sinism' (see Sects. 13, 20, 26 *h*).

[d] The 27th, 28th and 29th.

[e] Takakusu (1), pp. 127, 222: (*a*) sores; (*b*) acupuncture; (*c*) general medicine; (*d*) casting out of demons; (*e*) paediatrics; (*f*) pharmaceutics; (*g*) alchemy for longevity; (*h*) 'invigorating the legs and body'.

[f] It is interesting, in view of what we saw about theriaca above (p. 204), that I-Ching says 'rich people may buy the valuable glue that comes from Ta-Chhin'.

cure any disease, and control the temperament. In the healing arts of acupuncture and cautery, and the skill of feeling the pulse, China has never been surpassed by any country of Jambūdvīpa (the *oikoumene*); the drug for prolonging life is found only in China.... Is there anyone, in the five parts of India, who does not admire China?[a]

In treatment I-Ching relied mainly on abstention from food; and wrote strongly against the use of excreta in therapy.[b]

(2) Ambassadors, Alchemists and Mathematicians

One should, however, be cautious of concluding that even these Chinese who went to India were too proud of the scientific achievements of their own country to be interested in Indian science. We recall the description given on p. 128 above of the numerous books of 'Brahmin' astronomy, mathematics, etc., which appeared in Chinese between the time of Fa-Hsien and that of I-Ching. Even though they were soon lost, their mere existence is significant. Some more details of Chinese-Indian scientific relations are found in the extraordinary stories[c] of Wang Hsüan-Tshê[1] and Hsüan-Chhao.[2]

Wang Hsüan-Tshê was an official who left China (for the second time) in +648 as ambassador to the court of Magadha (modern Patna), where at that time Harsha Vardhana, the friend of Hsüan-Chuang, was reigning. But at this time Harsha died and a usurping minister (A-Lo-Na-Shun[3] in the Chinese records) thought fit to attack the Chinese party, plunder their goods, and kill most of Wang's retinue. Wang, however, was a man of resource; he escaped to the mountains, made contact with the Kings of Nepal and Tibet, who were at that time allied with China, and descending again with an army of considerable size, gave battle to the usurper and completely overthrew him. It was at this time that he helped Hsüan-Chhao, who had been studying Sanskrit and Buddhism in India for some years, to return to China. The ambassador then himself returned home by another route, taking with him the usurping Indian and other prisoners, whom he presented to the emperor at Chhang-an with an account of his proceedings.

A version of this, written somewhat over a century later, is of the greatest interest, as it preserves what may be one of the earliest passages on mineral acids.[d] Hitherto it has been generally believed that the mineral acids were first known in Europe in the +13th century, and J. R. Partington, our greatest authority, finds the first mention of

[a] Tr. Takakusu (1), mod.
[b] Although a northerner from Hopei, he was a great enemy of onions.
[c] Pelliot (8); Lévi (1); Chavannes (4); Grousset (3), pp. 246 ff.; V. A. Smith (1), p. 169; *TH*, p. 1342.
[d] This passage was noted by Pauthier (1), Julien (2), Pelliot (8) and Yule (2), vol. 1, p. 69, but none of them appreciated its scientific significance.

[1] 王玄策 [2] 玄超 [3] 阿羅那順

them in the *Pro Conservanda Sanitate* of the French Franciscan Vital du Four about +1295.[a] But in the *Yu-Yang Tsa Tsu* of Tuan Chhêng-Shih (*c.* +860) we read:

> Wang Hsüan-Tshê captured an Indian prince named A-Lo-Na-Shun. He had with him a scholar versed in arts and gramarye named[b] Na-Lo-Mi-So-Pho,[1] who said he could make people live for two hundred years. (The Emperor) Thai Tsung was very astonished and invited him to live in the Chin Yen Mên (Palace), to make the drugs for prolonging life. The Emperor asked the Minister of War, Mr Tshui Tun-Li, to be in charge of it. The Indian said 'In India there is a substance called Pan Chha Cho Shui[2] (Pan-chha-cho water) which is produced from minerals in the mountains, has seven varieties of different colours, is sometimes hot, sometimes cold, can dissolve herbs, wood, metals, and iron—and if it is put into a person's hand, it will melt and destroy it. The skulls of camels have to be used as recipients for it....' Finally the Indian died in Chhang-an.[c]

Foreshadowing perhaps the later 'alkahest' or universal solvent of Arabic chemistry, this passage at any rate strongly suggests that a mineral acid was known in the +7th century. It gives colour to the hints about strong acids in Ray's (usually so unconvincing) *History of Hindu Chemistry*. The *Rasārṇava Tantra* (dated by Renou (1) as of the +12th century) speaks of the 'killing' of iron and other metals by a *viḍa* (solvent?) which is prepared from green vitriol (*kāsīsa*), pyrites, etc.[d] From the *Rasaratna-samuccaya* (which according to Renou may go back to the +8th century), the process of 'killing' certainly seems to be the formation of salts from metals.[e]

In the meantime the monk Hsüan-Chhao had hoped to be left in peace to translate Buddhist sūtras. But in +664 he received an imperial order to return to India to search for certain famous physicians and to collect medicinal plants. Interest in alchemy and the elixir of life was of course at this time very strong at the Chinese capital. Hsüan-Chhao succeeded in finding the physician or alchemist (who may have been the So-Po-Mei[3] or the Lu-Chia-I-To[4] recorded[f] in *Thung Chien Kang Mu*) and sending him back to China; then he continued his search for drugs and herbs. Finally, his way home was blocked by the Tibetans in revolt, and by the newly appearing Arab power, so that he died at the age of sixty in an Indian monastery.

In spite of these and many other evidences of Indian-Chinese scientific contact, definite proof of Indian influence on Chinese science is hard to adduce. A careful comparison of the traditional pharmacopoeias of China and India, which would trace out pharmacological borrowings, is greatly to be desired. It is probable that drugs such as chaulmoogra oil, used for leprosy, which have been for many centuries in the Chinese pharmacopoeia, were of Indian origin. Ta-fêng-tzu,[5] the oil of chaulmoogra[g]

[a] Private communication to the author. See Sarton (1), vol. 2, p. 408 and vol. 3, p. 531.

[b] Pelliot (8) conjectures Skr. Nārāyaṇasvāmin; Waley (16) agrees. Pauthier (1) conjectured Skr. Panjāb water; Pelliot (8) thinks *phāṇṭa* water. This latter term meant a liquid prepared by filtration.

[c] Ch. 7, p. 7*a* (tr. auct.). There is a similar passage in *Chiu Thang Shu*, ch. 198, p. 12*b*.

[d] Ray (1), vol. 1, pp. 65, 72.

[e] Ray (1), vol. 1, pp. 118 ff.

[f] *TH*, pp. 1371, 1372, 1374.

[g] R258. Cf. Burkill (1), vol. 1, p. 1204.

1 那羅邇娑婆 2 畔茶佉水 3 娑婆寐 4 盧迦逸多 5 大風子

(*Hydnocarpus anthelmintica*), first appears in the *Pên Tshao Yen I Pu I*[1] of about +1380, i.e. rather late. The materials for making such a comparison are now largely available in Western languages. But all such studies are rendered excessively difficult by the absence of a firm chronology for India, and the great uncertainty which exists in the dating of even the most important Indian scientific texts.

Influences of Chinese mathematics on Hindu mathematics are, however, unmistakable.[a] The proof of the Pythagoras Theorem used by Chao Chün-Chhing[2] in his +2nd-century commentary on the *Chou Pei*[3] (the oldest mathematical classic) appears again in the work of Bhāskara (+1150). The rule for the area of the segment of a circle given in the *Chiu Chang Suan Shu*[4] (Nine Chapters on the Mathematical Art) of the +1st century appears again in the +9th-century work of Mahāvīra. Indeterminate problems of the *Sun Tzu Suan Ching*[5] (Master Sun's Mathematical Manual) of the +3rd century are found in Brahmagupta (+7th century). Āryabhaṭa (+5th century) has geometrical survey material very like that of Liu Hui of the +3rd. These intervals of time are often so long that independent development would have seemed equally likely, if we were not able to point to many more cases of the same kind.

At this stage it is difficult, without anticipating later expositions of complex subjects, to attempt any balance-sheet of Indian influences on Chinese science and technology. The question will arise in almost every specialised field. It seems extremely probable that for some of the most fundamental ideas such as the equatorial system of lunar mansions in astronomy,[b] and the general theory of pneumatic physiology in medicine,[c] there was an outflow of seminal concepts from ancient Mesopotamia which were developed in quite different ways by the Indians, the Chinese and the Greeks. We shall see in one instance at least, that of the arithmetic cycle of notes in the musical scale,[d] that the Chinese believed they had had it from the west while the Greeks were equally sure that they had received it from the east. One could thus speak of a community of descent, with little or no diffusion between the secondary centres. As we have seen,[e] and shall see again,[f] atomism, whether originating in India or elsewhere, never took root in China.

Diffusion of course there was, even if it concerned mainly matters less important for us. Indian grammar undoubtedly stimulated Chinese philological study in the +5th and +6th centuries.[g] Indian and Chinese painting was known in both cultures, and it is said that the six great principles of painting enunciated by Hsieh Ho[6] about +480 were identical with those contained in earlier Indian writings.[h] Indian music

[a] Kaye (3) has listed them. He thinks that the mathematical book of Chang Chhiu-Chien (*c.* +500) (see on, Sect. 19*c*) is the nearest equivalent to the Indian works of contemporary and later date.

[b] See below, Sect. 20*e*.

[c] See below, Sect. 44.

[d] See below, Sect. 26*h*.

[e] See above, pp. 154ff.

[f] Below, Sect. 26*b*.

[g] Kroeber (2; 3; 4, p. 234).

[h] Bagchi (1), p. 163. Others, e.g. Rowland (2), p. 145, deny this.

1 本草衍義補遺 2 趙君卿 3 周髀 4 九章算術 5 孫子算經
6 謝赫

came through Kucha to China just before the Sui period and had a great vogue there in the hands of exponents such as Tshao Miao-Ta, a man of Brahminical origin. Conversely, when Hsüan-Chuang visited Bhāskara Kumāra, King of Assam (Kāmarūpa), in +638, they played in his honour a Chinese musical piece which had been composed to celebrate a Thang victory in +619, only nineteen years before. The Chinese pagoda is without question a development of the Indian stūpa through stages which have been traced.[a] Yet the Indian contribution in engineering technology was extremely small, apart from the machinery associated with cotton, which came late; this is a point to which we shall return.

Although the contact between China and India was mostly carried on through Buddhist channels, some Hindu influence is detectable, for instance, in the sculptured columns, possibly of Thang date, still extant in a temple at Chhüanchow (old Zayton) in Fukien, which have been described by Coomaraswamy (3). They are Singhalese in style, however, and may be connected with the enforced residence, in the +15th century in that city, of the King of Ceylon,[b] who left descendants there. In the other direction, as we shall see,[c] certain Hindu cults were strongly influenced, perhaps even called into being, by the passage of Taoism to India through Assam in the +7th century.

Roughly we may think of the +3rd to the end of the +7th centuries as the great Chinese-Indian period, though occasional embassies from Indian States continued until +1015. From the +8th to the +13th was the great Chinese-Arab period. A brief consideration of it will terminate this discussion.

(j) CHINESE-ARAB CULTURAL AND SCIENTIFIC CONTACTS

Islam, that great new power which was to affect so profoundly east-west relations had arisen just before the departure of Hsüan-Chuang for India—the Hejira of the Prophet, from which the Muslim year-reckoning still dates, had taken place in +622.[d] The complete conquest of Iran in +652 had brought Arab domination rather quickly to the frontiers of Chinese influence; it is not, therefore, surprising that contact should have been established early both by land and sea.[e] But for a period, contact with non-Islamic Persia was intensified. In +638 the last Sassanid king, Yazdagard III, sent an embassy to Thang Thai Tsung to beg for aid, but this was declined, as the king learned when retreating into Turkestan after the Battle of Nahawand (+642).[f] Yazdagard's son Fīrūz (Pi-Lu-Ssu) established himself in Tokharestan in some degree

[a] Boerschmann (1); Combaz (5).
[b] Cf. Sects. 22, 29 below.
[c] Cf. Sect. 15*f* below.
[d] As this is the first place where it becomes necessary to refer to many Arabic and Persian names, I should like to mention the help which may be derived (apart from the larger works of Sarton; Mieli; Hitti; and Brockelmann) from the handy guides of 'Abd al-Jalīl and Sauvaget. But so long as no two Arabists agree on the indexing of the numerous names borne by the same individual, any contact with this field will continue to cause irksome travail.
[e] Hitti (1), p. 344.
[f] These events provided dramatic material for the pen of Gibbon (vol. 9, ch. 51, pp. 375ff.).

of dependence on China, but could not hold it, and fled to the court at Chhang-an in +670, where he built a Zoroastrian temple and died soon afterwards. Nine years later the son of Fīrūz, Narses (Ni-Nieh-Ssu), made an attempt to recover his domains with the help of a Chinese expeditionary force, but fruitlessly, for he was back again at Chhang-an in +707. Small Persian border-States such as Ṭabaristān remained independent of Islam for many years, and ten of their embassies reached China between +713 and +755. Some Persians settled in Hainan Island.[a] The son of the last non-Islamic ruler of Tabaristān came to the Chinese capital and never returned home. It was like the end of the Bactrian-Indian Greeks all over again, and while one is tempted to think that in such periods purely political anxieties predominated, it can well be imagined that the desire to save certain deposits of learning or skill could have led to transmissions.

The first direct contacts of the Chinese with Islam soon gathered an accretion of legends, among which it is difficult to disentangle the truth.[b] Chinese historians speak clearly, however,[c] of an ambassador, one Sulaimān, from Arabia (Ta-Shih[1]),[d] in +726. This was just at the end of the Umayyad period, during the reign of the Caliph Hishām, so it is possible that Sulaimān was sent to seek help for the decaying house.

In the +8th century relations were very close. The Battle of the Talas River in +751, which established Muslim supremacy in western Central Asia, was the high-tide mark of Islamic expansion, for though the great general and governor al-Ḥajjāj al-Thaqafī had promised the governorship of China to whichever of his commanders, Muḥammad or Qutaibah, should first set foot on its soil,[e] no hostile Arab army ever did so. All this did not prevent the second Abbasid Caliph, Abū Ja'far al-Manṣūr, only a few years later, from sending a contingent of Arab troops (probably largely Persians if not Iraqis) to help the young emperor Su Tsung to regain control of the empire after the defeat of his father Hsüan Tsung in the An Lu-Shan rebellion.[f] This was in +756. Two years later, for reasons unknown, there were serious disturbances among the Arab maritime colony at Canton. It will be seen that from the +7th century onwards there were ample opportunities of intellectual contact between Arabs and Chinese. At the end of the +8th century, in +798, Hārūn al-Rashīd sent

[a] They were originally the crews and passengers of Persian ships pirated by the inhabitants apparently under the direction of a Chinese official Phêng Jo-Fang; the monk Chien-Chen saw settlements of them in +749 (Takakusu (3), p. 462).

[b] See Pelliot (5); Mason (1); Broomhall (1); and the summary in *TH*, pp. 1355 ff. The alleged tomb of Sa'd ibn abū Waqqāṣ, one of the prophet's uncles and a famous military commander in the first Arab conquests, exists to this day in Canton, but it is certain that he never went to China (Brockelmann (1), p. 15).

[c] E.g. *Thung Chien Kang Mu* (*TH*, p. 1400), taken from *Hsin Thang Shu*, ch. 221B, p. 12a; cf. *Chiu Thang Shu*, ch. 198, p. 17a.

[d] Perhaps from Persian Tazi and Tadjikistan (Hirth (9); Yule (2), vol. 1, p. 88).

[e] Hitti (1), p. 212.

[f] *TH*, pp. 1395, 1402, 1436, 1438. It is thought that this was the beginning of the spread of Islam in China; see Sarton (1), vol. 3, p. 1587; d'Ollone *et al.* (1); Drake (2).

[1] 大食

a mission to Chhang-An to arrange for coordinated strategy against the Tibetans, who were annoying both parties.[a] Not until +878 were Arab-Chinese trade relations seriously interfered with, by the sacking of Canton in the rebellion of Huang Chhao,[b] of which accounts exist by both Chinese and Arabic authors. But such trade relations recovered, and during the time of the Chin and Sung dynasties in the +11th century, there are records of more than twenty semi-commercial semi-diplomatic Arab missions; this was in the period of the Buwaihid (Persian) domination of the caliphate.

From the beginning of Islam,[c] the Arabs had been very conscious of China and India, as befitted the descendants of the people of Ta-Chhin and the middlemen of Abyssinian Axum. One of the most famous *ḥadīths* (orally transmitted utterances of the Prophet) ran 'Seek for Learning, though it be as far away as China'.[d] When the foundations of the new capital, Baghdad, were laid by al-Manṣūr in +762, he said (according to Abū Ja'far al-Tabarī): 'We have the Tigris to put us in touch with lands as far as China, and bring us all that the seas yield as well as the foods of Mesopotamia and Armenia. Then there is the Euphrates to carry for us all that Syria, al-Raqqah, and adjacent lands have to offer.'[e] And this eastward orientation runs through all the science of Islam.

In the middle of the +9th century (under the caliphate of al-Mutawakkil) Abū'l Ḥasan 'Alī al-Ṭabarī[f] wrote his great medical work *Firdaus al-Ḥikmah* (The Paradise of Wisdom), and it is striking that he quoted from Indian physicians such as Caraka, Suśruta and Vāgbhaṭa II, no less than from Hippocrates, Galen and Dioscorides.[g] In +771 an Indian scholar had brought the astronomical work *Sūrya Siddhānta*[h] to Baghdad, where it was translated by Muḥammad ibn Ibrāhīm al-Fazārī,[i] and this was probably one way in which the Hindu numerals reached the Mediterranean region.[j] The transmission was continued by the great Ibn Mūsa al-Khwārizmī, whose work *Ḥisāb al-Jabr wa'l-Muqābalah* (The Calculation of Integration and Equation), written about +820, was the greatest medieval work on algebra.[k]

Representative of this trend stands above all the great al-Bīrūnī (Abū al-Raiḥān al-Bīrūnī, +973 to +1048),[l] who having followed Mahmūd of Ghaznah in his conquest

[a] *TH*, p. 1465; cf. Gibb (1). [b] *TH*, p. 1506; cf. Schafer (2).

[c] The oldest name for the Islamic faith in Chinese was *Hui*,[1] perhaps derived from the Uighur people who first represented it on Chinese borders, but it later became known as the Chhing Chen Chiao[2] (Pure and True Teaching); cf. d'Ollone *et al.* (1); Broomhall (1). This is still its name, and I myself have had the privilege of visiting its mosques, as notably at Huihsien in Kansu.

[d] Hitti (1), p. 393. [e] Tr. Hitti (1), p. 292.

[f] Sarton (1), vol. 1, p. 574; Hitti (1), p. 365; Mieli (1), pp. 71, 73. Not to be confused with the historian just quoted, Abū Ja'far al-Ṭabarī, of about fifty years later (Sarton (1), vol. 1, p. 642; Hitti (1), p. 390).

[g] The parts relevant to Indian knowledge have been translated by Siggel (1).

[h] Tr. Burgess (1); see Sarton (1), vol. 1, p. 386.

[i] Sarton (1), vol. 1, p. 530; Hitti (1), p. 373; Mieli (1), p. 69.

[j] Reference will be made again to this and cognate subjects in Section 19 on Mathematics.

[k] Sarton (1), vol. 1, p. 563; Hitti (1), p. 379; Mieli (1), p. 82; tr. Karpinski (1).

[l] Sarton (1), vol. 1, p. 707; Hitti (1), p. 376 Mieli (1), p. 98; tr. Sachau (1).

[1] 回 [2] 清眞教

of India, wrote about +1030 his admirable work *Ta'rīkh al-Hind*. This is not only a history and geography of India in the ordinary sense, but a wide-ranging examination of all the sciences of the Hindus, together with original observations of the author. In the +13th century Muslim geographers gave many similar accounts of China and Chinese science, for example, the encyclopaedists Muḥammad ibn Ibrāhīm al-Anṣārī al-Dimashqī (+1256 to +1326)[a] and Aḥmad ibn 'Abd al-Wahhāb al-Nuwairī (+1279 to +1332).[b] Further information came through the *Taqwīm al-Buldān* of Abū'l-Fidā al-Aiyūbī[c] written in +1321, and by the work of the Persian geographer Ḥamdallah al-Mustaufī al-Qazwīnī (+1281 to +1340).[d] And there were not a few writers who had travelled to China and seen things for themselves, just as the Chinese Buddhists had visited India. The first was Sulaimān al-Tājir (Sulaimān the Merchant) whose stay in China took place a little before +851.[e] Another was Ibn Wahb al-Baṣrī, who had an interview with the Thang emperor, Hsi Tsung, in +876, and gave an account of it afterwards.[f] The greatest was the lovable Ibn Baṭṭūṭah (+1304 to +1377),[g] called by Sarton the greatest traveller of Islam, and the greatest, not excepting Marco Polo, of all medieval times. Ibn Baṭṭūṭah describes the construction of Chinese ships, the making of porcelain, water-raising machinery, paper money, coal, trade inspections, a system of old-age pensions and so on.

Much has already been said (pp. 140, 187) about the unity of all the east and west which came about after the Mongol conquests. But a few striking details may be given to show how much greater the personal contacts were between scientists from the two ends of the Old World at that time than is generally realised.[h] After the Mongol leader Hūlāgu Khan sacked Baghdad in +1258 and put an end to the Abbasid caliphate, he entrusted the illustrious Naṣīr al-Dīn al-Tūsī[i] (+1201 to +1274) with

[a] Sarton (1), vol. 3, p. 800; Mieli (1), p. 275; tr. Mehren (1).

[b] Sarton (1), vol. 3, p. 620; Mieli (1), p. 262.

[c] Sarton (1), vol. 3, p. 793; Mieli (1), p. 268; tr. Reinaud & Guyard (1).

[d] Sarton (1), vol. 3, p. 630; Mieli (1), p. 262, tr. le Strange (1) and J. Stephenson (1).

[e] *TH*, p. 1503; Sarton (1), vol. 1, p. 571; Hitti (1), p. 343; Mieli (1), p. 79; tr. Renaudot (1); J. T. Reinaud (1); Ferrand (2); Sauvaget (2). There is doubt whether Sulaimān was more than one of the informants whose experiences were collected in the *Akhbār al-Ṣīn wa'l-Hind*.

[f] Renaudot (1); Reinaud (1). Cf. Sect. 27*c* (5) below.

[g] Sarton (1), vol. 3, p. 1614; Hitti (1), p. 569; Mieli (1), p. 275; tr. Defrémery & Sanguinetti (1); Yule (2), vol. 4, pp. 80 ff.

[h] Yule (2), vol. 1, p. 167, says, with a flourish: 'Chinese engineers were employed on the banks of the Tigris, and Chinese astronomers, physicians, and theologians, could be consulted at Tabriz.' This is the kind of remark that gets remembered; unfortunately, apart from the fact that the idea of a Chinese theologian is odd in itself, one cannot substantiate the details which he and his editor, Cordier, give. D'Ohsson ((1), vol. 2, p. 611) is responsible for the engineers, but quotes no contemporary source; his second reference (vol. 3, p. 265) seems to refer only to Marāghah. The references to Quatremère are also useless, save that Hūlāgu had trouble getting rid of Tibetan lamas and was attended by Chinese physicians. The hydraulic engineering contact could be really important, for example, in connection with the invention of the lock-gate for canal traffic, coming as it does just at the right time for a westward transmission (cf. Sect. 28*f* below). But what authority did d'Ohsson have?

[i] Sarton (1), vol. 2, p. 1001; Hitti (1), p. 378; Mieli (1), p. 150. Not to be confused with other Islamic scientists who share the last name. Naṣīr al-Dīn al-Tūsī was a Persian, and as personal adviser to a Mongol conqueror occupies a historical position somewhat similar to Yehlü Chhu-Tshai (see p. 140); unfortunately for Iraq he did not meet with the same success in commending irrigated agriculture to the nomad leaders.

the formation of an astronomical observatory at Marāghah in Azerbaidjan, south of Tabriz. The observatory was equipped with the best instruments constructed up to that time, and the library is said to have contained over 400,000 volumes.[a] Hūlāgu sent for astronomers from China to collaborate in the work, and we even know the name of one of them, Fu Mêng-Chi (though not its characters).[b] At Marāghah they met with men from as far west as Spain, for instance, Yaḥyā ibn Muḥammad ibn abū'l-Shukr al-Maghribī al-Andalusī,[c] who published astronomical tables and other books from the observatory, including his *Risālat al-Khiṭā wa' l-Īghūr* (On the Calendar of the Chinese and Uighurs). None of the astronomical instruments of Marāghah, used throughout the last half of the +13th century, survived, though we know a good deal about them owing to the detailed description given by al-'Urḍī al-Dimashqī, a Syrian.[d] Later, in its proper place, we shall have occasion to compare them with the contemporary astronomical instruments made by Kuo Shou-Ching, under the Yuan dynasty in China about +1279, for the observatory which still exists near the south-east corner of the Tartar city wall at Peking.

Another example of this kind of contact was the astronomical treatise with lunar tables written in +1362 by 'Aṭā ibn Aḥmad al-Samarqandi for a Mongol prince of the Yuan dynasty, Chen-Hsi-Wu-Ching.[e] The autograph MS is in Paris and, as illustrated by Sarton, shows a title-page bearing both Chinese and Arabic writing; the content, however, has not yet been investigated.

Returning from the field of astronomy to that of the medical and biological sciences, we come upon the remarkable work of Rashīd al-Dīn al-Hamdānī (+1247 to +1318),[f] Persian physician and patron of learning, prime minister under the greatest of the Mongol rulers of Persia, Ghāzān Maḥmūd Khān. His *Jāmi'al-Tawārīkh* (Collection of Histories) contains a great quantity of information on China, especially on the Mongols, Buddhism, and the use of paper money.[g] About +1313 he caused to be prepared an encyclopaedia of Chinese medicine, the *Tanksuq-nāmah-i Īlkhān dar funūn-i 'ulūm-i Khiṭāi* (Treasures of the Ilkhan on the Sciences of Cathay). Sphygmology (pulse lore), anatomy, embryology, gynaecology, pharmaceutics and other medical subjects are dealt with, and—very notably—the Chinese ideographic language is considered as superior to alphabetic ones for science because the meaning of a character is independent of its pronunciation.[h] In 'Wank Shu Khu' we can recognise, as Sarton says, Wang Shu-Ho[1] the famous Chin dynasty physician (+265 to +317) who wrote the principal manual on the pulse, the *Mo Ching*.[2] This interesting outpost of Chinese thought in Persian literature has been rather fully described by Adnan

[a] See Minorsky (1). [b] Sarton (1), vol. 2, p. 1005.

[c] Sarton (1), vol. 2, p. 1015; Mieli (1), p. 153; (2) vol. 2, p. 172; Brockelmann (2), vol. 1, p. 474, suppl. vol. 1, p. 868. Other Spanish Muslims travelled 'as far as China in quest of learning', as we know from the list of Ibn Muḥammad al-Tālimsānī al-Maqqarī (Mieli (1), p. 272; Hitti (1), p. 578).

[d] Sarton (1), vol. 2, p. 1013; tr. Jourdain (1); Seemann (1).

[e] Sarton (1), vol. 3, p. 1529. We have not been able to identify the prince; no characters are available.

[f] Sarton (1), vol. 3, p. 969. [g] Tr. Quatremère (1). [h] See Sect. 49 below.

[1] 王叔和 [2] 脈經

PLATE XII

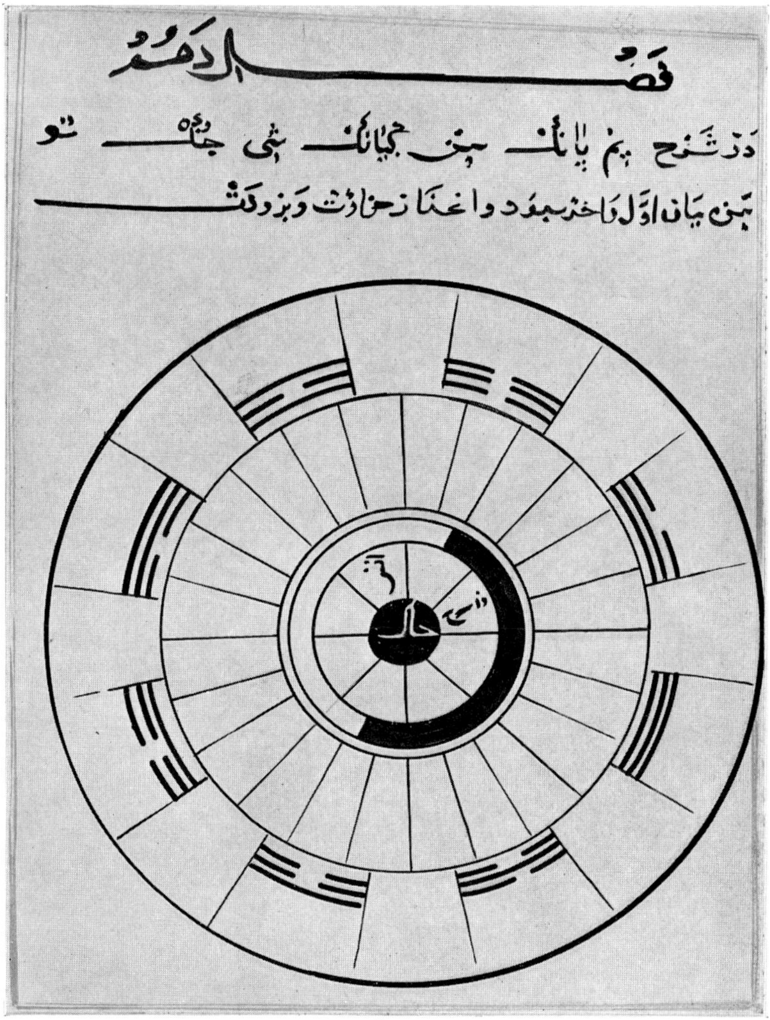

(*a*)

Fig. 34 (with Pl. XIII). Three medical drawings from the *Tanksuq-nāmah-Īlkhān dar funūn-i 'ulūm-i Khiṭāi* (Treasures of the Ilkhan on the Sciences of Cathay), prepared by Rashīd al-Dīn al-Hamdānī about +1313. In (*a*) the aegis of the Eight *Kua* over the day and night of the patient is related to the rise and fall of fever; (*b*) shows an anatomical sketch of the viscera (heart, diaphragm, liver and kidney can be made out); and (*c*) is a pulse-lore diagram. Though the text is in Persian, the Chinese origin of the illustrations is quite clear. From Süheyl Ünver (1, 3).

PLATE XIII

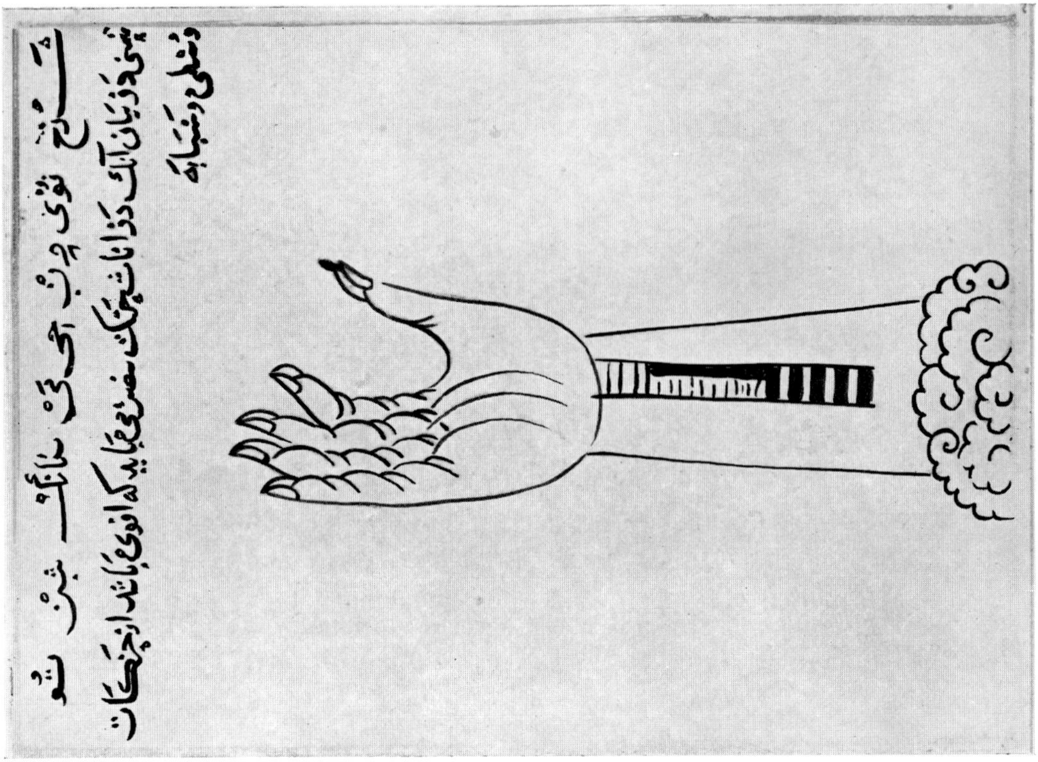

(*c*)

(*b*)

Adivar (1), and by Süheyl Ünver (1, 2), who has published the illustrations with his Turkish translation (see Fig. 34).

Transmission in the other direction is attested in a remarkable account found in the *Fihrist al-ʿulūm* (Index of the Sciences) of Abū'l-Faraj ibn Abū Yaʿqūb al-Nadīm (d. +995, called al-Warrāq al-Baghdādī, the stationer of Baghdad).[a] In this bibliography,[b] finished in +988, there is a story concerning the great Rhazes, physician and alchemist (Muḥammad ibn Zakarīyā al-Rāzī, +850 to +925):[c]

al-Rāzī said, 'A Chinese scholar came to my house, and remained in the town [probably Baghdad] about a year. In five months he learnt to speak and write Arabic, attaining indeed eloquence in speech and calligraphy in writing. When he decided to return to his country, he said to me a month or so beforehand, "I am about to leave. I would be very glad if someone would dictate to me the sixteen books of Galen before I go." I told him that he had not sufficient time to copy more than a small part of it, but he said, "I beg you to give me all your time until I go, and to dictate to me as rapidly as possible. You will see that I shall write faster than you can dictate." So together with one of my students we read Galen to him as fast as we could, but he wrote still faster. We did not believe that he was getting it correctly until we made a collation and found it exact throughout. I asked him how this could be, and he said, "We have in our country a way of writing which we call shorthand, and this is what you see. When we wish to write very fast we use this style, and then afterwards transcribe it into the ordinary characters at will." But he added that an intelligent man who learns quickly cannot master this script in under twenty years.'[d]

From this fascinating glimpse of Arab-Chinese contact, it is clear that the Chinese scholar, whose name has unfortunately not been preserved, was using the cursive script known as 'grass-writing' (*tshao-shu*[1]). While the story was told by al-Nadīm *à propos* of the writing methods of the Chinese,[e] it strongly indicates, if it does not absolutely prove, that there was at least one translation of Galen into Chinese in the +10th century. Yet, as will be seen (Sect. 44), hardly any perceptible influence of Hellenistic medicine upon Chinese medicine can be found.[f] Chinese medicine, until modern times, remained faithful to its special concepts—the two principles (*yin-yang*[2]), the five elements (*wu hsing*[3]), stasis (*yü*[4]), pneuma (*chhi*[5]), pulse lore greatly elaborated (*mo hsüeh*[6]) (some of which may have reached the western world through Ibn Sīnā),[g] acupuncture (*pien chen*[7]), moxa (*chiu*[8]), the use of mineral drugs long before the West, and so on.

[a] Sarton (1), vol. 1, p. 662; Hitti (1), p. 414; Mieli (1), p. 95.

[b] 'Not one in a thousand', says Sarton, 'of the books quoted in the *Fihrist* is now extant. To realise the importance of this index, the reader need but ask himself what it would mean to the classical scholar to have such a catalogue (with biographical notes!) of the libraries of Alexandria or Pergamon. Unfortunately, the losses of the books listed in the *i-wên-chih* (bibliographies) of the Chinese historical annals have been almost as great.

[c] Sarton (1), vol. 1, p. 609; Hitti (1), p. 365; Mieli (1), p. 89.

[d] Tr. Ferrand (1), vol. 1, p. 135; eng. auct.

[e] Hitti hints that al-Nadīm possessed, or saw, many Chinese books.

[f] Sarton (1), vol. 3, p. 261, has also noted this in remarking upon the presence of Muslim and Nestorian physicians at the Chinese court.

[g] (+980 to +1037), Sarton (1), vol. 1, p. 709; Mieli (1), p. 102; Hitti (1), p. 367.

1 草書 2 陰陽 3 五行 4 鬱 5 氣 6 脈學 7 砭針 8 灸

(i) The Focal Character of Islamic Science; the Transmission of Chinese Technology, but not of Chinese Scientific Thought, to Medieval Europe

The science of Asia has a dividing line running north and south through Bactria and the opening of the Persian Gulf. The science and scientific thought of Arabic civilisation forms in a very real sense a unity with European science—not only because at the farthest extension of Islam the Mediterranean was a Muslim lake, and Spanish no less than Persian Muslims contributed to the progress of science; but also because, as everyone knows, Arabic was the channel through which the writings of the Greeks reached the medieval Europeans. Mention has already been made of some of the great translators from Arabic, Adelard of Bath, Robert of Chester and Michael Scot, and there are detailed accounts of a hundred more in the encyclopaedia of Sarton. All of the important, and most of the less important, Greek scientific texts were translated into Arabic between the +7th and +11th centuries, and then translated back later by such men as these into Latin. Direct translation from the Greek does not begin before the +12th century with scholars such as James the Venetian (*fl.* +1128), Robert Grosseteste of Suffolk (b. +1175), and William Moerbeke the Fleming (b. +1215). In this remarkable phenomenon of transmission through different languages in the same geographical area, other Near Eastern languages, such as Syriac and Hebrew, played a minor but still significant part.

As an example of the Syriac contribution, interesting because it has a bearing on east-west transmission, we may mention the learned bishop Severus Sebokht,[a] who flourished at Qen-neshre in the middle of the +7th century, wrote on geography and astronomy, and composed in +660 one of the earliest treatises on the astrolabe. In one of the fragments of his books still extant, written in +662, he refers to the Hindu numerals.[b] He seems to have been hurt, says D. E. Smith, by the arrogance of certain Greek (Byzantine) scholars, who looked down on the Syrians, and in defending the latter, claims for them descent from the Chaldeans and other earliest Mesopotamian astronomers. He says:

> I will omit all discussion of the science of the Hindus, a people not the same as the Syrians; their subtle discoveries in astronomy, discoveries that are more ingenious than those of the Greeks and the Babylonians; their valuable methods of calculation; and their computing, which surpasses description. I wish only to say that this computation is done by means of nine signs. If those who believe, because they speak Greek, that they have reached the limits of science, should know these things, they would be convinced that there are also others who know something.[c]

And he ends by saying that science is international, and no monopoly of the Greeks.

[a] Sarton (1), vol. 1, p. 493; D. E. Smith (1), vol. 1, p. 166.

[b] Nau (1).

[c] Tr. Nau (1); eng. D. E. Smith.

In connection with this, another text of interest may be mentioned. Marqos Bayniel (+1224 to +1317), the Nestorian Metropolitan of Cathay, Mar Yahbh-Allaha III, enthroned as Patriarch in +1281, was a Chinese.[a] He accomplished the overland journey of Marco Polo in the opposite direction, accompanied by his friend Rabban Bar Sauma, a native of Peking (Khanbaliq), who had been ordained priest there. At Marāghah in +1278 the Catholicos Mar Denha gave them letters for Palestine. Later Marqos wrote an account of his travels in Persian, of which the Syriac version alone remains.[b] Unfortunately, there is nothing of specifically scientific interest in it, but the two friends remain noteworthy as possible vectors, and their travels were even more remarkable than this, as we shall shortly see.

Into the body of Latin learning, however, East Asian science was not incorporated. Knowledge of what the Chinese and Indians had contributed to the scientific patrimony of mankind did not penetrate through to the Franks. For some reason or other—presumably because there had been an abiding tradition, almost of a folk-lore character, of the names of the greatest Greek and Roman writers—when translations were being made from Arabic into Latin, it was always the famous authors of Mediterranean antiquity who were chosen, and never the books of Islamic scholars concerning the science of India or China.[c] Of the sixteen Arabic and Persian books cited in the preceding paragraphs on Chinese-Arabic relations, only one had been translated into a European language before +1700 (al-Khwārizmī by Robert of Chester) and only one more before +1800 (Sulaimān al-Tājir by Renaudot (1) in +1718). Six are still entirely unavailable for those who do not read Arabic or Persian. The others have been translated since +1800.[d]

Here may be mentioned the pioneer work of Andreas Müller Greifenhagius (1),[e] *Historia Sinensis Abdallae cogn. Abu Said Beidawi*, of +1679. This had nothing to do with the Persian theologian 'Abdallah ibn 'Umar al-Baidāwī (d. +1286),[f] as its translator thought, but was part[g] of the *Raudat ūlī'l-albāb fī Tawārīkh al-akābir wa'l-ansāb* (Garden of the Intelligent on History and Genealogy) of Dāwūd ibn Muḥammad al-Banākitī (d. +1330);[h] which in its turn was derived from the *Jami'al-Tawārīkh* of Rashīd al-Dīn al-Hamdānī, just mentioned.[i] The Latin translation seems to have had little influence and the content was mostly non-scientific. Meanwhile the Islamic scholars continued to be better informed about East Asia than those of Western Europe. Kahle (1) has drawn attention to a Turkish work by Abū Bakr ibn Bahrām al-Dimashqī of almost exactly the same date as Müller's. Though at first sight

[a] Mieli (1), p. 169; Yule (2), vol. 1, p. 119; and especially Chabot (1). We do not know his Chinese name, but he was a native of Shansi and may have been of Uighur origin.

[b] Tr. Budge (2).

[c] One of the few who have appreciated this was Schrameier (1), writing in 1888 at Peking. Cf. Sarton (6).

[d] Al-Bīrūnī, English, 1888; al-Dimashqī, French, 1874; Abū'l-Fida, French, 1848; al-Qazwīnī, English, 1919; Ibn Baṭṭūṭah, French, 1874–9; al-'Urḍī al-Dimashqī, French, 1809; al-Hamdānī's *Jami'*, French, 1836; the *Tanksuq-nāmah*, Turkish, 1939.

[e] On him see A. Müller (1) and Lach (2).

[f] Sarton (1), vol. 2, pp. 870, 871.

[g] Part 8.

[h] Sarton (1), vol. 3, pp. 971, 976.

[i] See Weston (1).

a translation of Blaeuw's *World Atlas* of +1662, it actually contains much original information, besides material drawn from the +13th-century work of al-Maghribī, the +14th-century *Jami'* and *Raudat*, and the +15th-century tables of Ulūgh Beg. Ibn Bahrām al-Dimashqī gives an account of the *I Ching* (cf. Sect. 13*g* below) and the lists of inventors (cf. p. 51 above). Moreover, Schefer (1) and Kahle (2) have described the early +16th-century Turkish work, *Khiṭāi-nāmah* (Description of China), by 'Alī Ekber. Like Ibn Baṭṭūṭah two centuries earlier, he greatly admired the Chinese women, and had much to say on porcelain, music and military matters.

It is quite clear, therefore, that throughout the formative period of modern European science there was no appreciation of Chinese and Indian contributions. This fact is not affected by the great work of the Jesuits in bringing Chinese sciences and techniques to the notice of the West, since that was all done after +1600 and mostly after +1650. Hitti saw this in part when he said[a] that 'literary Islamic geography' left no direct impression on European thought. As the works of the geographers found no translators into Latin, most of their contributions failed to pass on. The Far West took no interest in the descriptive geography of the Far East, nor in accurate cartography, nor in Chinese or Indian pharmaceutics or medicine.

But if East Asian science did not filter through to the Franks and Latins (i.e. to precisely that part of the world where, by a series of historical 'accidents', the geographical and social determinism of which has yet to be worked out, modern science and technology were later to develop)—the case was quite otherwise for East Asian technology. Only for the more abstract sciences was this barrier or filter effective. Technical inventions, on the contrary, show a slow but massive infiltration from east to west throughout the first fourteen centuries of the Christian era.[b] And this confirms the suspicion, which arose from quite different grounds at the beginning of this section (p. 154), namely, that while diffusion (in the anthropological sense) was very important in the case of techniques, it was much less operative for science proper or scientific thought.

For example (to take only the better-known themes), let it be assumed that the Mongols and other Central Asian peoples conveyed to Europe the efficient animal harnesses and the knowledge of gunpowder, while the Arabs transmitted in the same direction paper, printing and the magnetic compass. One may well ask what they could have been expected to take back? Until the +15th century West European technology may be said to have been less advanced than that of any other Old World region. The excessive rigidity of Aristotelian logic as seen in the scholastics could have had no appeal to Asian minds, and indeed modern science had to break that shell in order to be born.[c] The Greek and Hellenistic scientific classics would indeed have

[a] (1), p. 387.

[b] We shall see many examples (cf. pp. 240, 242), and may later tabulate them. But the basic papers of Lynn White (1) and C. Stephenson (1), in which this is beginning to be recognised, may here be mentioned.

[c] As in the attacks of men such as Joseph Glanvill and Robert Boyle on Aristotelianism, recounted in the memorable paper of Francis Gotch (1).

been a treasure greater even than the sūtras which the devoted Buddhists went to seek in India, but by the +13th century not all of them were available, and none of them had been fully assimilated into European thought before the Renaissance and the time of Vesalius and Galileo. They could of course have been obtained from the Arabic, and the failure of the Chinese so to obtain them probably argues a lack of theoretical interest on the part of Chinese scholars complementary to the technological poverty of Europe.

All in all, one may conclude that while the opportunities for transmission of thought between China and the West were greater than would at first sight appear, the use that was made of them was less. At the very beginning of the first Christian millennium a Greek-speaking Scyth acquainted with the Greek cities of the Black Sea and the books which were read in them, or a Greek-speaking Alan, *might* have conversed in some intermediate language of the steppes with a Chinese-speaking Hun. And a Romano-Syrian merchant *might* have learnt enough Chinese to give or receive ideas while in port at Canton. But on the whole Chinese and Western science seem to have influenced each other so little that the probability is against the existence of such fruitful conversations. For technology the matter stands otherwise.

(*k*) TYPES AND MOTIVES OF TRAVELLERS; THE POLITICAL URGE FOR STRATEGIC FLANKING MOVEMENTS

Before leaving the subject of culture contacts one thing more must be said. We have drawn attention to some of the artistic and literary evidence, we have traced the trade-routes, and we have cited some of the more oustanding facts known about the travels of individuals who could have been the carriers of ideas and techniques. It remains to refer to the types of these travellers and their motives.

A great deal has been said above concerning merchants, from the prehistoric exchangers of bronze and furs through the Romano-Syrians and Parthians to Marco Polo and his colleagues in the Mongol age. But one does not often find discussions of the part which they may really have played in transmission. Presumably they would be more likely to be interested in techniques than in abstract ideas. Waley[a] has well pointed out, however, that this assumption may be dangerous, and reminds us that Lü Pu-Wei, the compiler, or at least the patron, of the encyclopaedia of natural philosophy, the *Lü Shih Chhun Chhiu* (p. 98), was a famous merchant. Legend even makes a merchant of Kuan I-Wu.[b] So also Kennedy (2), in his interesting comparison of the *Gospels of the Infancy* with the *Lalita Vistara* and the *Vishṇu Purāṇa*, in which he works out parallels in Indian and early Christian mythology, though very hard on the sailors as such, grants to the merchants a considerable share in transmission.[c] But

[a] (4), p. 114.

[b] I.e. Kuan Chung, minister of the feudal State of Chhi in the −7th century, after whom is named the *Kuan Tzu* book (cf. p. 150 above).

[c] Kennedy gives many interesting quotations from classical authors on the character of sailors and merchants.

he lays the main burden on those communities where Christians and Indians lived side by side, as in Babylonia in the +1st century, and especially in Bactria and the Kushan empire, where there was certainly daily association between Christians, Jews, Buddhists and Hindus. It would be well worth while to push further this line of study.[a]

Almost as much has been said about the political and diplomatic travellers, Megasthenes, Chang Chhien and so on. One cannot overlook the strategic needs which prompted their voyages. The pattern for this was set already in the −2nd century. Lattimore[b] has shown how the Chinese, penetrating into the northern arid lands of the Gobi and its margins, could set up oasis 'cells' of Chinese culture, could dominate and assimilate, but never integrate their culture with that of the nomads, hence felt continually the need for ever further outflanking positions. Hence the Kansu pan-handle and the Wall, hence imperialism and conquest rather than culture synthesis. As we saw, the mission of Chang Chhien, which had such great consequences, was undertaken primarily to obtain the Yüeh-chih as allies against the Huns.

Exactly similar was the mission from Byzantium to the second Thang emperor in +643. During the preceding seven years, as Yule[c] says, the Muslim Arabs had wrested Syria from the Roman empire and Persia from the Sassanian kings, while Yazdagard, the last of that line, had sent for help to China. The envoys of Fu-Lin might discuss drugs and silk, but their real mission was to harass the rear of Islam. Rather more than a century later the Caliphate in its turn sent to Chhang-an (+798) with the object of obtaining allies against the Tibetans. And after the Mongol invasions, this situation constantly repeated itself in the triangular struggle between the Mongol power, Islam, and the crusading Franks. The detailed story has been told by Pelliot (10), Minorsky (2), and Grousset in his brilliant book *Bilan de l'Histoire*.[d] Quite consciously to obtain Nestorian allies against the Saracens did King Louis send the Franciscans John of Plano Carpini,[e] Andrew of Longjumeau,[f] and William Ruysbroeck[g] to Chinghiz Khan about +1250. They were unsuccessful, but a Christian-Mongol alliance was arranged by the Armenian king Haython I four years later,[h] and the sack of Baghdad by Hūlāgu in +1258 must have seemed in a sense a Christian victory, for there were many Nestorians in the Mongol army. But the crusader cities of the Levant rejected the Mongol alliance, preferring to join with the Muslim Mamluks of Egypt under Baybars,[i] who shortly afterwards put an end to the Mongol expansion in Islamic lands, and then turned against the Crusaders, expelling them by +1291.

[a] Cf. Lévi (5). [b] (1), p. 170. [c] (2), vol. 1, p. 55.

[d] Especially the essay 'Europe et Asie'.

[e] See Beazley (1), vol. 2, pp. 279 ff.

[f] See Beazley (1), vol. 2, pp. 317 ff. [g] See Beazley (1), vol. 2, pp. 320 ff.

[h] Sarton (1), vol. 3, p. 953. An excellent account of him is to be found in Beazley (1), vol. 2, pp. 382 ff. Armenian relations with China had sometimes been close, and certain families claimed Chinese origins (Yule (2), vol. 1, p. 94) which could well have been the case through exchanges of minor princesses and the movements of exiles.

[i] Hitti (1), pp. 655 ff.

It was in this final period of unsettlement, before the definitive Islamisation of the Mongols in Iran and the liquidation of the Levantine crusader strongholds, that the last effort at a Christian-Mongol alliance was made, by that very Marqos the Chinese, Mar Yahbh-Allaha III, whom we have met with above. The Ilkhan Arghun sent his friend Rabban Bar Sauma in +1287 to Byzantium and Rome to negotiate an alliance for the encirclement of the Mamluks on two fronts. Bar Sauma got as far as Paris and Gascony (where he called upon the English king), but the Europeans of the West were too divided to be able to undertake anything important. Minorsky (2) has continued the story by showing how just as Islam was at one moment caught between the Mongols and the Franks; so later on, in the +15th century, the Turcomans[a] found themselves between the Ottomans and the Venetian Republic; while in the +16th the Ṣafawids[b] were caught between the Sultans and Western Europe. Similar situations led repeatedly to similar diplomatic contacts.

Many see the apotheosis of these flanking movements in the correspondence between the final failure of the Crusades and the great voyages of discovery.[c] The fall of the Latin colonies in Syria, by giving the Egyptian Mamluks exclusive control of Indian and Asian commerce, spurred on the navigators of the Far West to find the Cape route to the Indies. Grousset makes an impressive juxtaposition of dates. In +1498 the Ottoman advance-guards reached Vicenza, but Vasco da Gama landed at Calicut. In +1503 the Turks scored great victories over the Venetians, but six years later Francisco d'Almeida, in a decisive naval engagement, obtained mastery of the Indian Ocean. In +1522 Rhodes fell to Sulaimān the Magnificent, but before five years were out the Portuguese dispersed the Muslim fleet at Gujerat. In the thirties the Turks were able to ravage Austria, but could not dislodge Joaõ da Castro from India. To some extent the decline of Islamic civilisation was due to the destruction of its central position in the trade (especially of spices) between Asia and the West. But we are already in sight of the ship carrying Matthew Ricci to China and the end of the Ages of Isolation. It is a proper point at which to end the story.

(*l*) GENERAL OBSERVATIONS

Having arrived so far, the reader may well ask whether any general principles are discernible among the mass of geographical and historical facts which we have been turning over.[d] It is, of course, hard to present the relationships between the several cultures of the Old World without at the same time sketching their individual characteristics (Iranian, Indian, Greek, Muslim and the like), but so great a task can obviously not be attempted in the present work. We may, however, remember at the outset that the position of China was unique in its isolation among the river-valley cultures of

[a] Brockelmann (1), p. 272.
[b] Hitti (1), p. 703.
[c] E.g. Grousset (4); Panikkar (1), pp. 30, 36, 49.
[d] It had been intended to postpone the writing of this piece until the completion of the whole of the Third Part of the work, but such a plan proved impracticable; nevertheless, the writer believes that no basic revision will be necessary. He is much indebted to Dr. G. Weltfish of Columbia University for valuable suggestions concerning the following paragraphs.

antiquity While the Fertile Crescent had Egypt not too far away on the west, and the Indus Valley civilisations comparatively near on the east, the Yellow River was very remote to the north and east across the vast barrier of the Himalayas and the Tibetan plateau.

In considering the contacts which interest us in the historical development of science and technology, we may make one deduction at any rate from the foregoing account, namely, that the period involved is of vital importance. For example, our estimate of the probabilities of transmission of specific ideas or detailed techniques before the time of Chang Chhien and the opening of the Old Silk Road towards the end of the −2nd century, must differ from estimates which impose themselves for later times. Earlier than that, in the bronze age, there had indeed been a remarkable cultural continuity between China and northern Europe across the Siberian steppes, even if it extended only to certain techniques of which we still have specimen products. Again, there were great differences in the extent to which China was 'open' as between the settled order of the Han period (−2nd to +2nd centuries) and the disturbances of the Nan Pei Chhao (+4th to +6th); and as between the Yuan dynasty and the Ming.

(1) Diffusion and Convergence

In seeking to distinguish between a contribution which a particular people made to the sum of human knowledge and power because it spread and was taken up by other different peoples, and an invention made truly independently in several places, hence the contribution of several peoples, we become involved (doubtless against our will) in those controversial questions which so stirred discussion in the twenties of this century among anthropologists and archaeologists. Although the exaggerations of Elliott Smith and W. J. Perry, in which all civilisation throughout the world was supposed to have emanated from ancient Egypt, or the earlier efforts of Terrien de Lacouperie and Ball to derive Chinese culture piece by piece from Mesopotamia, are now no longer seriously entertained, extreme diffusionist doctrine still finds supporters.[a] Views more generally held seem more reasonable.

> That diffusion is responsible [writes Dixon[b]] for a large number of apparently disconnected similar traits is probable, but there remains a considerable residuum for which independent origin is the only rational explanation. For common sense and the laws of probability must be applied to all cases, and when an explanation by diffusion requires us to assume that the extremely improbable or almost impossible has occurred, the *onus probandi* becomes very heavy. Where the physical difficulties are serious we must refuse to be carried away by vague generalities, and demand concrete proof. Until such proof is forthcoming the alternative of independent invention or convergence must be preferred. That diffusion has been responsible for cultural development to a far greater extent than independent invention is certain but occasional independent invention cannot, in the face of the evidence, be denied.

[a] For example Raglan (1).

[b] R. B. Dixon (1), p. 223.

Here the anthropologist is using the word convergence in a sense similar to that which it has in discussions of biological evolution (or so, at least, one supposes). There we define it as the process whereby analogous but not homologous structures have come to resemble one another as adaptation to environment has proceeded. Homologous organs are those which have the same essential structure, inherited from common ancestors of the organisms in which they exist, though they may be very differently modified in carrying out different functions. For example, the pentadactyl limbs of all air-breathing vertebrates are described as homologous, though the paddles of the whale are far removed from what ordinary people would regard as forelimbs. Analogous organs, on the other hand, though they resemble each other superficially, have fundamentally different structures; for example, the wings of birds and insects, or the ambulatory appendages of arthropods and vertebrates. This distinction probably also applies to biochemical functions. For instance, terrestrial gastropods, insects, reptiles and birds, all excrete uric acid as the main end-product of their nitrogenous metabolism, and they do so as an adaptation to the terrestrial environment with the cleidoic ('closed-box') type of embryonic development which that has implied, but it is probable that the reaction systems whereby the uric acid is produced are not identical in all these phyla.

When applied to social evolution, the concept of convergence need not necessarily mean independent invention of a high order. It may only mean that when presented with the same rather simple problems, people in different parts of the world solved them in the same way. Leroi-Gourhan,[a] in an excellent discussion of the diffusion of primitive inventions, points out that the environment and nature of man imposes considerable limitations on what is possible—e.g. an ornament *can* only be suspended from the ears or the nose, fibres *can* only be spun by twisting, there are only about ten possible ways in which hatchets *can* be fitted with handles, and so on. And he says, 'Convergence is the burden from which the ethnologist can never free himself, the trap in which every theory leaves some of its fur behind.'[b] In other words, a good deal of parallel development in isolated cultures has to be expected, especially in the earlier stages.

At the intellectual level the analogue of these simple compulsions exists, perhaps, in certain examples which we have already discussed, or shall before long, e.g. the contemporaneous appearance of theories of the type of the 'ladder of souls' in Aristotle and Hsün Tzu,[c] and the simultaneity of the Eleatic paradoxes with those of Hui Shih.[d] Moreover, since the heavens are the same from whatever part of the world they may be observed, it is not perhaps so very extraordinary to find a parallel march of astronomical knowledge at the two ends of the Old World. Chatley (9) has emphasised many correspondences. Just about the time, he says, when eclipses began

[a] (1), vol. 2, pp. 321 ff. [b] (1), vol. 2, p. 464.

[c] See Sect. 9*e* below. The properties of plants and animals are the same everywhere.

[d] See Sect. 11*c* below. Continuity and discontinuity are universal concepts. The suggestion has already been made (p. 154 above) that atomic theories could have arisen quite independently in India, Greece and China.

to be regularly recorded in Babylonia (Nabonassar, −8th century), we find them carefully noted in Chinese writings. While the Pythagorean school flourished (−600 to −300) the scholars and diviners of China were developing the *I Ching* (Book of Changes)[a] into a universal repository of concepts which included tables of antinomies (Yin and Yang) and a cosmic numerology; all this was systematised in the Han. While Naburiannu and Timocharis were beginning to make observations of star positions in Babylonia and Greece (all lost), Shih Shen and Kan Tê were registering their own in China (and these are still preserved).[b] The soli-lunar cycles of Meton and Callippos appear in China under different names at the same time, where indeed they may be much older, but quite probably the Greeks derived them from the Babylonians also.[c] The cycles of Plato and Berossos have their analogues in China from −300 onwards at least. After Chang Chhien's time, and contemporary with the first applications of water-power known to us in China, between the Chi-Pin embassies and the visit of Kan Ying to the borders of Iraq and Roman Syria, Liu Hsin produced (+25) his *Three Sequences Calendar*, which far excelled, as Chatley says, all predecessors for accuracy and system, and antedated Ptolemy's *Almagest* by more than a century. Yet rules and tables for the periods of the sun, moon and planets were bound to be closely similar in whatever part of the world they originated; so that originality and independence necessarily reduce to the rather technical assessment of the degree of precision attained, and this may vary according to the assumptions adopted, so that even priority may be hard to establish. A good case of this kind is the history of the successive attempts in East and West to calculate the exact value of π; for many centuries the Chinese were far ahead on this, but what was the point of it?

(2) Simplicity and Complexity

Few will contest that in social evolution, just as in biological evolution, there has occurred a passage from the simple to the complex and from the lesser to the higher degree of organisation. It will also be readily admitted that the simpler the idea, invention, technique or machine, the more likely it may be that it originated independently in different parts of the world. But this principle is surely conditioned by the degree of complexity which the culture producing the invention has attained. The most ancient techniques were naturally all comparatively simple, and yet, as is well known, the most probable supposition is that they originated only in one place, radiating outwards in all directions from there. Techniques subsequently invented were naturally more complicated, and in due course it becomes almost impossible to believe that any of them could have been invented more than once. But at the same time the simpler techniques were still needed and used in these more developed cultures, and there is now less difficulty in imagining that they could have sprung up independently; for the degree of sophistication of mankind (if the expression is

[a] See below, Sect. 13*g*.
[b] See below, Sect. 20*f*.
[c] See below, Sect. 20*h*.

permissible) was continually growing, and skills, both manual and mental, were becoming ever subtler and more widespread. The logical conclusion of this argument would be that in recent times, after the rise of modern science and technology, simultaneous discoveries and inventions of a very high order of sophistication would be likely to occur. This is exactly what has happened, as appears from the study made by Ogburn,[a] who has listed a number of such correspondences.[b] To take a concrete illustration, the simple combination of gear-wheels required to make a hodometer seems not very likely to have been invented twice in the −3rd century, but today, if it did not already exist, it could be put together by any young mechanic in almost any country in the world. In sum, the degrees of complexity of inventions likely to diffuse, should follow up the rising curve of general cultural attainment, simple at first and then increasingly complicated. And they should be followed at a due (and probably narrowing) interval by the degrees of complexity of inventions likely to prove capable of being made independently.

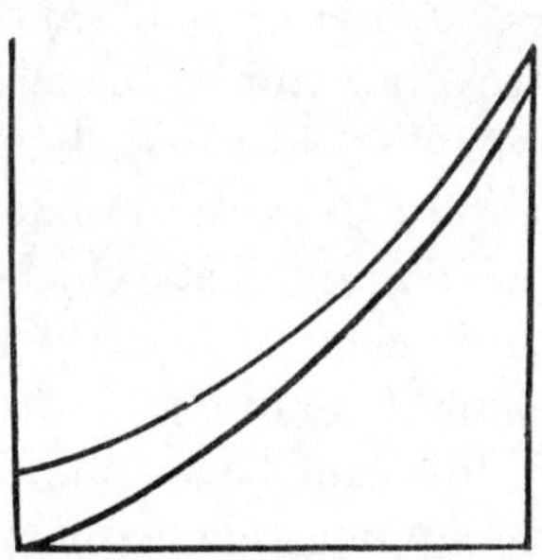

Indeed, the generally accepted opinion, as set forth, for example, in the books of Gordon Childe, and outlined in an elegant passage by Vidal de la Blache,[c] is that all the oldest and most fundamental inventions, such as those of fire, the wheel, the plough, textile weaving, the domestication of animals and so on, can only be supposed to have been made at one centre and to have spread outwards from there. The earliest civilisations of the Mesopotamian basin are held largely responsible. The maps of Bishop (2), one of which, showing the area which had the use of bronze in antiquity, is here reproduced (Fig. 11), well illustrate this. It is certainly very difficult to believe that bronze metallurgy can have been invented more than once. But when one comes to much later dates, e.g. during the +1st millennium, one feels the same certainty about more complex inventions, for instance the rotary quern, the water-wheel, the windmill, the draw-loom, the magnetic compass and the camera obscura. A dual origin of any of these can hardly be envisaged.

[a] (2), especially pp. 90 ff. See also the interesting paper of Harrison (2).

[b] One which occurs to me as I write is the first successful reception of radio-wave echoes from the moon, which was accomplished quite independently in America and in Hungary during the second world war. While this section was being prepared, I was much impressed by the remark of a scientific colleague. During an after-dinner discussion on the principles which should govern the release of confidential scientific documents, someone suggested that they should be 'declassified' after fifty years. He then said that that would help nobody, since 'everything which had not, within twenty years, been discovered again by other scientific workers, would probably be either unimportant or wrong'. The contrast between this conception and that of indispensable diffusion, apparently required for earlier ages, was quite startling.

[c] (1), p. 285.

(3) Priority and Transmission

The historian of science and technology who wishes to evaluate the contributions of the different cultures cannot therefore avoid the seemingly endless task of trying to establish clear priorities. His conclusions, like those of the natural sciences themselves, are ever subject to revision. The arch, once thought to be a comparatively late Etruscan invention, now turns out to have been familiar to the Sumerians.[a] Techniques of medical gymnastics, generally supposed to have been a gift of 18th-century Sweden to Europe, can be shown to arise in fact directly from the practices of early medieval Taoists in China.[b] The gimbals of the navigator, and still more important the mounting of the gyroscope of the automatic navigator, go back, not only to the charcoal-stoves which warmed the hands of medieval prelates in chilly cathedrals, but to the self-righting perfume-burners with which a +2nd-century Chinese mechanic furnished the beds of the princes and officials of Chhang-an.[c] The automatic clock-drive of the astronomical telescope first appears, not as usually believed, in early 19th-century Europe, but in +2nd-century China;[d] and its equatorial mounting was first made, not in the workshops of Uraniborg or Vienna, but in Mongol Khanbaliq, though Kuo Shou-Ching in +1276 had nothing more than a sighting-tube with cross-wires to put in it.

In some of these cases there is nothing whatever to suggest that there was any connection between the inventions. When Fraunhofer invented his clock-drive in 1824 he certainly did not know that the Chinese had for many centuries caused their equatorial armillary spheres to rotate by water-power, for true Chinese astronomy had gone down into a temporary limbo after the Jesuits had 'reformed' it, and practically nothing was known of these machines until the work of modern historians.[e] In other cases, other circumstances prevail. Sometimes so long a period elapses between two events that there must have been many opportunities of transmission, yet we have no evidence of it. Sometimes the period is very short, yet we have clear evidence that a transmission took place. For instance, the first great segmental arch bridge was constructed by Li Chhun in China shortly after +600, but no such structures were built elsewhere until Italy followed with several of the kind shortly after +1300, and the design has of course flourished since.[f] Here the dates correspond well with reports which the Italian travellers in Mongol China could have brought back. And again we would like to know what those Chinese engineers on the Tigris in Mongol Iraq were doing in the +13th century. But there is no hint; conviction arises by implication only. Similarly, the first suspension-bridges with iron chains

[a] See Sect. 28*d* below.
[b] See Sects. 10*i*, 44 below.
[c] See Sect. 27*c* below.
[d] See Sect. 20*g* below.
[e] Of course one can never absolutely exclude that Fraunhofer might have picked up the old volumes of Gaubil (like the Cambridge Observatory, most observatory libraries possess them), and read what seemed to be a legend about turning armillary spheres. He might then very likely have forgotten that he had ever derived his idea from that source.
[f] See Sect. 28*e* below.

were constructed in China at least as early as the +6th century,[a] yet though they had many successors in that part of the world, especially among the Tibetans and other Himalayan peoples, the principle was not suggested in Europe until the end of the +16th and the first bridge not built until the end of the +18th. In this case there is some reason to suspect conscious imitation, though no proof. Elsewhere we know of transmission, though the intervening period was very short. Take the parachute. De la Loubère saw this used by acrobats in Siam in +1688, and his description was read a century later by Lenormand, who made some successful experiments and introduced the device to Montgolfier.[b] This is not to deny that the idea of the parachute had been proposed in Europe at the time of the Renaissance, but there are Asian references to it much earlier still.

In some instances, as for paper,[c] we have a veritable time-table, constructed by Carter, for its travel from +2nd-century China to +13th-century Europe. Similar charts could now be drawn up both for printing[d] and gunpowder.[e] Finally, we sometimes meet with extraordinary temporal coincidences. Thus the development of quantitative cartography in China takes its rise just about the time when Ptolemy finished his work in Europe, yet there is not a trace of concrete evidence that Phei Hsiu had ever heard of attempts in other parts of the world to lay a network of coordinates over the earth's surface so as to determine as correctly as possible the positions of particular places. This suggests a kind of diffusion different from what we have so far considered, and to which we shall presently return.

(4) Simultaneous Developments

The simultaneities are sometimes quite troubling. Some ideas, some sciences, and some complicated inventions have a disagreeable habit of appearing (and even disappearing) almost at the same time at both ends of the Old World. At the beginning of this section we noted the remarkable case of contemporaneous −4th-century knowledge of the lunar periodicity in the reproduction of marine invertebrates; and also the curious parallelism in the rise and fall of anatomical dissection. The explanation just suggested for the simultaneous appearance of theories of the type of the 'ladder of souls', and for philosophical paradoxes of the Eleatic sort, will not so easily cover examples such as these.

And then there are the mechanical inventions. Toothed wheels, with the many applications of gearing which immediately result, appear and flourish almost exactly contemporaneously in Hellenistic and Chinese civilisation (−2nd to +1st centuries).

[a] See Sect. 28*e* below.
[b] See Sect. 27*j* below.
[c] See Sect. 32 below.
[d] See Sect. 32 below. There is no dispute that China has the priority also for the invention of movable-type printing in the late +11th century, but it did not become general there since the large number of characters required in the fount made it simpler to continue to use wood blocks, which amounted to 'going directly into stereotype'. The first Chinese type faces were made of earthenware; later, the Koreans, nearly a century before Gutenberg, used founts of Chinese characters cast in bronze.
[e] See Sects. 30, 33 below.

Correspondingly, the hodometer appears in Europe between the −1st and +1st centuries, and in China at some time between the −2nd and +2nd centuries.[a] For the water-wheel (i.e. the ex-aqueous wheel, as we shall later call it, deriving power from the flow or fall of water), the present position is that we find it on the north coast of Asia Minor about −60, and in China about +20, but since in the latter case the wheel was used not to grind cereal grain but to operate metallurgical bellows by a necessarily complicated system, the pre-existence of at least several decades of trial and use must be admitted. Hence we may say that the appearance of the water-wheel occurred simultaneously at both ends of the Old World.[b] There is here, however, a complication in that we may be dealing with two separate inventions, the horizontally mounted type of water-wheel and the vertically mounted type. Though we do not know for certain what was the nature of the earliest 'European' water-wheel, the vertical one—which came to be known as the Vitruvian—became common later, while in China, on the contrary, all the evidence goes to show that the earliest water-wheels were horizontally mounted.[c] Since the vertical type may have been a derivative of the noria (peripheral pot water-raising wheel) and hence probably of Indian origin, while the horizontal type originated rather as an extension of the rotary quern, it may conceivably be that some of the problems of simultaneous invention may ultimately be resolved by finding that we are really dealing with two inventions and not one. Still, it is striking that the period of the inventions of the water-wheel coincides with Chinese-Greek contacts in Central Asia.

This type of explanation, however, shows no sign of accounting for the strange circumstance in the history of optics, that the camera obscura appears both among the Arabs and the Chinese in the +10th century.[d] In cases such as this, the only hope remaining is that the progress of research will redeem the inadequacy of our present information, and give priority to one side or the other, or conceivably to some intermediate focal area.

We always come back to the same questions. Who was talking about gear-wheels in −1st-century Bactria? Did the Roman-Syrian merchant Chhin Lun, who visited China in +226, happen to take an active interest in cartography? Could a copy of a tractate of al-Haitham's on optics have reached Canton or Hangchow during his lifetime? What drawings of textile machinery could have been found in the baggage of Polo or Pegolotti?

[a] See Sect. 27*c* below.
[b] See Sect. 27*f* below.
[c] It will later be shown that this distinction runs through the whole of Chinese and Western ancient and medieval engineering. Chinese technicians always tended to mount wheels horizontally, and Western technicians vertically.
[d] See Sect. 26*g* below.

(5) The Watcher of the Tides and the Man from Chhin

Such considerations endow with vivid interest the few scraps of information which we have about scientific men and technologists in the Central Asian sphere. First, it should be remembered that Hellenistic science reached further east than is often realised. The Greek cities in Parthia (An-Hsi) produced a remarkable astronomer in Seleucos the Chaldean,[a] a native of Seleuceia-on-the-Erythraean-Sea,[b] whose *floruit* was in the neighbourhood of −140. This place was somewhere on the north-east coast of the Persian Gulf. He spent his life, says Tarn (1), maintaining, against Hipparchus, the theory of Aristarchus of Samos that the earth goes round the sun, and explained the tides in the Gulf by the resistance opposed by the moon to the diurnal rotation of the atmosphere. He discovered periodical inequalities in the tides which he connected with the position of the moon in the zodiac.[c] As we shall later find,[d] the dependence of the sea tides upon the moon was distinctly stated by Wang Chhung about +80, and the possibility naturally presents itself that some breath of the work of Seleucos might have reached him. It would have been the bare idea, doubtless, with no name attached to it, but this can only be pure speculation, and, moreover, in China the idea had been 'in the air' earlier.

That such ideas might travel rather fast is suggested by the undeniable fact that the only approximately correct account of silk and the silkworm in all classical Western literature was given by Pausanias (d. +180) in his *Description of Greece*, i.e. exactly at the time when the merchants of the 'embassy' from Marcus Aurelius to China (+166) must have just returned.[e] After him, down to the +12th century, writers continued to repeat all the old nonsense about silk fibres growing like a kind of wool on trees, or being spun from multicoloured wild flowers.[f]

That Greek Bactria was in fairly close contact with the Hellenistic world is shown by many evidences. Propertius (−50 to −10) has a poem about a girl at Alexandria called Arethusa whose husband was twice in Bactria (and stayed too long there) and had 'seen the Seres'.[g] Dio Chrysostom the rhetorician (b. +50) describes the Bactrian merchants frequenting the commercial quarter of Alexandria.[h] Great interest therefore attaches to a curious incident which took place during the campaigns between the Chinese and the people of Ferghana towards the end of the −2nd century. Details will be given later on[i] about the efforts of the Chinese to improve their breed of horses by importing chargers from Ta-Yuan (Ferghana); this was one of Han Wu Ti's

[a] Tarn (1), p. 43, thinks that if he had really been a Chaldean, the Greek tradition would have preserved his original name, as it did for Kidinnu, Naburiannu and Berossos. Apparently the word 'Chaldean' was a title of honour assumed by Greek astronomers who had studied in the schools of Babylonia.

[b] The Erythraean Sea, of course, as in the *Periplus* later, meant the Persian Gulf and the Indian Ocean as well as what we now call the Red Sea.

[c] Cf. Sarton (1), vol. 1, p. 183.

[d] Below, Sect. 21 *i*.

[e] This was first pointed out by Reinaud (2), p. 185.

[f] See below, Sect. 31.

[g] *Elegies*, IV, 3.

[h] *Discourses*, 32.

[i] See Sect. 30 below.

major interests. After the death of Chang Chhien in −114 he obtained information about the excellence of these horses and sent one mission after another to purchase them, but with limited success. Then some young men who had been attached to one of the missions informed him that there were particularly fine horses kept in the city[a] of Erh-Shih[1] by a prince called Mu Kua,[2] whereupon he sent an envoy to obtain them, but the prince thought himself sufficiently far from China to be safe, and murdered the envoy. Several fruitless expeditions across the Tarim basin (Sinkiang) followed, but in −101, after Li Kuang-Li[3] had been appointed to command the expeditionary force, an army of 30,000 men did succeed in reaching Erh-Shih and captured the outer city.

It was this siege which involved matters of technical interest. The *Shih Chi* says:

In the city of the king of Ta-Yuan there were no wells, and the people had to obtain water from a river outside. Experts in hydraulic engineering (*shui kung*[4]) were therefore despatched to divert the course of the river so as to deprive the city of water, or to effect an opening (by sapping the foundations of the walls) through which the city might be laid open of access.[b]

In due course the water-supply was in fact cut off, and the Ferghanese, who had retired to the inner city, assassinated Mu Kua and proclaimed their willingness to give up the horses. Li Kuang-Li, debating with his advisers whether to accept this offer or not, had to take the following facts into consideration:

It was reported that the (Ta-)Yuan (people) had recently secured the services of a man of Chhin who knew how to bore wells (*Chhin jen chih chhuan ching*[5]); and that the city was still well supplied with provisions....[c]

Eventually the terms were accepted, and General Li and his army left with the horses.

Everything here depends upon what Ssuma Chhien meant by the 'Man from Chhin'. Tarn,[d] in a judicious discussion, has declined on various grounds to accept the usual view that the well-boring engineer was a Chinese deserter. By a curious coincidence, deserters are mentioned in an adjacent passage in the same chapter; the expression used is different, though what they did is incidentally of much interest to us in itself. Ssuma Chhien says, talking of the people occupying the tracts from Ta-Yuan westwards as far as the country of An-Hsi:

These countries produced no silk or varnish (lacquer), and they did not know the casting (*chu*[6]) of iron utensils.[e]...When some deserters from the retinue of a Chinese embassy

[a] Perhaps modern Urutepe.
[b] Ch. 123, p. 17*b* (tr. Hirth (2), mod.).
[c] *Shih Chi*, ch. 123, p. 19*a* (tr. Hirth, 2).
[d] (1) pp. 309 ff.
[e] Some texts have money (*chhien*[7]) instead of iron (*thieh*[8]) here, but the text seems to read more smoothly if the latter is preferred. *Chhien Han Shu*, ch. 96A, p. 18*b*, the same passage, always has 'iron'.

1 貳師 2 毋寡 3 李廣利 4 水工 5 秦人知穿井 6 鑄
7 錢 8 鐵

(*Han shih wang tsu*[1]) had settled there, they taught them to cast weapons and utensils. And when they secured Chinese yellow and white metal (gold and silver) they used it for utensils and not for coining money.[a]

Here then pejorative words are used which are not applied to the man from Chhin. If he was not from Chhin in the sense of China, then he came from Ta-Chhin, which, as we saw earlier on, meant Roman Syria. Tarn, indeed, discounts the possibility that there could have been a Syrian engineer in a Bactrian frontier-State at the time of the final collapse of the Seleucid power, and prefers the view that he was a Bactrian Greek of some kind, since there are authorities for believing that the term Ta-Chhin was sometimes loosely applied to include Bactrian regions. What was apparently unknown to Tarn, however, was the evidence which we shall later be able to adduce[b] that the art of boring deep wells[c] goes back to the Han period in Szechuan, and −100 would not seem too early for the appearance of an expert in it. Moreover, Szechuan, the home of the technique, had been one of the most important parts of the Chhin State before the First Unification. Perhaps therefore he was Chinese after all. The mystery remains—but from the point of view of the present argument, the most interesting possibility would of course be that his origin was Syrian. Whatever sort of Greek he was, he could have discussed Ctesibios and Aristotle with the Chinese metallurgical deserters, even if only through an interpreter. We should remember them when considering the spread of the invention of cast iron.[d]

Greek skill in well-digging is attested by the author of the *Periplus*,[e] who was impressed by those at Barygaza, which were supposed to go back to the time of Alexander the Great. It is true that Alexander had a well-boring engineer with him;[f] we even know his name[g]—but the Barygaza wells were certainly not so old, and dated from the conquests of the Bactrian Euthydemid Apollodotus in the region,[h] about −177. In Persia the Seleucid Greeks had continued to keep up the old irrigation systems, which included those underground tunnels marked by a line of well-shafts which are still used to this day (*qanāt* or *kārīz*; the *hyponomoi* (ὑπόνομοι) of

[a] *Shih Chi*, ch. 123, p. 15*b* (tr. Hirth, 2).

[b] See Sect. 37 below.

[c] I mean here specifically bore-holes of the artesian type and not simply substantial vertical shafts lined with masonry.

[d] A still earlier spread of it is to be surmised by the fact that according to Pan Ku the people of Kucha were good at founding and casting (*Chhien Han Shu*, ch. 96B, p. 9*a*).

[e] § 41.

[f] And numerous other engineering and technical specialists (Tarn (2), vol. 1, p. 12; vol. 2, p. 39). There was Diades of Thessaly, expert in siege machines; Deinocrates the town-planner; bematists (surveyors and geographers) such as Bacton, Diognetos and Philonides. Nearchos and Onesicratos were nautical experts. The architect and engineer Aristobulos was a particular friend of Alexander's and afterwards wrote about him. But all this happened before the great period of the study of mechanics by the Alexandrians, so that eastward transmission of their knowledge by these men cannot have taken place.

[g] Gorgos the Metal-Worker (Γόργος ὁ μεταλλευτής), Strabo XV, 700; Tarn (1), p. 149.

[h] Tarn (1), p. 148.

[1] 漢使亡卒

Polybius).[a] It is thus of particular interest that the *Shih Chi* contains[b] an account of an elaborate engineering work constructed in China by Chuang Hsiung-Phi[1] in −120 (the 'Dragon-Head Canal') which involved tunnels and well-shafts at regular intervals.[c] This was twenty years before the hydraulic engineers of Li Kuang-Li found themselves in the army besieging Erh-Shih, but the resemblance with Persian methods does seem to indicate some contacts during the −2nd century. The *qanāt* system, however, did not become widespread in China, presumably because the local conditions did not call for it. Obviously there is a great deal that we do not yet know about the beginnings of hydraulic engineering and the mutual borrowings of Persian, Greek and Chinese technologists.

The story of the deserters reminds us of another episode in Central Asia eight centuries later, when after the Battle of the Talas River (+751), Chinese paper-makers were captured and induced to continue their craft at Samarqand.[d] This is known to have been the earliest passage of paper-making from East to West, and it occurred no less than six hundred years after the first invention. There does not seem to be any record of the names of these men, but by a fortunate chance certain details have been preserved of other technicians also captured in the same battle. One of the prisoners taken was an officer, Tu Huan,[2] and when he returned to China eleven years later he told his family about the Chinese artisans who had settled at the Abbasid capital (Kūfah). It so happened that his brother was the great scholar Tu Yu,[3] and the information was thus written down. In Tu Yu's *Thung Tien*, we find:[e]

> As for the weavers who make light silks (in the Arab capital), the goldsmiths (who work) gold and silver (there), and the painters; (the arts which they practise) were started by Chinese technicians. For example—for painting Fan Shu[4] and Liu Tzhu[5] from the capital (Sian), and for silk throwing and weaving (*chi lo*[6]) Yüeh Huan[7] and Lü Li[8] from Shensi.

The importance of this goes far beyond the well-known mutual influences of styles in Chinese and Muslim art (cf. Hornblower, 1). It gives us confidence that if we find good reason to suspect a transmission of some device or technique westwards in the +8th century, we need not hesitate to believe in the existence of men capable of transmitting it, and circumstances in which they could do so.

By a strange coincidence, it had been also on the Talas River, in −36, that a Chinese army, victoriously storming the fortified capital of a Hun leader (Shanyu), had come upon some very peculiar mercenary soldiers. The punitive expedition which thus killed the Shanyu (who, after the murder of a Chinese diplomatic envoy seven years

[a] x, 28, 3ff. The fact that Polybius mentions them must mean that they go back well beyond −146. Cf. Stein (8); B. Fisher (1); A. Smith (1).

[b] Ch. 29, p. 5*b*.

[c] See below, Sect. 28*f*.

[d] Carter (1), p. 97.

[e] *Thung Tien*, chs. 191–3, tr. Pelliot (32), eng. auct.; parallel passages in *Thai-Phing Huan Yü Chi*, ch. 186, p. 15*b*; *Wên Hsien Thung Khao*, ch. 339, p. 7*a*.

1 莊熊羆 2 杜環 3 杜佑 4 樊淑 5 劉泚 6 機絡
7 樂隈 8 呂禮

earlier, had settled in Eastern Sogdiana) was commanded by Kan Yen-Shou,[1] Protector-General of Central Asia, and his assistant, Chhen Thang.[2] According to the careful description of the battle in the biography of the former, the Chinese, upon setting up their camp and beginning the siege of the city, could see from afar 'more than a hundred foot-soldiers, lined up on either side of the gate in a fish-scale formation (*yü lin chhen*[3]), practising military drill....'[a] After the sack of the place, 145 soldiers were captured alive.

It was at first thought (Dubs, 28) that this was the Macedonian phalanx, and that the soldiers were Greek mercenaries, but this seems both historically and technically[b] impossible. The singularity of the expression, however, for which Chinese literature affords no exact parallel, has led to the view that the soldiers were no other than Roman legionaries, exercising in the *testudo* formation of interlocked shields.[c] It is accordingly urged (Dubs, 6, 29) that they were the remnants of the captives made by the Parthians on the occasion of their great victory over Crassus at Carrhae in −54.[d] All this may be accepted, but it is not certain that we are right in identifying the captured soldiers with the Romans seen drilling before the fight. One argument in favour of doing so has been provided by Dubs (30), who points out that in +5 there was a city in Kansu province, newly-founded, called Li-kan[4] or Li-chien,[4] its name being thus almost identical with that which, as we saw above (p. 173), was applied to Media, Syria, and perhaps all the eastern Roman empire. Still more remarkable, the name was changed by Wang Mang in +9 to Chieh-lu,[5] which at any rate could mean 'prisoners captured in the storming of a city'. The evidence points, therefore, to a settlement of the remaining Romans as a military colony on the Old Silk Road, where they married Chinese women and spent the rest of their days.[e]

Perhaps we should not expect too much of these soldiers, but if any of them were at all skilled in military engineering, some fragments of knowledge might have been exchanged between the cultures of Rome and China.[f]

[a] *Chhien Han Shu*, ch. 70, pp. 9*b* ff., tr. Duyvendak (16).

[b] Technically, because the shields of the Macedonians were round, and did not overlap like fish-scales, as those of the Roman *testudo*-forming infantry did.

[c] Additional evidence that the soldiers were Romans lies in the fact that the Shanyu's city was defended, not only by earth walls, but by a double palisade of wooden stakes. This was a typical piece of Roman military technique.

[d] It is true that in the twenty years between the battles, the Romans had 1500 miles to move as far east as Antioch in Margiana, but the pace seems rather slow. One wonders, therefore, whether the men whom the Chinese saw, and perhaps captured, were not Persians of some kind who had been trained by Romans after Carrhae.

[e] The colony can be identified as having been near Yungchhang,[6] a *hsien* city on the motor-road of the present day (about half-way between Chiu-chhüan and Lanchow). Though it does not count for much, I cannot help remarking that when I passed that way in November 1943, I wrote in my log-book 'Country people in the Yungchhang upper valley have very red cheeks.' This is most unusual in a Chinese population. Yet it is hard to believe that any local stock could have perpetuated its physical characters 'under the drums and tramplings' of twenty centuries.

[f] The whole story is of interest both in connection with military technology (cf. Sect. 30 below), and map-making (cf. Sect. 22*d* below). In the latter place reasons will be given for viewing without enthusiasm a concrete instance of culture transmission proposed by Dubs (30).

[1] 甘延壽 [2] 陳湯 [3] 魚鱗陣 [4] 驪靬 [5] 揭虜 [6] 永昌

Literary analogues of the 'Man from Chhin' (if he was really from anywhere west of Bactria) are still to seek. The books of Indian mathematics and astronomy with the 'Brahmin' titles, in the Sui and Thang, we have already noted, and of course there were many Buddhist sūtras with medical and astronomical-astrological content,[a] but apparently texts from the West never got through. The only story of such a transmission turns out on examination[b] to be baseless. Yet further work may reveal something; for example, a Nestorian bibliography found at Tunhuang contains a title *Hun Yuan Ching*,[1] which one might venture to translate 'On the First Cause of the Universe'.[c] But if the Nestorians and Manichees brought anything, it was probably all theology.

(6) Spreading Technique and Localised Science; the Acceptance or Rejection of the Strange

So far we have considered chiefly simplicity and complexity as criteria for the probability of diffusion of an idea or a technique at different stages of civilisation. But we must remember the distinction between scientific theories and observations as such, on the one hand, and technological inventions on the other. One can only by courtesy call the invention of the wheel, for example, applied science, since the science was hardly there beforehand to be applied. The arguments about the value of π seem to come long after the successful attempts of wood-workers to make wheels—judging at least from Chinese evidence. I am far from suggesting that the history of science and the history of technology do not have to be written together, but when we come to consider travel, was there not perhaps a kind of filtration process? Was it not the inventions of immediate practical use which tended to travel, rather than the scientific and pre-scientific observations, speculations and theories? Would it not therefore be in the field of technology that diffusion would be supreme,[d] while in that of scientific thought and observation we might expect to find quite often independent development or convergence?

For techniques, a sort of catch-and-ratchet mechanism perhaps operated. 'People may accept', said Leroi-Gourhan,[e] 'a language less subtle than what they had before, or a less developed religion, but except for temporary regressions due to war's devastation, they never go back from the plough to the hoe.' Moreover, there is some *a priori* improbability in the diffusion of whole systems of ideas, which tended no doubt to be limited by specific ethnic characteristics, and the travel of fragments of

[a] Pp. 128, 211 above; see also Sect. 19*c* below. [b] Sect. 20*a*.

[c] Saeki (1), pp. 67 ff.

[d] P. F. Cressey (1) has partly seen this, for he emphasises that 'during the centuries of limited and indirect contact, only items of a material nature reached Europe. They came as isolated traits, divorced from any related complex of Chinese usage or custom.' This he is easily able to illustrate—gunpowder but no temple fire-crackers; paper but no paper money. Cf. pp. 154, 222 above.

[e] (1), vol. 2, p. 322.

[1] 渾元經

these idea-systems would rather be expected. There was certainly no all-or-none process at work.

Definite limits to the validity of this point of view there certainly are, for we must accept such powerful evidence as that assembled by Filliozat (1), which shows the close resemblance of Greek pneumatic medicine to ancient Indian medicine, and this has parallels astonishingly close[a] in the ancient Chinese physiological concepts of *chhi*[1] and *fêng*,[2] so that all might find a Mesopotamian origin. Besides, Babylonian astrological ideas travelled to China,[b] and Babylonian knowledge of the arithmetic cycle of musical notes travelled both east and west,[c] while the only plausible explanation of the prominence of the lunar mansions in several Asian astronomical systems is a Babylonian origin.[d] But what travelled in this way seems to have been the exception rather than the rule. Broadly speaking, Chinese science, for two millennia before the coming of the Jesuits, and in spite of opportunities of intellectual intercourse much greater than has often been pictured, had very little in common with that of the West—hence the interest of the subject of the present book. But Chinese technological inventions poured into Europe in a continuous stream during the first thirteen centuries of the Christian era, just as later on the technological current flowed the other way. This is now at last achieving a measure of recognition, as can be seen in such writings as those of Lynn White, C. Stephenson, Kroeber (3) and Thorndike (8).

It seems desirable here to give concrete examples of this, but first a few words may be devoted to the opposite phenomenon, the rejection by one culture of loan material from another—a question which has greatly interested ethnologists. The fact must be faced that by the time of the Thang, for example (+7th to +10th centuries), the patterns of Chinese and European thinking had become so fixed that the acceptance of any new element from outside was quite difficult if not impossible, although, paradoxically, as we have seen above, contact between China, Iran, Islam and India was then particularly close. Chinese medicine seems to have gone on its way largely independent of outside influences. Similarly, in the Yuan time, in the +13th century, when Jamāl al-Dīn came with his astronomical mission from Persia to Peking, the direct effect upon Chinese astronomical practice seems to have been nil,[e] for the systems were too different, though indirectly it stimulated Kuo Shou-Ching to his great invention of the equatorial mounting. Conversely, the Chinese had been observing and registering sun-spots since the −1st century, but even if any hint of this had reached medieval Europe, imitation or extension of the observations would not have been feasible, since in European ideas the sun was a perfect body and by

[a] Cf. Sects. 43, 44. [b] Cf. Sect. 14*a*. [c] Cf. Sect. 26*h*. [d] Cf. Sect. 20*e*.

[e] See Sect. 20*g* below. For example, the astrolabe was never adopted by the Chinese. Perhaps this was because they were not interested in the interconversion of altazimuth and ecliptic coordinates. The astrolabe was a manifestation of two characteristically European developments, Euclidean geometry and Ptolemaic planetary theory. The Chinese preferred algebraic methods and less speculation about celestial motions.

[1] 氣 [2] 風

definition could not have spots.[a] Still more important would have been the acceptance of the Chinese conception that the celestial bodies floated freely in a vast inane, or were driven by winds along their regular courses in empty space, but no medieval European mind would have been prepared to give up the belief in solid concentric crystalline spheres, with all the theological implications of geocentrism; and, indeed, when the Jesuits first learnt of these old Chinese theories of the emptiness of the heavens, they set them down as yet another of the absurdities of the mandarins and bonzes.[b] Instances could be multiplied indefinitely, from the question of the fixity of biological species[c] to the ideas of Plutonian geology,[d] but the point needs no labouring; every civilisation seems to have to come eventually to truth by the hard road, and often in history they could give each other very little help. Like Odoric and his Buddhist friend, they could only agree to differ, but Odoric's side had no monopoly of wisdom.

(7) The Westward Flow of Techniques

Without trespassing too much on the detailed treatment which will be given in succeeding volumes of this work, here are a few of the things which may be said about the transmission of mechanical and other techniques. A few fundamental ones diffused in all directions from ancient Mesopotamia, e.g. the wheeled vehicle, the windlass and the pulley. Ancient Egypt was responsible for the swape (counterbalanced bailing bucket), and perhaps also for the crank, though this is not certain. From somewhere in the Fertile Crescent, or possibly from farther west, came the rotary quern. The fundamentals of the locksmith's art took shape early in Babylonia and Egypt and then spread all over the Old World with comparatively little development until the Renaissance. The Indians contributed the flume-beamed swape and perhaps the important invention of the noria. The only Persian invention of the first rank was the windmill, which found its way out, in strangely differing forms, all over the world. Unless the rotary quern be attributed to them, the ancient Europeans of the Mediterranean basin launched only one valuable mechanical technique, namely, the pot chain pump, later so characteristic of Arabic lands as the *sāqīyah*—though, of course, in the persons of the Alexandrian theoreticians they went further than all others in the classification and description of machines. But China produced a profusion of developments which reached Europe and other regions at times varying between the +1st and the +18th centuries: (*a*) the square-pallet chain-pump; (*b*) the edge-runner mill and the application of water-power to it; (*c*) metallurgical blowing-engines operated by water-power; (*d*) the rotary fan and winnowing machine; (*e*) the piston-bellows; (*f*) the horizontal-warp loom (possibly also Indian), and the drawloom; (*g*) silk reeling, twisting and doubling machinery; (*h*) the wheelbarrow; (*i*) the sailing-carriage; (*j*) the wagon-mill; (*k*) the two efficient harnesses for draught-animals, i.e. the breast-strap or postilion

[a] See Sect. 20*i* below. Cf. Kanda (1).
[b] See Sect. 20*j* below.
[c] See Sects. 10*e*, 39 below.
[d] See Sect. 23*b* below.

harness, and the collar harness; (*l*) the cross-bow; (*m*) the kite; (*n*) the helicopter top and the zoetrope; (*o*) the technique of deep drilling; (*p*) the mastery of cast iron; (*q*) the 'Cardan' suspension; (*r*) the segmental arch bridge; (*s*) the iron-chain suspension-bridge; (*t*) canal lock-gates; (*u*) numerous inventions in nautical construction, including water-tight compartments, aerodynamically efficient sails, the fore-and-aft rig, and (*v*) the stern-post rudder; (*w*) gunpowder and some of its associated techniques; (*x*) the magnetic compass, used first for geomancy and then, also by the Chinese, for navigation; (*y*) paper, printing, and movable-type printing; and (*z*) porcelain. I come to a stop, having exhausted the alphabet, but many more instances, even important ones, could be given.

We must not suppose that the last word has been said on any of these developments, nor that adequate evidence exists in all cases to prove conclusively that the later European uses were derived from the earlier Chinese practice. The feature common to all examples is that firm evidence for their use in China antedates, and sometimes long antedates, the best evidence for their appearance in any other part of the world. This is summarised in Table 8.

Many factors operated, no doubt, other than facility of transmission in fixing these figures. Certain water-raising machines, for example, probably failed to travel because the custom of other cultures was to employ different types of machines, no less efficient, and only the omnivorous curiosity of the Renaissance appropriated them. Again, certain successes of Chinese culture, such as the early mastery of cast iron, may well have been due to the nature of the ores employed, permitting fusion at temperatures lower than those possible elsewhere, so that the art could not have been imitated even if its principles had spread. In any case, it is clear that certain favourite ideas of occidental historians will need far-reaching revision in the light of these and similar facts. 'However far it may or may not be possible', wrote Toynbee, 'to trace back our Western mechanical trend towards the origins of our Western history, there is no doubt that a mechanical penchant is as characteristic of the Western civilisation as an aesthetic penchant was of the Hellenic, or a religious penchant was of the Indic and the Hindu.'[a] It is to be feared that all such valuations of East and West are built on insecure foundations.

It may now be interesting to see what engineering principles the West was able to contribute to Chinese civilisation at the time of the Jesuits (+17th century), for which we have abundant information. The only really important mechanical elements which had been missing from Chinese culture were (*a*) the screw, the absence of which complemented the poverty of pedal and treadle motions in ancient European civilisation, though these had been extremely prominent in Chinese technique, e.g. in the square-pallet chain-pump. The screw in its simple form had reached the Chinese earlier through Arab contacts, but the Jesuits brought it in the form of the Archimedean water-raiser. They also brought (*b*) the Ctesibian double force-pump. Cylinder-pumps for liquids had not been characteristic of Chinese engineering tradition, though the

[a] (1), vol. 3, p. 386.

Table 8. *Transmission of mechanical and other techniques from China to the West*

		Approximate lag in centuries
(*a*)	Square-pallet chain-pump	15
(*b*)	Edge-runner mill	13
	Edge-runner mill with application of water-power	9
(*c*)	Metallurgical blowing-engines, water-power	11
(*d*)	Rotary fan and rotary winnowing machine	14
(*e*)	Piston-bellows	*c.* 14
(*f*)	Draw-loom	4
(*g*)	Silk-handling machinery (a form of flyer for laying thread evenly on reels appears in the +11th century, and water-power is applied to spinning mills in the +14th)	3–13
(*h*)	Wheelbarrow	9–10
(*i*)	Sailing-carriage	11
(*j*)	Wagon-mill	12
(*k*)	Efficient harness for draught-animals: Breast-strap (postilion)	8
	Collar	6
(*l*)	Cross-bow (as an individual arm)	13
(*m*)	Kite	*c.* 12
(*n*)	Helicopter top (spun by cord)	14
	Zoetrope (moved by ascending hot-air current)	*c.* 10
(*o*)	Deep drilling	11
(*p*)	Cast iron	10–12
(*q*)	'Cardan' suspension	8–9
(*r*)	Segmental arch bridge	7
(*s*)	Iron-chain suspension-bridge	10–13
(*t*)	Canal lock-gates	7–17
(*u*)	Nautical construction principles	>10
(*v*)	Stern-post rudder	*c.* 4
(*w*)	Gunpowder	5–6
	Gunpowder used as a war technique	4
(*x*)	Magnetic compass (lodestone spoon)	11
	Magnetic compass with needle	4
	Magnetic compass used for navigation	2
(*y*)	Paper	10
	Printing (block)	6
	Printing (movable type)	4
	Printing (metal movable type)	1
(*z*)	Porcelain	11–13

double-acting piston-bellows for air was ancient, and there had been an approximation to the suction lift-pump in the valved buckets used in the deep brine-field borings. Then there was (*c*) the crankshaft, which had not been used in Chinese machinery, though the crank itself had been employed for many centuries, probably since the Han. The last important introduction was (*d*) clockwork, a distinctively European invention of the early +14th century. Thirteen other devices or machines, introduced to China by the Jesuits, were quite redundant, that is to say, they had been known there already in some form or other for many generations. Table 9 gives the delays in the eastward transmissions.

Table 9. *Transmission of mechanical techniques from the West to China*

	Approximate lag in centuries
(*a*) Screw	14
(*b*) Force-pump for liquids	18
(*c*) Crankshaft	3
(*d*) Clockwork	3

Two Chinese devices at least there were which seem never to have passed to any other culture: (*a*) the south-pointing carriage; and (*b*) the repeating, or 'magazine', crossbow and catapult. The first of these has importance as one of the ancestors of all cybernetic machines, but presumably meant nothing to other peoples because they were not interested in the magic prestige of a vehicle with a figure which could keep pointing south irrespective of the course traversed. The failure of the second to spread is more curious, since the portable crossbow was mass-produced as the standard weapon of the Han armies.

A special category may be needed for certain techniques which the late medieval West knew were used in China and yet did not adopt. P. F. Cressey has instanced (*a*) water-tight compartments in shipbuilding; (*b*) paper money; and (*c*) the use of coal. Knowledge of these things came at least as early as the time of Marco Polo, but none of them fitted into contemporary European patterns, and sometimes it can be shown rather clearly why they did not do so.[a] Thus we always have to take into account not only the time required for knowledge of a technique to travel from one civilisation to another, but also the latent period which may follow before its practical adoption. Roughly speaking, for all three examples just given, this latent period was of the order of four centuries.

[a] Cf. Sect. 29*g*. They would be examples of Kroeber's 'ideational germs', p. 247 below.

(8) Stimulus Diffusion

In discussing these possibilities of influence passing from one civilisation to another in ancient and medieval times, it must be realised that the wholesale taking over of idea-systems or patterned structures is not a necessary supposition. A simple hint, a faint suggestion of an idea, might be sufficient to set off a train of development which would lead to roughly similar phenomena in later ages, apparently wholly independent in origin. A man might hear of writing as such, yet have no specimen of what had actually been written, and such an idea might stimulate the elaboration of an entirely new and distinct system of writing.[a] Or the news that some technical process had successfully been accomplished in some far-away part of the world, might encourage certain people to solve the problem anew entirely in their own way.[b] This process has been recognised for some time as an important possibility by scholars such as Chatley (23), Wheeler (2) and Kroeber (2), the last of whom has named it 'stimulus diffusion'.

The simplest thing to do is to give some examples of it (or what may be examples of it) from the material described in later sections of this book. Let us first take deep drilling.

The art of drilling deep wells or bore-holes, such as are used today for exploiting fields of petroleum, is specifically Chinese, for we have much evidence for it going back to the Han period (−1st to +1st centuries) in Szechuan. Furthermore, the method used for so long was essentially the same as that employed in California and Pennsylvania before the application of steam power. In ancient Szechuan, moreover, natural gas issuing from some of the bore-holes was already utilised to evaporate the brine drawn up or flowing from others. No trace of the radiation of the technique into other cultures appears for a millennium, after which there are one or two

[a] This is indeed known to have taken place. A syllabic 'alphabet' was invented for the Cherokee language in 1821 by a man of mixed race, Sequoya (John Gist), who knew no English but was impressed by the advantages of writing (Foreman). Some of the signs were like English letters but others were newly invented. A parallel case was that of Doalu Bukere, who in 1849 devised a syllabary for the Vei language of Liberia (Ellis). Yet a third syllabary was invented by an African prince of the Bamun tribe in the Cameroons in 1899, and this gradually turned into an alphabet (Métraux). The ideographic writings of tribal peoples along the south-western borders of China were devised by men who had a model ready at hand, and the same applies to the inventors of the Hsi-Hsia writing in the middle ages (cf. p. 133 above). More surprising is the case of the Chukcha herdsman, Tynnevil, in the Soviet Arctic, who in 1932 perfected an ideographic system; though this did not spread, it aroused much interest in Russian linguistic circles (Semushkin (1), p. 226). Chatley, Wheeler and Kroeber all suggest the more speculative view that this process, the imitation of 'writing as such', applies also to the spread of written languages in the ancient world, and could account both for the first invention of Chinese ideographs, and for the rise of the many alphabets following the first Phoenician one.

[b] This might cover the case of printing, for no one suggests that Gutenberg ever actually saw a Chinese printed book; what cannot be excluded is that he heard them talked about. But the invention would have been a re-invention and not something with a large element of originality. Tarn (1), p. 373, presents an argument that Tshai Lun's invention of paper was due to the diffusion of a stimulus. Chang Chhien had reported that the Parthians and other Westerners wrote on parchment in horizontal lines (*Shih Chi*, ch. 123), and this may, he thinks, have made the Chinese dissatisfied with silk and bamboo slips.

references in Arab writers, followed shortly by the successful borings of the first artesian wells in Europe in the +12th century. It has not yet been established that the method by which these were drilled was the same as the ancient Chinese method, but since we know of no other before modern times, it almost certainly was. Yet the early +12th century, before the Mongol conquests, was a most unpropitious time for the passage of any idea from China to Europe; it would have had to come through Arab sailors on the sea route, probably through Moorish Spain, and Arab sailors would not have been likely visitors to Szechuan. It seems reasonable to think, therefore, of the transference purely of an idea, indeed rather of an encouragement—'Certain men *have*, for many generations, in Cinasthan, gained wealth from the earth by means of drilled holes, which, with patience, and suitable tools, may be made.' The rest would follow, but in this case the technical answer may have been almost exactly identical. Such was more generally not the case.

The windmill, as has been said, is a Persian invention. But in +8th-century Seistan it was horizontally mounted, and this position it retained when it was introduced to China at the beginning of the Mongol period (the second half of the +13th century) and set to work all along the eastern coast pumping brine into innumerable salterns. From the first, however, the European windmill was mounted vertically, like a Vitruvian water-wheel requiring right-angled gearing; as we may see from the earliest +14th-century illustrations. The suggestion therefore presents itself that somebody (very probably someone associated with the crusaders in the Levant) said to someone else—'Certain Saracens *have*, indeed, harnessed the power of the wind to grind their corn.' Not being at all clear how this had been accomplished, and thinking in terms of vertically mounted wheels, the millwrights, probably in England or northern France, solved the problem in their own way.[a]

A case closely similar might be found in the more abstract realm of quantitative cartography, which we have several times touched upon.[b] Only a few decades after the death of Ptolemy, and just at the time when his meridians and parallels were falling into oblivion (except in a restricted Byzantine milieu), thus bringing to a close the great period of Greek geographical science, Phei Hsiu, at the other end of the Old World, is writing that essay on cartographic technique which firmly established the Chinese system of the rectangular grid, not to be superseded until modern times.[c] Here the basic principle was the same, but the actual system distinctly different, for the Chinese grid made no allowance for the curvature of the earth. The chief difficulty in the supposition of stimulus diffusion in this case lies perhaps in the fact that Chang Hêng, Ptolemy's own contemporary, seems to have been at work already along characteristic Chinese lines, but even so, this might not wholly invalidate it. The travel of individual persons, especially merchants, who might have said: 'In Ta-Chhin, learned men have thought of throwing a criss-cross network of parallel lines over the earth, to fix the positions of places,' at the appropriate time, is not in doubt.

[a] See Sect. 27*i* below.
[b] Pp. 198, 231.
[c] See Sect. 22*d* below.

The most complex and fascinating of all these chains of events is perhaps the history of the paddle-wheel boat. The idea of bladed wheels transmitting power *to* water (ad-aqueous wheels) for the purpose of motion over water, could readily have arisen at any time from the Vitruvian (vertically mounted) water-wheel, or from the current-operated noria. In fact it seems to have been first proposed by the anonymous Byzantine author whose military engineering projects have come down to us from the end of the +4th century, but this almost certainly remained a proposal on paper only. Then in the +6th century the Byzantine general Belisarius mounted corn-mills on moored boats during the siege of Rome by the Goths, and from the +8th century onwards (if not several centuries earlier) there were true paddle-boats in China, operated by treadmills and used time after time for naval purposes on lakes and canals. They have, indeed, continued until the present day, for those who visit Canton harbour and the Pearl River may still see them.[a] It may be, therefore, that at some time in the Thang dynasty or a little earlier the message was received that 'In the domains of Fu-Lin (Eastern Rome) men have seen boats with wheels in motion', whereupon the Chinese engineers, misunderstanding the purpose of the wheels, proceeded to construct true paddle-boats and not merely floating corn-mills. Such a message could well have been transmitted by any of those +7th-century embassies from Byzantium which have been mentioned above. And there may even have been two messages, one of which was not misunderstood, because ship-mills also existed in China, at least from the Sung dynasty onwards, and still do. Then after long years the transmission seems to have occurred again in the opposite direction, for treadmill paddle-boats are depicted in the technological MSS of late medieval Europe (e.g. Guido da Vigevano, +1335; Konrad Kyeser, +1407), though the first record of one actually constructed refers to Blasco de Garay in +1545. The reverse transmission would thus have to be fixed, like so many others, in the time of Marco Polo. And when the first steamships, with their paddle-wheels, churned the waters of the Chinese estuaries, their European crews, encountering the old treadmill junks, were absolutely unable to believe that they could be anything else than crude imitations of their own vessels.

To take one last example, it may well be that metal coinage, when its full story is known, will illustrate the theme of stimulus diffusion.[b] The classical statement is that the marking of small pieces of precious or standard metal with specific dies, or with designs produced in casting, first developed in Lydia, one of the eastern regions of Asia Minor, in the early −7th century.[c] These coins were stamped. And it is agreed that the making of metal money rapidly spread, not only to all the Greek city-States

[a] See Sect. 27*g* below.

[b] As quite a number of the examples given by Kroeber (2) seem to me not to do. Porcelain, for example, can hardly be adduced. There may have been some independence in the European researches of the +16th and +17th centuries which sought to imitate it, but their ultimate results could not but be identical with those which the Chinese had long before obtained. There was therefore no essentially new material in which the social-intellectual-practical pattern could manifest itself. Kroeber (3) took up the question of minted money, but assumed that the Lydian coins had a clear priority in time.

[c] Breasted (1), p. 313, illustrates examples.

but also eastwards to the Persian culture-area. It appears, however, from the recent comprehensive work of Wang Yü-Chhüan (1), that the earliest coined money in China goes back almost to the Shang. The knife coinage (associated with the feudal State of Chhi on the eastern coast) begins in the −9th century, and the spade coinage (associated with the imperial State of Chou) in the −8th.[a] Both these were cast, like all subsequent Chinese metal coinages. To this generalisation there was one exception, however, the development by the State of Chhu in the south-east, in the −8th or −7th century, of small pieces of gold stamped, like the coins of Lydia, with the imprint of a square seal. Disc coins do not appear in China until the period of the Warring States (−4th or −3rd century). Meanwhile, metal money had appeared in India[b] in the Magadha empire (Śaiśunāga and Nanda dynasties) late in the −7th century;[c] the coins were punch-marked, and some of the dies were square. In the light of the discussion at the beginning of this section about the routes across the central parts of the Old World, it seems hard to believe that there could have been much culture-contact between the Chou people, the Indians before Buddha's time, and the Lydians, yet the time-relations remain troubling. For the present, it can only be recorded (*a*) that when the first Lydian coins were struck, there had already been coined (cast) money in China for at least two centuries, (*b*) that the first stamped Indian coins followed the Lydian ones by rather less than a century, and conversely (*c*) that when the first Chinese (cast) disc coins were made, there had been (stamped) Greek disc coins for nearly three centuries.

Such possibilities have been envisaged by Kroeber (2). Stimulus diffusion, he said, 'occurs in situations where a system or pattern as such meets with no resistance to its spread, but there are difficulties with regard to the transmission of the concrete content of the system. In this case, it is the idea of the complex or system which is accepted, but it remains for the receiving culture to develop a new content.' What is really involved, he goes on, in every true example of stimulus diffusion is the birth of a pattern new to the culture in which it develops, though not completely new in human culture. There is historical connection and dependence, but there is also originality, because the pattern is executed in new materials by different methods.[d] Stimulus diffusion might be defined as 'new pattern growth initiated by precedent in a foreign culture'.[e] Moreover, as Kroeber[f] has pointed out, such stimuli might be more likely to slumber in the form of 'ideational germs' perhaps for centuries, till awakened to life by changes of internal environment, possibly brought about by the

[a] This dating, though conservative compared to some views, has still not won universal agreement; for a different opinion see Yang Lien-Shêng (3). But even if the knife and spade coins did not go back beyond the −5th century, the general argument of this paragraph would be unaffected.

[b] See Kosambi (1) on the Taxila hoard, and, of course, Marshall (1).

[c] A Persian stimulus is probable (Tarn (2), vol. 1, p. 86).

[d] The pattern is therefore in the biological sense 'analogous', not 'homologous' (cf. p. 227).

[e] Writing systems have already been mentioned. But of course within literature itself the same process could take place. Tarn (1) argues forcefully that there were such effects in the Graeco-Indian culture-contact zone, as regards drama (p. 384), and as regards the literary *genre* in which kings consult with philosophers (p. 414).

[f] (3), p. 9.

sprouting of other germs, or the conscious imitation of imported products. And another factor which should be borne in mind is the possibility that the effective stimulus might be transmitted not by a more or less garbled message, but by a very small number of human individuals. Aristeas, Kan Ying, Chhin Lun, as we have seen, seem to stand alone, but they may have had more nameless companions and competitors than we realise. The relations of the Old World and the New are outside the scope of this book, but although most Amerindian archaeologists have upheld a kind of Monroe Doctrine about the origins of Mexican, Mayan and Peruvian civilisations, the number of culture traits which they share with the East Asian continental groups is so suggestive that one cannot but wonder whether a few direct stimuli did not reach them across that great waste of waters.

Speaking of the spread of the idea of 'writing as such', Kroeber uses the words 'a stimulus towards an original but *induced* local invention in China'. This is interesting, for the concept of stimulus diffusion in social evolution has an exact parallel in the individual morphogenesis of vertebrates. Modern experimental embryology has shown that when xenoplastic transplantations are made (as between frogs and newts, for instance) a piece of inductor tissue which normally would stimulate the tissue on which it acts to produce horny jaws will induce teeth if teeth are the usual response made by the competent tissue receiving the stimulus. The stimulus is not species-specific, and the responding tissue responds 'according to its own tradition', i.e. its own genetic determinism. So also must it have been many times during the course of social evolution in the history of technology—an idea arrived, and the nature of the response which took place depended upon the genius of the local culture.

This seems all that can usefully be said at this stage about the problem of contacts between the civilisation of China and the other civilisations of the Old World. The time has now come to study as carefully as possible what that civilisation contributed to science and technology. Such a study will, as Robert Hooke wrote, in the words which have been chosen for the opening page of the present book, 'lay open to us an *Empire* of Learning, hitherto only fabulously described', and 'admit us to converse with the best and greatest of that *Empire*, that either are, or ever have been'.

BIBLIOGRAPHIES

A CHINESE BOOKS BEFORE +1800

B CHINESE AND JAPANESE BOOKS AND JOURNAL ARTICLES SINCE +1800

C BOOKS AND JOURNAL ARTICLES IN WESTERN LANGUAGES

For details of the plan of these Bibliographies see pp. 20ff.

In Bibliographies A and B there are two modifications of the Roman alphabetical sequence: transliterated *Chh* comes after all other entries under *Ch*, and transliterated *Hs* comes after all other entries under *H*. Thus *Chhen* comes after *Chung* and *Hsi* comes after *Huai*. This system applies only to the first words of titles. Moreover, where *Chh* and *Hs* occur in words used in Bibliography C, i.e. in a western language context, the normal sequence of the Roman alphabet is observed.

When obsolete or unusual romanisations of Chinese words occur in entries in Bibliography C, they are followed, wherever possible, by the romanisations adopted as standard in the present work. If inserted in the title, these are enclosed in square brackets; if they follow it, in round brackets. When Chinese words or phrases occur romanised according to the Wade-Giles system or related systems, they are assimilated to the system here adopted without indication of any change. Additional notes are added in round brackets. The reference numbers do not necessarily begin with (1), nor are they necessarily consecutive, because only those references required for this, the first volume of the series, are given.

ABBREVIATIONS

AA	*Artibus Asiae*
A/AIHS	*Archives Internationales d'Histoire des Sciences* (continuation of *Archeion*)
AAN	*American Anthropologist*
AB	*Art Bulletin* (New York)
AGNT	*Archiv f. d. Geschichte d. Naturwissenschaften u. d. Technik*
AH	*Asian Horizon*
AHR	*American Historical Review*
AJA	*American Journal of Archaeology*
AJP	*American Journal of Philology*
AKML	*Abhandlungen f. d. Kunde des Morgenlandes*
AM	*Asia Major*
AMSR	*American Sociological Review*
ANI	*Ancient India*
AO	*Acta Orientalia*
APAW/PH	*Abhandlungen d. preussischen Akademie d. Wissenschaften zu Berlin* (Phil.-hist. Klasse)
AQ	*Antiquity*
ARLC/DO	*Annual Reports of the Librarian of Congress* (Division of Orientalia)
ARSI	*Annual Reports of the Smithsonian Institution*
ARTHA	*Arethusa*
AS/BIHP	*Bulletin of the Institute of History and Philology, Academia Sinica*
ASEA	*Asiatische Studien; Etudes Asiatiques*
ASIA	*Asia*
B	*Byzantion*
BAFAO	*Bulletin de l'Association Française des Amis de l'Orient*
BCGS	*Bulletin of the Chinese Geological Society*
BE/AMG	*Bibliothèque d'Etudes* (*Annales du Musée Guimet*). See *RAA and RHR*
BEFEO	*Bulletin de l'Ecole Française del'Extrême Orient*
BGSC	*Bulletin of the Chinese Geological Survey*
BLSOAS	*Bulletin of the London School of Oriental and African Studies*
BMFEA	*Bulletin of the Museum of Far Eastern Antiquities* (Stockholm)
BVSAW/PH	*Berichte über d. Verhandlungen d. sächsischen Akademie d. Wissenschaften zu Leipzig* (Phil.-hist. Klasse); formerly *Ber. u. d. Verhdl. d. kgl. Sächsischen Gesellschaft d. Wiss.*
CB	*Centrablatt f. Bibliothekswesen*
CET	*Ciel et Terre*
CHER	*Chhing-Hua Engineering Reports* (*Kung-Chhêng Hsüeh Pao*)
CHESJ	*Chhing-Hua Engineering Society Journal* (*Kung-Chhêng Hsüeh Hui Hui Khan*)
CHI	*Cambridge History of India*
CHJ	*Chhing-Hua Hsüeh Pao* (Chhing-Hua (Ts'ing-Hua University) Journal)
CIB	*China Institute Bulletin* (New York)
CIBA/T	*Ciba Review* (Textile Technology)
CJ	*China Journal of Science and Arts*
CLR	*Classical Review*
CMH	*Cambridge Medieval History*
CNH	*Cambridge Natural History*
CNRS	Centre Nationale de la Recherche Scientifique
CP	*Classical Philology*
CQ	*Classical Quarterly*
CR	*China Review*
CSPSR	*Chinese Social and Political Science Review*
CTR	*Contemporary Review*
DVN	*Dan Viet-Nam*
DWAW/PH	*Denkschriften d. Akademie d. Wissenschaften, Wien* (Vienna) (Phil.-hist. Klasse)
E	*Ecology*
EI	*Encyclopaedia of Islam*
ENG	*Engineering*
ESA	*Eurasia Septentrionalis Antiqua*
ETH	*Ethnos*
FEQ	*Far Eastern Quarterly*
FF	*Forschungen und Fortschritte*
FL	*Folklore*
FMNHP/AS	*Field Museum of Natural History* (Chicago) *Publications; Anthropological Series*
G	*Geography*
GGM	*Geographical Magazine*
GHA	*Göteborgs Högskolas Årsskrift*
GJ	*Geographical Journal*
GM	*Geological Magazine*
GR	*Geographical Review*
GZ	*Geographische Zeitschrift*
HB	*Human Biology*
HJAS	*Harvard Journal of Asiatic Studies*
I	*L'Ingegnere*
IAQ	*Indian Antiquary*
IC	*Islamic Culture*
IPR	Institute of Pacific Relations
ISIS	*Isis*
JA	*Journal Asiatique*
JAFL	*Journal of American Folklore*
JAOS	*Journal of the American Oriental Society*
JBAA	*Journal of the British Archaeological Association*

JDZWT *Japanisch-deutsche Zeitschrift f. Wissenschaft u. Technik*
JEH *Journal of Economic History*
JMH *Journal of Modern History*
JPOS *Journal of the Peking Oriental Society*
JRAI *Journal of the Royal Anthropological Institute*
JRAS *Journal of the Royal Asiatic Society*
JRAS/B *Journal of the Royal Asiatic Society of Bengal* (cf. *TAS/B*)
JRAS/BOM *Journal of the Royal Asiatic Society, Bombay Branch*
JRAS/KB *Journal of the Royal Asiatic Society, Korea Branch*
JRAS/NCB *Journal of the Royal Asiatic Society, North China Branch*
JRCAS *Journal of the Royal Central Asian Society*
JRSA *Journal of the Royal Society of Arts*
JS *Journal des Savants*
JWCBRS *Journal of the West China Border Research Society*
JWKHS *Jen Wên Kho-Hsüeh Hsüeh-Pao* (Kunming)
KDVS/HFM *Kgl. Danske Videnskabernes Selskab* (Hist. Filol. Medd.)
KHCK *Kuo Hsüeh Chi-Khan*
KHS *Kho-Hsüeh*
KSP *Ku Shih Pien.* See Ku Chieh-Kang (*2*); Lo Kên-Tsê (*3*)
MAI/NEM *Mémoires de l'Académie des Inscriptions et Belles Lettres*, Paris (Notices et Extraits des MSS.)
MCB *Mélanges Chinois et Bouddhiques*
MCM *Macmillan's Magazine*
MCMU *Memoirs of the Carnegie Museum* (Pittsburgh)
MGGH *Mitteilungen d. geographischen Gesellschaft in Hamburg*
MGGW *Mitteilungen d. geographischen Gesellschaft in Wien*
MGSC *Memoirs of the Chinese Geological Survey*
MMI *Mariner's Mirror*
MQ *Modern Quarterly*
MRAI *Mémoires de l'Académie Royale des Inscriptions et Belles Lettres* (Paris).
MRC Medical Research Council (U.K.)
MRDTB *Mem. Research Dept. Toyo Bunko* (Tokyo)
MS *Monumenta Serica*
MSOS *Mitteilungen d. Seminar f. orientalischen Sprachen* (Berlin)
N *Nature*
NCR *New China Review*
NGM *National Geographic Magazine*
NJKA *Neue Jahrbücher f. d. klassische Altertum*
O *Observatory*
OAV *Orientalistisches Archiv.* (Leipzig)
OAZ *Ostasiatische Zeitschrift*
OMO *Österreiche Monatschrift f. d. Orient*
OR *Oriens*
ORT *Orient*
PA *Pacific Affairs*
PAAAS *Proceedings of the American Academy of Arts and Sciences*
PIAJ *Proceedings of the Imperial Academy of Japan*
PIS *Proceedings of the Iranian Society*
PL *Philologus*
PNHB *Peking Natural History Bulletin*
PNISI *Proc. Nat. Instit. of Sciences of India*
PRGS *Proceedings of the Royal Geographical Society*
PRSB *Proceedings of the Royal Society* (Series B)
PS *Palaeontologica Sinica*
PSBA *Proc. Soc. Biblical Archaeology*
PTRSB *Philosophical Transactions of the Royal Society* (Series B)
QBCB/C *Quarterly Bulletin of Chinese Bibliography (Chinese edition), Thu Shu Chi-Khan*
QJMS *Quarterly Journal of Microscopical Science*
RA *Revue Archéologique*
RAA/AMG *Revue des Arts Asiatiques (Annales du Musée Guimet).* See *BE/AMG* and *RHR/AMG*
RAS Royal Asiatic Society
RAS/ON *Occasional Notes of the Royal Astronomical Society*
REES *Revue des Etudes Ethnographiques et Sociologiques*
RGI *Rivista Geografica Italiana*
RH *Revue Historique*
RHR/AMG *Revue de l'Histoire des Religions (Annales du Musée Guimet).* See *BE/AMG* and *RAA/AMG*
RI *Revue Indochinoise*
RIIA Royal Institute of International Affairs
ROC *Revue de l'Orient Chrétien*
RP *Revue Philosophique*
RQS *Revue des Questions Scientifiques* (Brussels)
RSO *Rivista di Studi Orientali*
RTP *Revue des Traditions Populaires*
RTS Religious Tract Society
S *Sinologica*
SA *Sinica*
SBE *Sacred Books of the East* series
SCI *Scientia*
SCK *Smithsonian Contributions to Knowledge*
SOS *Semitic and Oriental Studies* (Univ. of California Publ. in Semitic Philology)
SP *Speculum*
SPAW/PH *Sitzungsberichte d. preussischen Akademie d. Wissenschaften* (Phil.-hist. Klasse)

SPCK Society for Promoting Christian Knowledge

SPMSE *Sitzungsberichte d. physikalisch-medizinischen Gesellschaft Erlangen*

SSQ *Social Science Quarterly*

SWAW/PH *Sitzungsberichte d. Akademie d. Wissenschaften, Wien* (Vienna) (Phil.-hist. Klasse)

TAPS *Transactions of the Amer. Philos. Society*

TAS/B *Transactions of the Asiatic Society of Bengal (Asiatick Researches)*

TITLV *Tijdschrift v. indische Taal-, Land- en Volkenkunde*

TNS *Transactions of the Newcomen Society*

TP *T'oung Pao*

UNN *United Nations News* (U.K.)

VAG *Vierteljahrsschrift d. astronomischen Gesellschaft*

VGEB *Verhandlungen d. Gesellschaft f. Erdkunde* (Berlin)

VKAWA/L *Verhandelingen d. Koninklijke Akademie v. Wetenschappen te Amsterdam* (Afd. Letterkunde)

VS *Variétés Sinologiques* series

WR *World Review*

YAHS *Yenching Shih-Hsüeh Nien-Pao* (*Yenching Annual of Historical Studies* or *Yenching Historical Annual*)

YCHP *Yenching Hsüeh-Pao* (*Yenching Journal of Chinese Studies*)

YJSS *Yenching Journal of Social Studies*

ZDMG *Zeitschrift d. deutschen morganländischen Gesellschaft*

ZGEB *Zeitschrift d. Gesellschaft f. Erdkunde* (Berlin)

ZOIAV *Zeitschrift d. österreichischen Ingenieur- u. Architekten-Vereines*

A. CHINESE BOOKS BEFORE +1800

Words which assist in the translation of titles are inserted in round brackets.
Alternative titles or explanatory additions to the titles are added in square brackets.

Abbreviations

C/Han	Former Han.	N/Sung	Northern Sung (before the removal of the capital to Hangchow).
H/Han	Later Han.	N/Wei	Northern Wei.
H/Shu	Later Shu (Wu Tai).	S/Sung	Southern Sung (after the removal of the capital to Hangchow).
H/Thang	Later Thang (Wu Tai).	*TT*	*Tao Tsang*.
J/Chin	Jurchen Chin.		
L/Sung	Liu Sung.		

Chêng Lei Pên Tshao 證類本草.
Reorganised Pharmacopoeia.
N/Sung, +1108.
Thang Shen-Wei 唐愼微.

Chi Chung Chou Shu 汲冢周書.
The Books of (the) Chou (Dynasty) found in the Tomb at Chi.
See *I Chou Shu*.

Chih Shêng Shih Lu 至聖實錄.
On Islam and its Prophet-Sage.
Chhing, +1721.
Liu Chih 劉智.

Chin Shih 金史.
History of the Chin (Jurchen) Dynasty [+1115 to +1234].
Yuan, *c.* +1345.
Tho-Tho (Toktaga) 脫脫 & Ouyang Hsüan 歐陽玄.
Yin-Tê Index, no. 35.

Chin Shu 晉書.
History of the Chin Dynasty [+265 to +419].
Thang, +635.
Fang Hsüan-Ling 房玄齡.
A few chs. tr. Pfizmaier (54–7).

Chiu Chang Suan Shu 九章算術.
Nine Chapters on the Mathematical Art.
H/Han, +1st century (containing material from C/Han).
Writer unknown.

Chiu Huang Pên Tshao 救荒本草.
Famine Herbal.
Ming, (+1395) +1406.
Chou Ting Wang (prince of the Ming) 周定王.
Preserved in chs. 46–59 of *Nung Chêng Chhüan Shu* (*q.v.*).

Chiu Thang Shu 舊唐書.
Old History of the Thang Dynasty [+618 to +906].
Wu Tai, +945.
Liu Hsü 劉昫.

Chiu Wu Tai Shih 舊五代史.
Old History of the Five Dynasties [+907 to +959].
Sung, +974.
Hsüeh Chü-Chêng 薛居正.

Chou Li 周禮.
Record of the Rites of (the) Chou (Dynasty) [descriptions of all government official posts and their duties.]
C/Han.
Compilers unknown.
Tr. E. Biot (1).

Chou Pei Suan Ching 周髀算經.
The Arithmetical Classic of the Gnomon and the Circular Paths (of Heaven).
Han, ±1st century (probably containing material from Chou).
Author unknown.

Chou Shu 周書.
History of the (Northern) Chou Dynasty [+557 to +581].
Thang, +625.
Linghu Tê-Fên 令狐德棻.

Chu Fan Chih 諸蕃志.
Records of Foreign Peoples.
Sung, +1225.
Chao Ju-Kua 趙汝适.
Tr. Hirth & Rockhill.

Chü Lu 橘錄.
Orange Record [citrus horticulture].
Sung, +1178.
Han Yen-Chih 韓彥直.
Tr. Hagerty (1).

Chu Shu Chi Nien 竹書紀年.
The Bamboo Books [annals].
Chou, −296 or before, such parts as are genuine (found in the tomb of An Li Wang, a prince of the Wei State; in +3rd century).

Chu Shu Chi Nien (*cont.*)
Writers unknown.
Tr. E. Biot (3).

Chu Tzu Chhüan Shu 朱子全書.
Collected Works of Chu Hsi.
Sung (ed. Ming; *editio princeps* +1713).
Chu Hsi 朱熹.
Partial tr. Bruce (1); le Gall (1).

Chu Tzu Yü Lei 朱子語類.
Classified Conversations of Chu Hsi.
Sung, *c.* +1270.
Ed. Li Ching-Tê 黎靖德.

Chuang Tzu [=*Nan Hua Chen Ching*] 莊子.
The Book of Master Chuang.
Chou, *c.* −290.
Chuang Chou 莊周.
Tr. Legge (5); Fêng Yu-Lan (5); Lin Yü-Thang (1); Wieger (7).
Yin-Tê Index no. (Suppl). 20.

Chung Yuan Yin Yün 中原音韻.
Tables of Sounds and Rhymes.
Sung, *c.* +1250.
Chou Tê-Chhing 周德清.

Chhen Shu 陳書.
History of the Chhen Dynasty [+556 to +589].
Thang, +630.
Yao Ssu-Lien 姚思廉 and his father Yao Chha 姚察.
A few chs. tr. Pfizmaier (59).

Chhi Min Yao Shu 齊民要術.
Important Arts for the People's Welfare.
N/Wei, probably +5th century.
Chia Ssu-Hsieh 賈思勰.

Chhieh Yün 切韻.
Dictionary of the Sounds of Characters [rhyming dictionary].
Sui, +601.
Lu Fa-Yen 陸法言.
See *Kuang Yün*.

Chhien Han Shu 前漢書.
History of the Former Han Dynasty [−206 to +24].
H/Han, *c.* +100.
Pan Ku 班固.
Partial trs. Dubs (2); Pfizmaier (32–4, 37–51); Wylie (2, 3, 10); Swann (1).
Yin-Tê Index no. 36.

Chhin Ting Ku Chin Thu Shu Chi Chhêng 欽定古今圖書集成.
See *Thu Shu Chi Chhêng*.

Chhin Ting Ssu Khu Chhüan Shu Chien Ming Mu Lu 欽定四庫全書簡明目錄.
Abridged Analytical Catalogue of the Books in the *Ssu Khu Chhüan Shu* Encyclopaedia, made by imperial order.
Chhing, +1782.
[There are two versions of this: (*a*) ed. Chi Yün, 紀昀, which contains mention of nearly all the books in the *Thi Yao*; (*b*) ed. Yü Min-Chung 于敏中, which contains only the books which were copied.]

Chhin Ting Ssu Khu Chhüan Shu Tsung Mu Thi Yao 欽定四庫全書總目提要.
Analytical Catalogue of the Books in the *Ssu Khu Chhüan Shu* Encyclopaedia, made by imperial order.
Chhing, +1782.
Ed. Chi Yün 紀昀.
Index by Yang Chia-Lo; Yü & Gillis.

Chhou Jen Chuan 疇人傳.
Biographies of (Chinese) Mathematicians (and Scientists).
Chhing, +1799.
Juan Yuan 阮元.
With continuations by Lo Shih-Lin 羅士琳, Chu Kho-Pao 諸可寶 and Huang Chung-Chün 黃鐘駿, in HCCC, ch. 159.

Chhu Hsüeh Chi 初學記.
Entry into Learning [encyclopaedia].
Thang, +700.
Hsü Chien 徐堅.

Chhun Chhiu 春秋.
Spring and Autumn Annals.
Chou.
Writers unknown.
Tr. Couvreur (1); Legge (11).

Chhun Chhiu Fan Lu 春秋繁露.
String of Pearls on the Spring and Autumn Annals.
C/Han, *c.* −135.
Tung Chung-Shu 董仲舒.
Partial trs. Wieger (2); Hughes (1); d'Hormon (ed.).
Chung-Fa Index, no. 4.

Chhung Hsü Chen Ching 冲虛眞經.
See *Lieh Tzu*.

Erh Ya 爾雅.
Literary Expositor [dictionary].
Chou material, stabilised in Chhin or C/Han.
Compiler unknown.
Enlarged and commented on *c.* +300 by Kuo Pho 郭璞.
Yin-Tê Index no. (Suppl.) 18.

Fa-Hsien Chuan.
See *Fo Kuo Chi*.

Fa-Hsien Hsing Chuan.
See *Fo Kuo Chi*.

Fêng Chhuang Hsiao Tu 楓窻小牘.
Maple-Tree Window Memories.
Sung, early +13th (after +1202).
Yuan Chhiung 袁褧.

Fêng Shen Yen I 封神演義.
Stories of the Promotions of the Martial Genii [novel].
Ming.
Hsü Chung-Lin 許仲琳.
Tr. Grube (1).

Fêng Su Thung I 風俗通義.
Popular Traditions and Customs.
H/Han, +175.
Ying Shao 應劭.
Chung-Fa Index no. 3.

Fo Kuo Chi 佛國記.
[= *Fa-Hsien Chuan* or *Fa-Hsien Hsing Chuan* 法顯行傳.]
Records of Buddhist Countries [also called Travels of Fa-Hsien].
Chin, *c.* +420.
Fa-Hsien (monk) 法顯.
Tr. Rémusat (1); Beal (1); Legge (4); H. A. Giles (3).

Hai Yao Pên Tshao 海藥本草.
Drugs of the Southern Countries beyond the Seas [or Pharmaceutical Codex of Marine Products].
Thang, *c.* +775 (or early +10th).
Li Hsün (acc. to Li Shih-Chen) 李珣.
Li Hsien (acc. to Huang Hsiu-Fu) 李玹.

Han Wu Ku Shih 漢武故事.
Tales of the Emperor Wu of the Han.
L/Sung and Chhi, +5th century.
Perhaps Wang Chien 王儉, based on Ko Hung 葛洪.
Tr. d'Hormon (ed.).

Han Wu Ti Nei Chuan 漢武帝內傳.
The Inside Story of Emperor Wu of the Han.
Chin, +4th century, or at least pre-Sui.
Anon. (ascr. Pan Ku), perhaps Ko Hung 葛洪.
TT/289.

Hou Han Shu 後漢書.
History of the later Han Dynasty [+25 to +220].
L/Sung, +450.
Fan Yeh 范曄.
A few chs. tr. Chavannes (6, 16); Pfizmaier (52, 53).
Yin-Tê Index, no. 41.

Hua Shu 化書.
Book of the Transformations (in Nature).
H/Thang, *c.* +940.
Attrib. Than Chhiao 譚峭.
TT/1032.

Hua Yang Kuo Chih 華陽國志.
Record of the Country South of Mount Hua [Historical Geography of Szechuan].
Chin, +347.
Chhang Chhü 常璩.

Huai Nan Tzu 淮南子.
[=*Huai Nan Hung Lieh Chieh* 淮南鴻烈解.
The Book of (the Prince of) Huai Nan. [compendium of natural philosophy.]
C/Han, *c.* −120.
Attrib. Liu An (prince of Huai Nan) 劉安.
Partial trs. Morgan (1); Erkes (1); Hughes (1); Chatley (1); Wieger (2).
Chung-Fa Index no. 5.
TT/1170.

Hsi Ching Tsa Chi 西京雜記.
Miscellaneous Records of the Western Capital.
Liang or Chhen, mid +6th century.
Attrib. to Liu Hsin 劉歆 (C/Han) and to Ko Hung 葛洪 (Chin) but prob. Wu Chün 吳均.

Hsi Yu Chi 西遊記.
Story of a Journey to the West [novel = *Monkey*].
Ming, +14th or +15th century.
Wu Chhêng-Ên 吳承恩.
Tr. Waley (17).

Hsia Hsiao Chêng 夏小正.
Lesser Annuary of the Hsia Dynasty.
Chou, between −7th and −4th.
Writers unknown.
Incorporated in *Ta Tai Li Chi* (*q.v.*).
Tr. R. Wilhelm (6); Soothill (5).

Hsiao Ching 孝經.
Filial Piety Classic.
Alleged Chou; perhaps Han, ±1st century.
Attrib. Tsêng Shen (pupil of Confucius) 曾參.
Tr. de Rosny (2); Legge (1).

Hsiao Tai Li Chi.
See *Li Chi*.

Hsin Hsiu Pên Tshao 新修本草.
Newly Reorganised Pharmacopoeia.
Thang, +659.
Li Chi (with 22 assistants) 李勣.
[This pharmacopoeia was probably very similar to the *Thang Pên Tshao*, *q.v.* It was lost in China but five chapters have been preserved as they were copied by a Japanese envoy in +732 and survived in Japan.]

Hsin Shu 新書.
New Book.
C/Han, −2nd century, but the present text may be partly Thang or pre-Thang.
Chia I 賈誼.

Hsin Thang Shu 新唐書.
New History of the Thang Dynasty [+618 to +906].
Sung, +1061.
Ouyang Hsiu & Sung Chhi 歐陽修, 宋祁.

Hsin Thang Shu (*cont.*)
Partial trs. des Rotours (1); Pfizmaier (66–74).
Yin-Tê Index, no. 16.

Hsin Wu Tai Shih 新五代史.
[= *Wu Tai Shih Chi.*]
New History of the Five Dynasties [+907 to +959].
Sung, *c.* +1070.
Ouyang Hsiu 歐陽修.

Hsin Yü 新語.
New Discourses.
C/Han, *c.* −196.
Lu Chia 陸賈.
Tr. v. Gabain (1).

Hsing Shui Chin Chien 行水金鑑.
Golden Mirror of the Flowing Waters.
Chhing, +1725.
Fu Tsê-Hung 傅澤洪.

Hsü Po Wu Chih 續博物志.
Supplement to the *Record of the Investigation of Things*. (Cf. *Po Wu Chih.*)
Sung, mid +12th century.
Li Shih 李石.

Hsü Shih Shih 續事始.
Supplement to the *Beginnings of All Affairs*. (Cf. *Shih Shih.*)
H/Shu, *c.* +960.
Ma Chien 馬鑑.

I Ching 易經.
The Classic of Changes [Book of Changes].
Chou with C/Han additions.
Compiler unknown.
Tr. R. Wilhelm (2); Legge (9); de Harlez (1).
Yin-Tê Index no. (Suppl.) 10.

I Chou Shu 逸周書.
[= *Chi Chung Chou Shu.*]
Lost Books of the Chou.
Chou, −296 or before, if genuine. (Found in the tomb of An Li Wang, a prince of Wei State; in +3rd century.)
Writers unknown.

I Wên Lei Chü 藝文類聚.
Literary Records Collected and Classified [encyclopaedia].
Thang, *c.* +620.
Ouyang Hsün 歐陽詢.

I Yin 易音.
Dictionary of the Original Sounds of Words in the *Book of Changes*.
Chhing, +1667 (in *Yin Hsüeh Wu Shu*).
Ku Yen-Wu 顧炎武.

Kêng Hsin Yü Tshê 庚辛玉册.
The Jade Book of the Kêng (-Tzu) and Hsin (-Chhou) Years (on Mineralogy, Alchemy and Pharmaceutics).
Ming, +1421.
Ning Hsien Wang (prince of the Ming) 寧獻王.

Khai Ho Chi 開河記.
Record of the Opening of the (Grand) Canal.
Sui.
Author unknown.

Khang-Hsi Tzu Tien 康熙字典.
Imperial Dictionary of the Khang-Hsi reign-period.
Chhing, +1716.
Ed. Chang Yü-Shu 張玉書.

Khao Kung Chi 考工記.
The Artificers' Record [a section of the *Chou Li* (*q.v.*)].
Han.
Compiler unknown.

Ko Chih Ching Yuan 格致鏡原.
Mirror of Scientific and Technological Origins.
Chhing, +1735.
Chhen Yuan-Lung 陳元龍.

Ku Chin Chu 古今註.
Commentary on Things Old and New.
Chin, mid +4th century.
Tshui Pao 崔豹.

Ku Yin Piao 古音表.
Catalogue of Ancient Pronunciations.
Chhing, +1667 (in *Yin Hsüeh Wu Shu*).
Ku Yen-Wu 顧炎武.

Kuan Tzu 管子.
The Writings of Master Kuan.
Chou and C/Han.
Attrib. Kuan Chung 管仲.

Kuang Yü Thu 廣輿圖.
Enlarged Terrestrial Atlas.
Yuan, +1320.
Chu Ssu-Pên 朱思本.
First printed, and the word *Kuang* added, by Lo Hung-Hsien (羅洪先), Ming *c.* +1555.

Kuang Yün 廣韻.
Enlargement of the *Chhieh Yün* Dictionary of the Sounds of Characters.
Sung (a completion by later Thang and Sung scholars, given its present name in +1011).

Lang Huan Chi 瑯環記.
On the Cyclical Recurrence of World Catastrophes.
Liao, +10th century.
I Shih-Chen 伊世珍.

Lei Phien 類篇.
Dictionary of Character Sounds.
Sung, +1067.
Ssuma Kuang 司馬光.

Li Chi 禮記.
[=*Hsiao Tai Li Chi.*]
Record of Rites [compiled by Tai the Younger].
C/Han, *c.* −50.
Ed. Tai Shêng 戴聖.
Tr. Legge (7); Couvreur (3); R. Wilhelm (6).
Yin-Tê Index no. 27.

Liang Shu 梁書.
History of the Liang Dynasty [+502 to +556].
Thang, +629.
Yao Chha 姚察, and his son Yao Ssu-Lien 姚思廉.

Liao Shih 遼史.
History of the Liao (Chhi-tan) Dynasty [+916 to +1125].
Yuan, *c.* +1350.
Tho-Tho (Toktaga) 脫脫 & Ouyang Hsüan 歐陽玄.
Partial tr. Wittfogel, Fêng Chia-Shêng *et al.*
Yin-Tê Index no. 35.

Lieh Tzu [=*Chhung Hsü Chen Ching*] 列子.
The Book of Master Lieh.
Chou and C/Han −5th to −1st century. (Ancient fragments of miscellaneous origin finally cemented together with much new material about +380.)
Attrib. Lieh Yü-Khou 列禦寇.
Tr. R. Wilhelm (4); L. Giles (4); Wieger (7).
TT/663.

Liu Shu Ku 六書故.
The Six Classes of Characters Explained.
Sung, *c.* +1275, printed +1320.
Tai Tung 戴侗.

Liu Thieh 六帖.
The Six Cards [encyclopaedia].
Thang, *c.* +800.
Pai Chü-I 白居易.
(Enlarged in Sung by Khung Chuan 孔傳.)

Loyang Chhieh Lan Chi 洛陽伽藍記.
Description of the Buddhist Temples of Loyang.
N/Wei, *c.* +530.
Yang Hsüan-Chih 楊衒之.

Lü Hsüeh Hsin Shuo 律學新說.
A New Account of the Science of the Pitch-Pipes.
Ming, +1584.
Chu Tsai-Yü (prince of the Ming) 朱載堉.

Lü Lü Ching I 律呂精義.
The Essential Meaning of the Standard Pitch-Pipes (in two parts).
[=The Pitch-Pipe Opus, *Lü Shu* 律書.]
Ming, +1596.
Chu Tsai-Yü (prince of the Ming) 朱載堉.

Lü Shih Chhun Chhiu 呂氏春秋.
Master Lü's Spring and Autumn Annals. [compendium of natural philosophy.]
Chou (end), *c.* −3rd century.
Attrib. Lü Pu-Wei 呂不韋.
Tr. R. Wilhelm (3).
Chung-Fa Index no. 2.

Lun Hêng 論衡.
Discourses Weighed in the Balance.
H/Han, +82 or +83.
Wang Chhung 王充.
Tr. Forke (4).
Chung-Fa Index no. 1.

Lun Yü 論語.
Conversations and Discourses (of Confucius).
Chou, late −5th or early −4th century.
Compiled by disciples of Confucius.
Tr. Legge (2); Waley (5); Ku Hung-Ming (1).
Yin-Tê Index no. (Suppl.) 16.

Mao Shih Ku Yin Khao 毛詩古音考.
Investigations on the Sounds in Mao's Version of the *Book of Odes*.
Ming, +1606.
Chhen Ti 陳第.

Mao Thing Kho Hua 茅亭客話.
Discourses with Guests in the Thatched Pavilion.
Sung.
Huang Hsiu-Fu 黃休復.

Mêng Chhi Pi Than 夢溪筆談.
Dream Pool Essays.
Sung, +1086.
Shen Kua 沈括.

Mêng Tzu 孟子.
Writings of Mencius.
Chou. −3rd century.
Mêng Kho 孟軻.
Tr. Legge (3).
Yin-Tê Index no. (Suppl.) 17.

Ming Ju Hsüeh An 明儒學案.
Schools of Philosophers of the Ming Dynasty.
Chhing, *c.* +1700.
Huang Tsung-Hsi & Wan Ssu-Thung 黃宗羲; 萬斯同.

Ming Shih 明史.
History of the Ming Dynasty.
Chhing, +1739.
Chang Thing-Yü 張廷玉 *et al.*

Mo Ching 墨經. See *Mo Tzu*.

Mo Tzu (incl. *Mo Ching*) 墨子.
The Book of Master Mo.
Chou. −4th century.
Mo Ti (and disciples) 墨翟.
Tr. Mei Yi-Pao (1); Forke (3).
Yin-Tê Index no. (Suppl.) 21; *TT*/1162.

Mu Thien Tzu Chuan 穆天子傳.
Account of the Travels of the Emperor Mu.
Chou, before −245 (found in the tomb of An Li Wang, a prince of Wei State; in +281.)
Writer unknown.
Tr. Eitel (1); Chêng Tê-Khun (2).

Nan Chhi Shu 南齊書.
History of the Southern Chhi Dynasty [+479 to +501].
Chhi and Liang, +520.
Hsiao Tzu-Hsien 蕭子顯.

Nan Chou I Wu Chih 南州異物志.
Strange Things of the South.
Chin, +3rd or +4th century.
Wan Chen 萬震.

Nan Fang Tshao Mu Chuang 南方草木狀.
Records of the Plants and Trees of the Southern Regions.
Chin, +3rd century.
Chi Han 稽含.

Nan Hai Chi Kuei Nei Fa Chuan 南海寄歸內法傳.
Record of Buddhist Practices Sent Home from the South Seas.
Thang, *c.* +689.
I-Ching (monk) 義淨.
Tr. Takakusu (1).

Nan Hua Chen Ching 南華眞經.
See *Chuang Tzu*.

Nan I I Wu Chih 南裔異物志.
Strange Things from the Southern Borders.
H/Han, end +2nd century.
Yang Fu 楊孚.

Nan Shih 南史.
History of the Southern Dynasties [Nan Pei Chhao period, +420 to +589].
Thang, *c.* +629.
Li Yen-Shou 李延壽.

Nung Chêng Chhüan Shu 農政全書.
Complete Treatise on Agriculture.
Ming. Composed +1625 to +1628; printed +1639.
Hsü Kuang-Chhi 徐光啓.
Ed. Chhen Tzu-Lung 陳子龍.

Pai Chhuan Hsüeh Hai 百川學海.
The Hundred Rivers Sea of Learning [a collection of separate books; the first *tshung-shu*].
Sung, late +12th or early +13th century.
Compiled and edited by Tso Kuei 左圭.

Pai Hu Thung Tê Lun 白虎通德論.
Universal Discussions at the White Tiger Lodge.
H/Han, *c.* +80.
Pan Ku 班固.
Tr. Tsêng Chu-Sên (1).

Pai Khung Liu Thieh. See *Liu Thieh*.

Pao Phu Tzu 抱樸(or 朴)子.
Book of the Preservation-of-Solidarity Master.
Chin, early +4th century.
Ko Hung 葛洪.
Partial trs. Feifel (1, 2); Wu & Davis (2); etc.
TT/1171–1173.

Pei Chhi Shu 北齊書.
History of the Northern Chhi Dynasty [+550 to 577].
Thang, +636.
Li Tê-Lin 李德林, and his son Li Pai-Yao 李百藥.
A few chs. tr. Pfizmaier (60).

Pei Chou Shu 北周書.
See *Chou Shu*.

Pei Hu Lu 北戶錄.
Northern Family Records.
Thang, +875.
Tuan Kung-Lu 段公路.

Pei Shih 北史.
History of the Northern Dynasties [Nan Pei Chhao period, +386 to +581].
Thang, *c.* +629.
Li Yen-Shou 李延壽.

Pei Thang Shu Chhao 北堂書鈔.
Book Records of the Northern Hall [encyclopaedia].
Thang, *c.* +630.
Yü Shih-Nan 虞世南.

Pên Tshao Kang Mu 本草綱目.
The Great Pharmacopoeia.
Ming, +1596.
Li Shih-Chen 李時珍.
Paraphrased and abridged tr. Read & collaborators (1–7) and Read & Pak (1) with indexes.

Pên Tshao Yen I 本草衍義.
The General Ideas of the Pharmacopoeia.
Sung, +1116.
Khou Tsung-Shih 寇宗奭.
TT/761.

Pên Tshao Yen I Pu I 本草衍義補遺.
Revision and Amplification of the *General Ideas of the Pharmacopoeia*.
Ming, *c.* +1380.
Chu Chen-Hêng 朱震亨.

Po Wu Chih 博物志.
Record of the Investigation of Things. (Cf. *Hsü Po Wu Chih*.)
Chin, *c.* +290.
Chang Hua 張華.

San Kuo Chih 三國志.
History of the Three Kingdoms [+220 to +280].
Chin, *c.* +290.
Chhen Shou 陳壽.
Yin-Tê Index no. 33.

San Kuo Chih Yen I 三國志演義.
The Three Kingdoms Story [novel].
Yuan.
Lo Kuan-Chung 羅貫中.
Tr. Brewitt-Taylor (1).

San Tshai Thu Hui 三才圖會.
Universal Encyclopaedia.
Ming, +1609.
Wang Chhi 王圻.

Sêng Hui-Sêng shih Hsi Yü Chi 僧惠生使西域記.
Record of Western Countries, by the monk Hui-Sêng.
N/Wei, *c.* +530.
Hui-Sêng (monk) 惠生.
In ch. 5 of *Loyang Chhieh Lan Chi* (*q.v.*).
Tr. Beal (1); Chavannes (3).

Shan Hai Ching 山海經.
Classic of the Mountains and Rivers.
Chou and C/Han.
Writers unknown.
Partial tr. de Rosny (1).
Chung-Fa Index no. 9.

Shih Chhü Li Lun 石渠禮論.
Report of the Discussions in the Stone Canal Pavilion.
C/Han, −51.
Attrib. Tai Shêng 戴聖.
Tr. Tsêng Chu-Sên (1).

Shih Chi 史記.
Historical Record (down to −99).
C/Han, *c.* −90.
Ssuma Chhien 司馬遷, and his father Ssuma Than 司馬談.
Partial trs. Chavannes (1); Pfizmaier (13–36); Hirth (2); Wu Khang (1); Swann (1); etc.
Yin-Tê Index no. 40.

Shih Ching 詩經.
Book of Odes, [ancient folksongs].
Chou, −9th to −5th century.
Writers and compilers unknown.
Tr. Legge (1, 8); Waley (1); Karlgren (14).

Shih Pên 世本.
Book of Origins [imperial genealogies, family names, and legendary inventors].
C/Han (incorporating Chou material).
Ed. Sung Chung (H/Han) 宋衷.

Shih Pên Yin 詩本音.
Dictionary of the Original Sounds of Words in the *Book of Odes*.
Chhing +1667 (in *Yin Hsüeh Wu Shu*).
Ku Yen-Wu 顧炎武.

Shih Shih 事始.
Beginnings of all Affairs.
Sui, +605 to +616.
Liu Tshun 劉存
or Liu Hsiao-Sun 劉孝孫.

Shih Wên Lei Chü 事文類聚.
Encyclopaedia of Events and Literature.
Sung, +1246.
Chu Mu 祝穆.

Shih Wu Chi Yuan 事物紀原.
Records of the Origins of Affairs and Things.
Sung, *c.* +1085.
Kao Chhêng 高承.

Shu Ching 書經.
Historical Classic [Book of Documents].
Chou, with later additions.
Writers unknown.
Tr. Medhurst (1); Legge (1, 10); Karlgren (12).

Shui Ching 水經.
The Waterways Classic [geographical account of rivers and canals].
Alleged C/Han, prob. San Kuo.
Attrib. Sang Chhin 桑欽.

Shui Ching Chu 水經注.
Commentary on the *Waterways Classic* [geographical account greatly extended].
N/Wei, late +5th or early +6th century.
Li Tao-Yuan 酈道元.

Shui Hu Chuan 水滸傳.
The Story of the Lake [novel = *All men are Brothers*].
Ming, *c.* +1380.
Ascr. Shih Nai-An 施耐庵.
Tr. Buck (1).

Shuo Wên Chieh Tzu 說文解字.
Analytical Dictionary of Characters.
H/Han, +121.
Hsü Shen 許慎.

Ssu Khu Chhüan Shu, etc.
See *Chhin Ting Ssu Khu Chhüan Shu*, etc.

Su Shen Liang Fang 蘇沈良方.
Beneficial Prescriptions collected by Su (Tung-Pho) and Shen (Kua).
Sung, *c.* +1100.
Su Tung-Pho & Shen Kua 蘇東坡; 沈括.

Suan Hsüeh Hsin Shuo 算學新說.
A New Account of the Science of Calculation (in Acoustics and Music).
Ming, +1603.
Chu Tsai-Yü (prince of the Ming) 朱載堉.

Sui Shu 隋書.
History of the Sui Dynasty [+581 to +617].
Thang, +636.
Wei Chêng 魏徵 *et al.*
Partial tr. Pfizmaier (61–5).

Sun Tzu Suan Ching 孫子算經.
Master Sun's Mathematical Manual.
San Kuo (+230 to +270, or more probably later).
Sun (given names unknown) 孫口口.

Sung Shih 宋史.
History of the Sung Dynasty [+960 to +1279].
Yuan, *c.* +1345.
Tho-Tho (Toktaga) 脫脫 & Ouyang Hsüan 歐陽玄.
Yin-Tê Index no. 34.

Sung Shu 宋書.
History of the (Liu) Sung Dynasty [+420 to +478].
S/Chhi, +500.
Shen Yo 沈約.
A few chs. tr. Pfizmaier (58).

Sung Yuan Hsüeh An 宋元學案.
Schools of Philosophers in the Sung and Yuan Dynasties.
Chhing, *c.* +1750.
Huang Tsung-Hsi & Chhüan Tsu-Wang 黃宗羲; 全祖望.

Ta Hsüeh 大學.
The Great Learning [The Learning of Greatness].
Chou, perhaps −4th century.
Prob. an unknown Confucian contemporary with Mencius.
Tr. Legge (2); Hughes (2).

Ta-Kuan Ching Shih Chêng Lei Pên Tshao 大觀經史證類本草.
Ta-Kuan reign-period Reorganised Pharmacopoeia.
See *Chêng Lei Pên Tshao*.

Ta Tai Li Chi 大戴禮記
Record of Rites [compiled by Tai the Elder].
C/Han, stabilised H/Han, +80/+100.
Ed. Tai Tê 戴德.
Tr. R. Wilhelm (6); cf. Legge (7).

Ta Thang Chhiu Fa Kao Sêng Chuan 大唐求法高僧傳.
Records of the High Monks (who went out to) seek for (the Books) of the Law in the time of the Thang.
Thang, *c.* +705.
I-Ching (monk) 義淨.
Tr. Chavannes (4).

Ta Thang Hsi Yü Chi 大唐西域記.
Records of the Western Countries in the time of the Thang.
Thang, +646.
Hsüan-Chuang (monk) 玄奘.
Ed. Pien Chi 辯機.
Tr. Julien (1); Beal (2).

Ta Thang Tzhu-Ên-Ssu San Tsang Fa-Shih Chuan 大唐慈恩寺三藏法師傳.
Life of the Master of the Law and the Tripiṭaka dwelling in the Great Loving-Kindness Temple in the time of the Thang.
Thang, *c.* +665.
Hui-Li (monk) 惠立.
Tr. Julien (1); Beal (3).

Ta-Yeh Tsa Chi 大業雜記.
Records of the Reign of Sui Yang Ti [the Ta-Yeh reign-period].
Sui.
Tu Pao 杜寶.

Tao Tê Ching 道德經.
Canon of the Virtue of the Tao.
Chou, before −300.
Attrib. Li Erh (Lao Tzu) 李耳(老子).
Tr. Waley (4); Chhu Ta-Kao (2); Lin Yü-Thang (1); Wieger (7); and very many others.

Tao Tsang 道藏.
Taoist Patrology [containing 1464 Taoist works].
All periods, but first collected and printed in the Sung. Also printed in J/Chin (+1186/+1191), Yuan, and Ming (+1445, +1598 and +1607).
Index by Wieger (6).
Yin-Tê Index no. 25.

Thai-Phing Huan Yü Chi 太平寰宇記.
Thai-Phing reign-period General Description of the World [Geographical Record].
Sung, +976 to +983.
Yüeh Shih 樂史.

Thai-Phing Kuang Chi 太平廣記.
Miscellaneous Records collected in the Thai-Phing reign-period.
Sung, +981.
Ed. Li Fang 李昉.

Thai-Phing Yü Lan 太平御覽.
Thai-Phing reign-period Imperial Encyclopaedia.
Sung, +983.
Ed. Li Fang 李昉.
Yin-Tê Index no. 23.

Thang Pên Tshao 唐本草.
Pharmacopoeia of the Thang Dynasty. (Cf. the *Hsin Hsiu Pên Tshao, q.v.*)
Thang, +660.
Ed. Su Kung 蘇恭.

Thang Shu.
See *Chiu Thang Shu* and *Hsin Thang Shu*.

Thang Yün Chêng 唐韻正.
Thang Dynasty Rhyme Sounds (compared with those of antiquity).
Chhing, +1667 (in *Yin Hsüeh Wu Shu*).
Ku Yen-Wu 顧炎武.

Thien Kung Khai Wu 天工開物.
The Exploitation of the Works of Nature.
Ming, +1637.
Sung Ying-Hsing 宋應星.

Thu Shu Chi Chhêng 圖書集成.
Imperial Encyclopaedia.
Chhing, +1726.
Ed. Chhen Mêng-Lei 陳夢雷 *et al.*
Index by L. Giles (2).

Thung Chien Kang Mu 通鑑綱目.
Essential Mirror of Universal History [the *Tzu Chih Thung Chien* condensed].
Sung, +1189.
Chu Hsi 朱熹 (and his school).
Partial tr. Wieger (1).

Thung Chih 通志.
Historical Collections.
Sung, *c.* +1150.
Chêng Chhiao 鄭樵.

Thung Chih Lüeh 通志略.
Compendium of Information [part of *Thung Chih* (*q.v.*)].

Thung Tien 通典.
Reservoir of Source Material on Political and Social History.
Thang, *c.* +812.
Tu Yu 杜佑.

Tshê Fu Yuan Kuei 册府元龜.
Collection of Material on the Lives of Emperors and Ministers.
Sung, +1013.
Ed. Wang Chhin-Jo & Yang I 王欽若; 楊億.

Tso Chuan 左傳.
Master Tsochhiu's Enlargement of the *Chhun Chhiu* (Spring and Autumn Annals), [dealing with the period −722/−468].
Chou, between −400 and −250, but with additions by Chhin and Han scholars.
Attrib. Tsochhiu Ming 左邱明.
See Karlgren (8); Maspero (1).
Tr. Couvreur (1); Legge (11).

Tu Shu Chi Shu Lüeh 讀書記數略.
Register of Numerical Categories.
Chhing, +1707.
Kung Mêng-Jen 宮夢仁.

Tzu Chih Thung Chien 資治通鑑.
Mirror of Universal History [−403 to +959].
Sung, +1084.
Ssuma Kuang 司馬光.

Tzu Shih Ching Hua 子史精華.
Essence of the Philosophers and Historians [dictionary of quotations].
Chhing, +1727.
Yün Lu 允祿 *et al.*

Wang Huai Lu 忘懷錄.
What Not to Forget to take with you (when Preparing for a Journey).
Sung, *c.* +1070.
Shen Kua 沈括.

Wei Lüeh 魏略.
Memorable Things of the Wei State (San Kuo).
San Kuo (Wei) or Chin, +3rd or +4th century.
Yü Huan 魚豢.

Wei Shu 魏書.
History of the (Northern) Wei Dynasty [+386 to +556].
N/Wei, +572.
Wei Shou 魏收.

Wên Hsien Thung Khao 文獻通考.
Historical Investigation of Public Affairs.
Sung, *c.* +1254 but not published until +1319.
Ma Tuan-Lin 馬端臨.

Wu Chhuan Lu 吳船錄.
Account of a Journey by boat to Wu [from Szechuan].
Sung, +1177.
Fan Chhêng-Ta 范成大.

Wu Ching Tsung Yao 武經總要.
Collection of the most important Military Techniques.
Sung (compiled by Imperial order), +1040 (+1044).
Ed. Tsêng Kung-Liang 曾公亮.

Wu Pei Chih 武備志.
Records of War Preparations.
Ming, +1628.
Mao Yuan-I 茅元儀.

Wu Tai Shih Chi. See *Hsin Wu Tai Shih.*

Yin Hsüeh Wu Shu 音學五書.
Five Works on Phonetics.
Chhing, +1667.
Ku Yen-Wu 顧炎武.

Yin Lun 音論.
Study of Ancient Pronunciations.
Chhing +1667 (in *Yin Hsüeh Wu Shu*).
Ku Yen-Wu 顧炎武.

Ying Tsao Fa Shih 營造法式.
Treatise on Architectural Methods.
Sung, +1097; printed +1103; revised +1141.
Li Chieh 李誡.

Yo Lü Chhüan Shu 樂律全書.
Collected Works on Music and Acoustics.
[Contains *Lü Hsüeh Hsin Shuo*, *Lü Lü Ching I*, and *Suan Hsüeh Hsin Shuo* (*q.v.*).]

Yo Lü Chhüan Shu (*cont.*)
Ming, *c.* +1610.
Chu Tsai-Yü (prince of the Ming) 朱載堉.

Yu-Yang Tsa Tsu 酉陽雜俎.
The Yu-Yang Miscellany [lit. Mixed Stew].
Thang, +863.
Tuan Chhêng-Shih 段成式.

Yü Hai 玉海.
Ocean of Jade [encyclopaedia].
Sung, +1267 (first pr. Yuan, +1351).
Wang Ying-Lin 王應麟.

Yuan Shih 元史.
History of the Yuan (Mongol) Dynasty [+1206 to +1367].
Ming, *c.* +1370.
Sung Lien 宋濂.
Yin-Tê Index no. 35.

Yuan Ying Chi 淵穎集.
The Vast and the Minute.
Liao, *c.* +1000.
Wu Lai 吳萊.

Yüeh Ling 月令.
Monthly Ordinances (of the Chou Dynasty).
Chou, between −7th and −3rd.
Writers unknown.
Incorporated in the *Hsiao Tai Li Chi* and the *Lü Shih Chhun Chhiu*, *q.v.*
Tr. Legge (7); R. Wilhelm (3).

B. CHINESE AND JAPANESE BOOKS AND JOURNAL ARTICLES SINCE +1800

Chang Hsi-Chhen (*1*) 章錫琛.
Erh-shih-wu Shih Jen Ming So Yin 二十五史人名索引.
Index to Persons Mentioned in the Twenty-five Dynastic Histories.
Kaiming, Shanghai, 1935; 2nd ed. 1946.

Chang Hsing-Lang (*1*) 張星烺.
Chung Hsi Chiao-Thung Shih-Liao Hui Phien 中西交通史料匯篇.
Materials for the Study of the Intercourse of China with Other Countries.
6 vols.:
(1) Ancient Times, and Europeans;
(2) Europeans;
(3) Africans and Arabs;
(4) Armenians, Jews and Persians;
(5) Turks and Central Asians;
(6) Indians.
Fu-Jen Univ. Press, Peiping, 1930.

Chang Hung-Chao (*1*) 章鴻釗.
Shih Ya 石雅.
Lapidarium Sinicum; A Study of the Rocks, Fossils and Minerals as known in Chinese Literature.
Chinese Geol. Survey, Peiping: 1st ed. 1921, 2nd ed. 1927. *MGSC* (ser. B), no. 2 (rev. P. Demiéville, *BEFEO*, 1924, **24**, 276).

Chang Tzu-Kao (*1*) 張子高.
Kho-Hsüeh Fa Ta Shih 科學發達史.
History of Scientific Discoveries (general).
Shanghai, *c.* 1930.

Chang Yin-Lin (*1*) 張蔭麟.
Ming Chhing chih Chi Hsi Hsüeh Shu ju Chung-Kuo Khao-lüeh 明清之際西學輸入中國考略.
History of the Introduction of Western Science and Technology into China in the Ming and Chhing Dynasties.
CHJ, 1924, **1** (no. 1), 38.

Chang Yin-Lin (*2*) 張蔭麟.
Chung-Kuo Li-Shih shang chih 'Chhi Chhi' chi chhi Tso-chê 中國歷史上之「奇器」及其作者.
Scientific Inventions and Inventors in Chinese History.
YCHP, 1928, **1** (no. 3), 359.

Chêng Chao-Ching (*1*) 鄭肇經.
Chung-Kuo Shui-Li Shih 中國水利史.
History of River Conservancy and Irrigation Engineering in China.
Com. Press, Chhangsha, 1939.

Chien Po-Tsan (*1*) 翦伯贊.
Chung-Kuo Shih Kang 中國史綱.
Outline of Chinese History (2 vols.).
Fifties Publishing Co., Chungking, 1944.

Chin Yü-Fu (*1*) 金毓黻.
Chung-Kuo Shih-Hsüeh Shih 中國史學史.
History of Chinese Historiography.
Pien I Kuan and Com. Press, Chungking, 1944.

Chou Ku-Chhêng (*1*) 周谷城.
Chung-Kuo Thung Shih 中國通史.
General History of China (2 vols.).
Kaiming, Chungking, 1941.

Chu Hsi-Tsu (*1*) 朱希祖.
Chung-Kuo Shih-Hsüeh chih Chhi Yuan 中國史學之起源.
Origins of Chinese Historical Science.
SSQ, 1.

Chung-Fa Series, no. 1 中法.
Lun Hêng Thung-Chien 論衡通檢.
Index to Words and Phrases in the *Lun Hêng*.
Centre Franco-Chinois d'Etudes Sinologiques. Peiping, 1943.

Chung-Fa Series, no. 2 中法.
Lü Shih Chhun Chhiu Thung-Chien 呂氏春秋通檢.
Index to Words and Phrases in the *Lü Shih Chhun Chhiu*.
Centre Franco-Chinois d'Etudes Sinologiques. Peiping, 1943.

Chung-Fa Series, no. 3 中法.
Fêng Su Thung I Thung-Chien 風俗通義通檢.
Index to Words and Phrases in the *Fêng Su Thung I*.
Centre Franco-Chinois d'Etudes Sinologiques. Peiping, 1943.

Chung-Fa Series, no. 4 中法.
Chhun Chhiu Fan Lu Thung-Chien 春秋繁露通檢.
Index to Words and Phrases in the *Chhun Chhiu Fan Lu*.
Centre Franco-Chinois d'Etudes Sinologiques. Peiping, 1944.

Chung-Fa Series, no. 5 中法.
Huai Nan Tzu Thung-Chien 淮南子通檢.
Index to Words and Phrases in *Huai Nan Tzu*.
Centre Franco-Chinois d'Etudes Sinologiques. Peiping, 1944.

Chung-Fa Series, no. 9 中法.
Shan Hai Ching Thung-Chien 山海經通檢.
Index to Words and Phrases in the *Shan Hai Ching*.
Centre Franco-Chinois d'Etudes Sinologiques. Peiping, 1948.

Chhen Nai-Chhien (*1*) 陳乃乾.
Pieh Hao So Yin 別號索引.
Synonymy of Chinese Personal Names.
Kaiming, Shanghai, 1936.

Chhen Tê-Yün (*1*) 陳德芸.
Ku Chin Jen Wu Pieh Ming So Yin
古今人物別名索引.
Synonymy of the Names of Distinguished Chinese, Ancient and Modern.
Lingnan University, Canton, 1937.

Chhen Wên-Thao (*1*) 陳文濤.
Hsien Chhin Tzu-Jan Hsüeh Kai Lun
先秦自然學概論.
History of Science in China during the Chou and Chhin periods.
Com. Press, Shanghai, 1934.

Chhen Yuan (*1*) 陳垣.
Huo-Hsien Chiao ju Chung-Kuo Khao
火祆教入中國考.
History of the Coming of Zoroastrianism to China.
KHCK, 1923, 1, 27.

Chhen Yuan (*2*) 陳垣.
Mo-Ni Chiao ju Chung-Kuo Khao
摩尼教入中國考.
History of the Coming of Manichaeism to China.
KHCK, 1923, 1, 203.

Chhen Yuan (*3*) 陳垣.
Yuan Hsi-Yü Jen Hua Hua Khao
元西域人華化考.
On the Sinisation of 'Western People' during the Yuan Dynasty.
Pt. 1. *KHCK*, 1923, 1, 573.
Pt. 2. *YCHP*, 1927, 2, 171.

Chhi Ssu-Ho (*1*) 齊思和.
Stories of the Inventions of Huang Ti.
YAHS, 1934, 2, 21.

Chhien Pao-Tsung (*1*) 錢寶琮.
Chung-Kuo Suan-Hsüeh Shih 中國算學史.
A History of Chinese Mathematics.
National Research Institute of History and Philology Monographs, ser. A, no. 6.
Academia Sinica, Peiping, 1932. Pt. I (Pt. II never published).

Chhüan Han-Shêng (*1*) 全漢昇.
Thang Sung Ti Kuo yü Yün-Ho
唐宋帝國與運河.
The Thang and Sung Empires and the Grand Canal.
Com. Press, Chungking, 1944.

Fan Shih 范適.
Ming chi Hsi-Yang Chhuan ju chih I-Hsüeh
明季西洋傳入之醫學.
On the Coming of Western Medicine to China during the Ming Dynasty.
Chinese Medical History Society, Shanghai, 1946.

Fan Wên-Lan (*1*) 范文瀾.
Chung-Kuo Thung Shih Chien Pien
中國通史簡編.
General History of China (2 vols.).
Peiping, 1948, and several later single-volume editions.

Fêng Chhêng-Chün 馮承鈞.
Chung-Kuo Nan-Yang Chiao-Thung Shih
中國南洋交通史.
History of the Contacts of China with the South Sea Regions.
Com. Press, Shanghai, 1937.

Fêng Yu-Lan (*1*) 馮友蘭.
Chung-Kuo Chê-Hsüeh Shih, 2 vols.
中國哲學史.
History of Chinese Philosophy.
Com. Press, Chhangsha, 1934, 2nd ed. 1941.

Fêng Yün-Phêng (*1*) 馮雲鵬.
Chin Shih So 金石索.
Collection of Carvings, Reliefs and Inscriptions.
[This was the first modern publication of the Han tomb-shrine reliefs.] 1821.

Hu Shih (*1*) 胡適.
Hu Shih Wên Tshun 胡適文存.
Collected Works.
Yatung, Shanghai, 1930.

Hsiang Ta (*3*) 向達.
Thang Tai Chhang-an yü Hsi-Yü Wên Ming
唐代長安與西域文明.
Western Cultures at the Chinese Capital (Chhang-an) during the Thang Dynasty.
YCHP Monograph Series, no. 2, Peiping, 1933.

Hsü Chien-Yen (*1*) 徐建寅.
Ko Chih Tshung-Shu 格致叢書.
Compendium of General Science.
1901.

Hsü Chung-Shu (*1*) 徐中舒.
In *An-yang Fa Chüeh Pao Kao*
安陽發掘報告.
Excavations at Anyang.
Academia Sinica, Peiping, 1929–31.

Ku Chieh-Kang (*1*) 顧頡剛.
Han Tai Hsüeh Shu Shih-Lüeh
漢代學術史略.
Outline History of Learning in the Han Dynasty.
Tungfang, Chungking, 1944.

Ku Chieh-Kang (*2*) (ed.) 顧頡剛.
Ku Shih Pien 古史辨.
Discussions on Ancient History and Philosophy (a collective work).
Vols. 1 to 3, and 5, Phu Shih, Peiping, 1916–31, 1935.

Ku Chieh-Kang (*5*) 顧頡剛.
Yü Chhien Hsüan-Thung hsiensêng Lun Ku Shih Shu 與錢玄同先生論古史書.
KSP, 1926, **1**, 59.
Eng. abstr. *CIB*, 1938, **3**, 67.

Kuo Mo-Jo (*2*) 郭沫若.
Ku Tai Shih-Hui chih Yen-Chiu
古代社會之硏究.
Studies in Ancient Chinese Society.
Shanghai, *c.* 1927.

Kuo Mo-Jo (*3*) 郭沫若.
Chia Ku Wên Tzu Yen-Chiu
甲骨文字硏究.
Researches on the Characters on the Oracle Bones (incl. astronomical and calendrical data).
2 vols., Peiping, 1931. Germ. abstr. by W. Eberhard, *OAZ*, 1932, **8**, 225.

Kuo Po-Kung (*1*) 郭伯恭.
Yung-Lo Ta Tien Khao 永樂大典考.
A Study of the Yung-Lo reign-period Encyclopaedia.
Com. Press, Shanghai, 1938.

Li Chao-Lo (*1*) 李兆洛.
Li Tai Yü Ti Yen Ko Hsien Yao Thu
歷代輿地沿革險要圖.
Historical Atlas of China.
1838, repr. 1879.

Li Chhang-Chih (*1*) 李長之.
Tao-Chiao Thu ti Shih-Jen Li Pai chi chhi Thung Khu
道敎徒的詩人李白及其痛苦.
The Taoist Poet Li Pai and his sufferings.
Com. Press, Chungking, 1943.

Liu Hsien-Chou (*1*) 劉仙洲.
Chung-Kuo Chi-Hsieh Kung-Chhêng Shih-Liao 中國機械工程史料.
Materials for the History of Engineering in China.
CHESJ, 1935, **3**, and **4** (no. 2), 27.
Reprinted Chhinghua Univ. Press, Peiping, 1935. With supplement, *CHER*, 1948, **3**, 135.

Lo Chen-Yü (*1*) 羅振玉.
Yin Shang Chen Pu Wên Tzu Khao
殷商貞卜文字考.
Investigations on the Characters written on the Shang Divination Bones.
1910.

Lo Kên-Tsê (*1*) 羅根澤.
Chan-Kuo Chhien Wu Ssu Chia Chu Tso Shuo
戰國前無私家著作說.
Absence of Books by Individual Writers before the Warring States Period.
KSP, 1933, **4**, 8.
Eng. abstr. *CIB*, 1938, **3**, 82.

Lo Kên-Tsê (*3*) (ed.) 羅根澤.
Ku Shih Pien 古史辨.
Discussions on Ancient History and Philosophy (a collective work), vol. 4.
Phu Shih, Peiping, 1933.

Lü Chen-Yü (*1*) 呂振羽.
Chien Ming Chung-Kuo Thung Shih
簡明中國通史.
General History of China, 2 vols.
Sêng-Huo, Peking, 1951.

Ma Kuo-Han (*1*) (ed.) 馬國翰.
Yü Han Shan Fang Chi I Shu 玉函山房輯佚書.
Jade Box Mountain Studio Collection of (Reconstituted) Lost Books.
1853.

Shih Chang-Ju (*3*) 石璋如.
Chhuan-Shuo chung Chou Tu ti Shih-Ti Khao-Chha
傳說中周都的實地考察.
A Field Investigation of the Traditional Sites of Settlements of the Chou People.
AS/BIHP, 1949, **20** B, 91.

Shih Thing-Yung (*1*) 施廷鏞.
Tshung-Shu Tzu Mu Shu Ming So Yin
叢書子目書名索引.
Index to the Tshung-Shu Collections.
Chhinghua Univ. Peiping, 1936.

Su Yuan-Lei (*1*) 蘇淵雷.
Hsüan-Chuang 玄奘.
Biography of Hsüan-Chuang.
Victory Press, Chungking, 1944.

Sun I-Jang (*1*) 孫詒讓.
Chhi Wên Chü Li 契文擧例.
Examples of Oracle-Bone Writing.
1904, printed 1917.

Sun Yü-Thang (*1*) 孫毓棠.
Chung-Kuo Ku Tai Shih-Hui Ching-Chi Lun-Tshung 中國古代社會經濟論叢.
Discussion of ancient Chinese Social and Economic History.
Yunnan Economic Council, Kunming, 1943

Têng Chhu-Min (*1*) 鄧初民.
Shê-Hui Shih Chien Ming Chiao Chhêng 社會史簡明教程.
Handbook of Chinese Social History.
Sêng-Huo, Chungking, 1943.

Than Chêng-Pi (*1*) 譚正璧.
Chung-Kuo Wên-Hsüeh Chia Ta Tzu-Tien 中國文學家大字典.
Biographical Dictionary.
Kuangming, Shanghai, 1941.

Ting Hsü-Hsien (*1*) 丁緒賢.
Hua-Hsüeh Shih 化學史.
History of Chemistry (in the West).
Shanghai, *c.* 1936.

Ting Wên-Chiang, Ong Wên-Hao & Tsêng Shih-Ying (*1*) (ed.) 丁文江, 翁文灝, 曾世英.
Chung-Hua Min-Kuo Hsin Ti-Thu 中華民國新地圖.
New Atlas of the Chinese Republic.
Shenpao, Shanghai, 1934.

Tung Tso-Pin (*1*) 董作賓.
Yin Li Phu 殷曆譜.
On the Calendar of the Shang Dynasty.
Academia Sinica, Lichuang, 1945.

Wang Ching-Ju (*1*) 王靜如.
Hsi-Hsia Wên chih Yen-Chiu 西夏文之研究.
Hsi-Hsia Studies, Pts. 1 and 2.
National Research Institute of History and Philology, Monographs, Ser. A, nos. 8 and 11.
Academia Sinica, Peiping, 1932, 1933.

Wang Jen-Chün (*1*) 王仁俊.
Ko Chih Ku Wei 格致古微.
Scientific Traces in Olden Times.
1896.

Wang Kuo-Wei (*1*) 王國維.
Wang Chung-Chhio Kung I Shu 王忠愨公遺書.
Collected Works. (48 vols.)
Com. Press, Chhangsha, 1940.

Wang Kuo-Wei (*2*) 王國維.
Yin Pu Tzhu so chien Hsien Kung Hsien Wang Khao 殷卜辭所見先公先王考.
Information about the Ancient Kings and Rulers on the Oracle-Bones.
In *Kuan Thang Chi Lin* 觀堂集林.
(also part of Wang Kuo-Wei, *1*).

Wang Yung (*1*) 王庸.
Chung-Kuo Ti-Li Hsüeh Shih 中國地理學史.
History of Geography in China.
Com. Press, Chhangsha, 1938.

Wên I-To (*1*) 聞一多.
Tshung Jen-Shou Shê-Shen Hsiang Than tao Lung yü Thu-Thêng 從人首蛇身像談到龍與圖騰.
Beings with human heads and serpent bodies considered as Dragon Totems.
JWKHS, 1942, **1**, 1.

Wu Chhêng-Lo (*1*) 吳成洛.
Chung Hsi Kho-Hsüeh I-Shu Wên-Hua Li-Shih Pien Nien Tui Chao 中西科學藝術文化歷史編年對照.
Comparative Tables of Scientific, Technological and Scholarly Achievements in China and Europe.
KHS, 1925, **10**, 1.

Wu Yü (*1*) 吳虞.
Wu Yü Wên Lu 吳虞文錄.
Collected Works.
Chhêngtu, 1936.

Yang Chia-Lo (*1*) (ed.) 楊家駱.
Ssu Khu Chhüan Shu Hsüeh Tien 四庫全書學典.
Bibliographical Index of the *Ssu Khu Chhüan Shu* Encyclopaedia.
World Book Co., Shanghai, 1946.

Yen Kho-Chün (*1*) (ed.) 嚴可均.
Chhüan Shang-Ku San-Tai Chhin Han San-Kuo Liu Chhao Wên 全上古三代秦漢三國六朝文.
Complete Collection of Prose Literature (including fragments) from Remote Antiquity through the Chhin and Han Dynasties, the Three Kingdoms and the Six Dynasties.
Finished 1836; published 1887–93.

Yin-Tê Series, no. 10 引得.
I Wên Chih Erh-shih Chung Tsung Ho Yin-Tê 藝文志二十種綜合引得.
Combined Indexes to Titles of Books in Twenty of the Bibliographies in the Official Histories.
Harvard-Yenching Institute, Peiping, 1932.

Yin-Tê Series, no. 23 引得.
Thai-Phing Yü Lan Yin-Tê 太平御覽引得.
Index to the *Thai-Phing Yü Lan* Encyclopaedia (Subjects, and Books quoted).
Harvard-Yenching Institute, Peiping, 1935.

Yin-Tê Series, no. 25 引得.
Tao Tsang Tzu Mu Yin-Tê 道藏子目引得.
Combined Indexes to the Authors and Titles of Books in Two Collections of Taoist Literature, by Ong Tu-Chien 翁獨健.
Harvard-Yenching Institute, Peiping, 1935.

Yin-Tê Series, no. 27 引得.
Li Chi Yin-Tê 禮記引得.
Index to Words in the *Li Chi* (a Concordance).
Harvard-Yenching Institute, Peiping, 1937.

Yin-Tê Series, no. 32 引得.
Shih Huo Chih Shih-wu Chung Tsung Ho Yin-Tê 食貨志十五種綜合引得.
Combined Indexes to the Economic Sections of Fifteen of the Official Histories.
Harvard-Yenching Institute, Peiping, 1938.

Yin-Tê Series, no. 40 引得.
Shih Chi chi Chu Shih Tsung Ho Yin-Tê 史記及注釋綜合引得.
Index to Words in the *Shih Chi* and its Collected Commentaries (a Concordance).
Harvard-Yenching Institute, Peiping, 1947.

Yin-Tê Series, Suppl. no. 10 引得.
Chou I Yin-Tê 周易引得.
Index to Words in the *I Ching* (a Concordance).
Harvard-Yenching Institute, Peiping, 1935.

Yin-Tê Series, Suppl. no. 16 引得.
Lun Yü Yin-Tê 論語引得.
Index to Words in the Confucian Analects (a Concordance).
Harvard-Yenching Institute, Peiping, 1940.

Yin-Tê Series, Suppl. no. 17 引得.
Mêng Tzu Yin-Tê 孟子引得.
Index to Words in Mencius (a Concordance).
Harvard-Yenching Institute, Peiping, 1941.

Yin-Tê Series, Suppl. no. 18 引得.
Erh Ya Yin-Tê 爾雅引得.
Index to Words in the *Erh Ya* Dictionary (a Concordance).
Harvard-Yenching Institute, Peiping, 1941.

Yin-Tê Series, Suppl. no. 20 引得.
Chuang Tzu Yin-Tê 莊子引得.
Index to Words in *Chuang Tzu* (a Concordance).
Harvard-Yenching Institute, Peiping, 1947.

Yin-Tê Series, Suppl. no. 21 引得.
Mo Tzu Yin-Tê 墨子引得.
Index to Words in *Mo Tzu* (a Concordance).
Harvard-Yenching Institute, Peiping, 1948.

Yuan Thung-Li 袁同禮.
Yung-Lo Ta Tien hsien-tshun Chhüan Mu Piao 永樂大典現存卷目表.
Census of the Locations and Contents of the still existing Volumes of the Yung-Lo reign-period Encyclopaedia.
QBCB/C, 1939 (n.s.), **1**, 246.

ADDENDA TO BIBLIOGRAPHY B

Chang Chêng-Lang (*1*) 張政烺.
Liu Shu Ku I 六書古義.
Investigation of the Ancient Meaning of the term 'Six Writings'.
AS/BIHP, 1948, **10**, 1.

Nagasawa, Kikuya (*2*) 長澤規矩也.
Shina-shoshi Kaidai 支那書籍解題.
Bibliography of Bibliographies of Chinese Books.
Tokyo, 1940.

C. BOOKS AND JOURNAL ARTICLES IN WESTERN LANGUAGES

ADNAN ADIVAR (1). 'On the *Tanksuq-nāmah-i Īlkhān dar Funūn-i 'Ulūm-i Khiṭāi*.' *ISIS*, 1940 (appeared 1947), **32**, 44.

AL-AIYŪBĪ, ABŪ'L-FIDĀ'. *Taqwīm al-Buldān* (Universal Geography). See Reinaud & Guyard.

ANDERSON, G. F. (1). 'The Wonderful Canals of China.' *NGM*, 1905, **16**, 68.

ANDERSON, JOHN (1). *A Report on an Expedition to Western Yunnan via Bhamo.* Calcutta, 1871.

ANDERSSON, J. G. (1). *Children of the Yellow Earth.* Kegan Paul, London, 1934.

ANDERSSON, J. G. (2). 'The Cave Deposit at Sha-Kuo-Thun in Fêngtien.' *PS* (ser. D), 1923, **1**, 1.

ANDERSSON, J. G. (3). 'An Early Chinese Culture.' *BGSC*, 1923, **5**, 1.

ANDERSSON, J. G. (4). 'Preliminary Report on Archaeological Research in Kansu.' *MGSC* (ser. A), 1925, **5**, 1.

ANDERSSON, J. G. (5). 'Researches into the Prehistory of the Chinese.' *BMFEA*, 1943, **15**, 7.

ANDERSSON, J. G. (6). 'Prehistoric Sites in Honan.' *BMFEA*, 1947, **19**, 1.

ANDERSSON, J. G. (7). 'Hunting Magic in the Animal Style.' *BMFEA*, 1932, **4**, 221.

ANDRÉE, R. (1). 'Scapulimantia.' In Boas Memorial Volume: *Anthropological Papers in honour of Franz Boas.* Stechert, New York, 1906, p. 143.

ANON. (1). *Manuel du Sinologue* (Bibl. Sinico-Japonaise no. 6). Maisonneuve, Paris, 1889.

APPLETON, W. W. (1). *A Cycle of Cathay; the Chinese Vogue in England during the 17th and 18th Centuries.* Columbia Univ. Press, New York, 1951.

ARENDT, C. (1). 'Synchronistische Regententabellen z. Geschichte d. chinesischen Dynastien.' *MSOS*, 1899, **2**, 152–250; 1900, **3**, 1–164; 1901, **4**, 114–70.

ARENDT, C. (2). 'Parallels in Greek and Chinese Literature.' *JPOS*, 1886, **1**, 29.

ARMSTRONG, E. A. (1). 'The Crane Dance in East and West.' *AQ*, 1943, **17**, 71.

ARMSTRONG, E. A. (2). 'The Ritual of the Plough.' *FL*, 1943, **54**, 250.

ARMSTRONG, E. A. (3). 'The Triple-Furrowed Field.' *CLR*, 1943, **57**, 3.

ARMSTRONG, E. A. (4). 'Mugwort Lore.' *FL*, 1944, **55**, 22.

ARMSTRONG, E. A. (5). 'Chinese Bull Ritual and its Affinities.' *FL*, 1944, **56**, 200.

ARMSTRONG, E. A. (6). 'Symbolism of the Swan and the Goose.' *FL*, 1944, **55**, 54.

AUBOYER, J. (1). 'L'Influence Chinoise sur le Paysage dans la Peinture de l'Orient et dans la Sculpture de l'Insulinde.' *RAA/AMG*, 1935, **9**, 228.

BACON, FRANCIS (1). *Philosophical Works*, ed. Ellis & Spedding. Routledge, London, 1905.

BAGCHI, P. C. (1). *India and China; a thousand years of Sino-Indian Cultural Relations.* Hind Kitab, Bombay, 1944. 2nd ed. 1950.

BALAZS, E. (=S.) (1). 'La Crise Sociale et la Philosophie Politique à la Fin des Han.' *TP*, 1949, **39**, 83.

BALAZS, E. (=S.) (2). 'Entre Révolte Nihiliste et Evasion Mystique' (the Seven Sages of the Bamboo Grove, and Pao Ching-Yen). *ASEA*, 1948, **1**, 27 (sequel to Balazs, 1).

BALAZS, E. (=S.) (4). 'Ein Vorlaüfer von Wang An-Schi' [Li Kou]. *SA*, 1933, **8**, 165.

BALL, C. J. (1). *Chinese and Sumerian.* Oxford, 1913. (crit. L. C. Hopkins, *PSBA*, 1914, **36**, 269; 1915, **37**, 24, 50 & 75. Cf. H. H. White, *NCR*, 1920, **2**, 37.)

BARBOUR, G. B. (1). (*a*) 'The Loess of China.' *CJ*, 1925, **3**, 454, 509. *ARSI*, 1926, 279. (*b*) 'The Loess Problem of China.' *GM*, 1930, **67**, 458.

BARBOUR, G. B. (2). 'Physiographic History of the Yangtze.' *GJ*, 1936, **87**, 17.

BARBOUR, G. B. (3). 'Recent Observations on the Loess of North China.' *GJ*, 1935, **86**, 54.

BARKER, SIR ERNEST (1). 'Italy and the West 410–476 A.D.', in *CMH*, 1911, vol 1, p. 393.

BARNETT, L. D. (1). 'Commercial and Political Connections of Ancient India with the West.' *BLSOAS*, 1917, **1** (1), 101.

BARTH, A. (1). Critique of Chavannes (4) and of Takakusu (1). *JS*, 1898, 261, 425, 522.

BEAL, S. (1) (tr.). *Travels of Fah-Hian [Fa-Hsien] and Sung Yün, Buddhist Pilgrims from China to India* (+400 and +518). Trübner, London, 1869.

BEAL, S. (2) (tr.). *Si Yu Ki [Hsi Yü Chi], Buddhist Records of the Western World, transl. from the Chinese of Hiuen Tsiang [Hsüan-Chuang].* 2 vols. Trübner, London, 1884. 2nd ed. 1906.

BEAL, S. (3) (tr.). *The Life of Hiuen Tsiang [Hsüan-Chuang] by the Shaman Hwui Li [Hui-Li], with an introduction containing an account of the Works of I-Tsing [I-Ching].* Kegan Paul, London, 1911.

BEAZLEY, C. R. (1). *The Dawn of Modern Geography.* 3 vols. (vol. 1, +300 to +900; vol. 2, +900 to +1260; vol. 3, +1260 to +1420). Vols. 1 and 2, Murray, London, 1897 and 1901. Vol. 3, Oxford, 1906.

BELL, M. S. (1). 'The Great Central Asian Trade Route from Peking to Kashgaria.' *PRGS* (n.s.), 1890, **12**, 57.

BELPAIRE, B. (1). 'Le Taoisme et Li Thai-Pai.' *MCB*, 1931, **1**, 1.

BERGMAN, FOLKE (1). *Archaeological Researches in Sinkiang.* Reports of the Sino-Swedish [Scientific] Expedition [to Northwest China]. 1939, vol. 7 (1).

BERNARD-MAÎTRE, H. (1). *Matteo Ricci's Scientific Contribution to China*, tr. E. T. C. Werner. Vetch, Peiping, 1935.

BERNARD-MAÎTRE, H. (2). *Sagesse Chinoise et Philosophie Chrétienne; Essai sur leurs Relations Historiques.* Mission Press, Tientsin, 1935.

BERNARD-MAÎTRE, H. (3). 'Le Père le Chéron d'Incarville, missionaire français à Pékin.' *A/AIHS*, 1949, **28**, 333.

BERNARD-MAÎTRE, H. (4). 'Notes on the Introduction of the Natural Sciences into the Chinese Empire.' *YJSS*, 1941, **3**, 220.

BERNARD-MAÎTRE, H. (5). *Le Père Matthieu Ricci et la Société Chinoise de son Temps (1552 à 1610).* 2 vols. Hsienhsien, Tientsin, 1937.

BERTHELOT, ANDRÉ (1). *L'Asie Centrale et Sud-Orientale d'après Ptolemée.* Payot, Paris, 1930.

BEVAN, E. R. (1). 'India in Early Greek and Latin Literature', ch. 16 in *CHI*, vol. 1, p. 391.

BIOT, E. (1) (tr.). *Le Tcheou-Li ou Rites des Tcheou* [*Chou*]. 3 vols. Imp. Nat., Paris, 1851. (Photographically reproduced, Wêntienko, Peiping, 1940.)

BIOT, E. (2). *Essai sur l'histoire de l'Instruction Publique en Chine.* Duprat, Paris, 1845.

BIOT, E. (3) (tr.). *Chu Shu Chi Nien* (Bamboo Books). *JA*, 1841 (3e sér.), **12**, 537; 1842, **13**, 381.

BIOT, E. (18). 'Mémoire sur les Colonies Militaires et Agricoles des Chinois.' *JA*, 1850 (4e sér.), **15**, 338, 529.

BISHOP, C. W. (1). 'Chronology of Ancient China.' *JAOS*, 1932, **52**, 232.

BISHOP, C. W. (2). 'Beginnings of Civilisation in Eastern Asia.' *AQ*, 1940, **14**, 301; *JAOS*, 1939, Suppl. no. 4, p. 35.

BISHOP, C. W. (3). 'The Rise of Civilisation in China with reference to its Geographical Aspects.' *GR*, 1932, **22**, 617.

BISHOP, C. W. (4). 'The Neolithic Age in Northern China.' *AQ*, 1933, **7**, 389.

BISHOP, C. W. (5). 'The Geographical Factor in the Rise of Chinese Civilisation.' *GR*, 1922, **12**, 19.

DE LA BLACHE, P. VIDAL (1). *Principes de Géographie Humaine.* Colin, Paris, 1921 and 1948.

BLACK, J. DAVIDSON & DE CHARDIN, TEILHARD (1). 'On the Discovery of a Palaeolithic Industry in Northern China.' *BCGS*, 1924, **3**, 45.

BLACK, J. DAVIDSON, LICENT, E. & DE CHARDIN, TEILHARD (1). 'On a presumably Pleistocene Tooth from the Sjara Osso Gol.' *BCGS*, 1927, **5**, 285.

BLOCH, MARC (1). *La Société Féodale.* Vol. 1. *La Formation des Liens de Dépendance.* Albin Michel, Paris, 1939. Vol. 2. *Les Classes et le Gouvernement des Hommes.* Albin Michel, Paris, 1940. (Evol. de l'Hum. Sér.; Moyen-Age, nos. 34 and 34 bis.)

BODDE, D. (1). *China's First Unifier, a study of the Chhin Dynasty as seen in the life of Li Ssu (−280 to −208).* Brill, Leiden, 1938. (Sinica Leidensia ser. no. 3.)

BODDE, D. (9). 'Some Chinese Tales of the Supernatural; Kan Pao and his *Sou Shen Chi*.' *HJAS*, 1942, **6**, 338.

BODDE, D. (10). 'Again Some Chinese Tales of the Supernatural; Further Remarks on Kan Pao and his *Sou Shen Chi*.' *JAOS*, 1942, **62**, 305.

BOERSCHMANN, E. (1). *Chinesische Pagoden.* de Gruyter, Berlin, 1931.

BOHLIN, B. (1). 'Newly Visited Western Caves at Tunhuang.' *HJAS*, 1936, **1**, 163.

BOUVAT, L. (1). *L'Empire Mongol (2e Phase).* Boccard, Paris, 1927. (*Histoire du Monde*, ed. E. Cavaignac, vol. 8 (3).) See Grousset (2).

BREASTED, J. H. (1). *The Conquest of Civilisation.* Harper, New York, 1926.

BRÉHIER, L. (1). *La Philosophie de Plotin.* Paris, 1928.

BRETSCHNEIDER, E. (1). *Botanicon Sinicum; Notes on Chinese Botany from Native and Western Sources.* 3 vols. Trübner, London, 1882 (printed in Japan). (Reprinted from *JRAS/NCB*, 1881, **16**.)

BRETSCHNEIDER, E. (2). *Medieval Researches from Eastern Asiatic Sources; Fragments towards the Knowledge of the Geography and History of Central and Western Asia from the thirteenth to the seventeenth century.* 2 vols. Trübner, London, 1888.

BRETSCHNEIDER, E. (3). 'Chinese Intercourse with the Countries of Central and Western Asia during the 15th century. I. Accounts of Foreign Countries and especially those of Central and Western Asia, drawn from the *Ming Shih* and the *Ta Ming I Thung Chih.*' *CR*, 1876, **5**, 132, 165. Reprinted in Bretschneider (2), vol. 2, p. 157.

BRETSCHNEIDER, E. (4). 'Chinese Intercourse with the Countries of Central and Western Asia during the 15th century. II. A Chinese Itinerary of the Ming period from the Chinese Northwest Frontier to the Mediterranean Sea.' *CR*, 1876, **5**, 227. Reprinted (abridged) in Bretschneider (2), vol. 2, p. 329.

BREWITT-TAYLOR, C. H. (1) (tr.). *San Kuo, or the Romance of the Three Kingdoms.* Kelly & Walsh, Shanghai, 1926.

BROCKELMANN, C. (1). *History of the Islamic Peoples.* Eng. tr. J. Carmichael & M. Perlmann. Putnam, New York, 1947.

BROCKELMANN, C. (2). *Geschichte d. arabischen Literatur.* Felber, Weimar, 1898. Supplementary volumes. Brill, Leiden, 1937.

BROOMHALL, M. (1). *Islam in China.* Morgan & Scott, London, 1910.

BROWNE, E. G. (1). *Arabian Medicine.* Cambridge, 1921. (French tr. H. J. P. Renaud; Larose, Paris, 1933.)

BRUCE, J. P. (1) (tr.). *The Philosophy of Human Nature, translated from the Chinese, with notes.* Probsthain, London, 1922. (Chs. 42-8, inclusive, of *Chu Tzu Chhüan Shu.*)

BRUNET, P. & MIELI, A. (1). *L'Histoire des Sciences (Antiquité).* Payot, Paris, 1935.

BUCK, P. (1) (tr.). *All Men are Brothers.* New York, 1933. (The *Shui Hu Chuan.*)

BUDGE, E. A. WALLIS (1). *Guide to the Egyptian Collections in the British Museum.* Brit. Mus. Trustees, London, 1909, and subsequent editions.

BUDGE, E. A. WALLIS (2) (tr.). *The Monks of Kublai Khan, or the History of the Life and Travels of Rabban Sauma and Marqos.* (Travels of Marqos Bayniel. Tr. from Syriac.) RTS, London, 1928.

BULLOCH, W. (1). 'History of Bacteriology.' In *A System of Bacteriology in Relation to Medicine*, vol. 1, p. 15. MRC, London, 1930.

BURDSELL, R. L. (1). 'Altitude and Location of Minya Konka.' *GR*, 1934, **24**, 118.

BURGESS, E. (1) (tr.). *Sūrya Siddhānta; a Translation of a Textbook of Hindu Astronomy, with notes and an appendix.* Ed. P. Gangooly, introd. by P. Sengupta. Calcutta Univ., Calcutta, 1860; repr. 1935.

BURKILL, I. H. (1). *A Dictionary of the Economic Products of the Malay Peninsula* (with contributions by W. Birtwhistle, F. W. Foxworthy, J. B. Scrivenor & J. G. Watson). 2 vols. Crown Agents for the Colonies, London, 1935.

BURKITT, F. C. (1). *The Religion of the Manichees.* Cambridge, 1925.

BURLING, J. & BURLING, A. H. (1). 'Western Influences on Chinese Art.' *CJ*, 1941, **34**, 149.

BURY, J. B. (1). *The Idea of Progress.* Macmillan, London, 1920.

BURY, J. B. (2). *The Invasion of Europe by the Barbarians.* Macmillan, London, 1928.

BUSHELL, S. W. (1). 'Ancient Roman Coins from Shansi.' *JPOS*, 1886, **1**, 17.

BUXTON, L. H. DUDLEY (1). *China, a Human Geography.* Oxford, 1929.

CABLE, M. & FRENCH, F. (1). *The Gobi Desert.* Hodder & Stoughton, London, 1942.

CARLES, W. R. (1). 'The Yangtze Kiang.' *GJ*, 1898, **12**, 225.

CARR-SAUNDERS, A. M. (1). *The Population Problem.* Oxford, 1922.

CARTER, T. F. (1). (*a*) *The Invention of Printing in China and its spread Westward.* Columbia, New York, 1925; 2nd ed. 1931. (*b*) 'The Westward Movement of the Art of Printing,' in *Yearbook of Oriental Art and Culture*, ed. A. Waley, vol. 1, p. 19. Benn, London, 1925. (rev. B. Laufer, *JAOS*, 1927, **47**, 71.)

CEDRENOS, GEORGIOS (*c.* +1057) (1). *Compendium Historiarum.* 2 vols. Bonn, 1838–9.

CHABOT, J. B. (1). *Histoire de Mar Jabalaha III.* Paris, 1895.

CHALFANT, F. H. (1). 'Early Chinese Writing.' *MCMU*, 1906, **4**, 1.

CHANG HSIN-HAI (1). 'Some Types of Chinese Historical Thought.' *CSPSR*, 1929, **13**, 321.

CHANG, MATTHIAS (1). *Synchronismes Chinois; Chronologie Complète et Concordance avec l'ère Chrétienne de toutes les dates concernant l'Histoire de l'Extrême-Orient; Chine, Japon, Corée, Annam, Mongolie, etc. VS*, no. 24, 1905.

CHAO YUAN-JÊN (1). 'Languages and Dialects in China' (with map). *GJ*, 1943, **102**, 63.

DE CHARDIN, TEILHARD & PHEI WÊN-CHUNG (1). *Le Néolithique de la Chine*, French Bookstore, Peiping, 1944. (Pub. Instit. de Géobiologie, Peiping, no. 10.)

DE CHARDIN, TEILHARD & YANG CHUNG-CHIEN (1). 'Preliminary Observations on the Pre-Loessic and Post-Pontian Formations in Western Shansi and Northern Shensi.' *MGSC*, 1930 (ser. A), no. 8, 34.

CHARLESWORTH, M. P. (1). 'Some Notes on the *Periplus Maris Erythraei.*' *CQ*, 1928, **22**, 92.

CHATLEY, H. (1). MS translation of the astronomical chapter (ch. 3, Thien Wên) of *Huai Nan Tzu.* Unpublished. (Cf. note in *O*, 1952, **72**, 84.)
CHATLEY, H. (2). 'The Development of Mechanisms in Ancient China.' *TNS*, 1942, **22**, 117. (Long abstr. without illustr. *ENG*, 1942, **153**, 175.)
CHATLEY, H. (3). 'Science in Old China.' *JRAS/NCB*, 1923, **54**, 65.
CHATLEY, H. (4). 'Did Ancient Chinese Culture come from Egypt?' *JRAS/NCB*, 1929, **60**, 79.
CHATLEY, H. (9). 'Ancient Chinese Astronomy.' *RAS/ON*, 1939, no. 5, 65.
CHATLEY, H. (23). *The Origin and Diffusion of Chinese Culture.* China Soc. London, Oct. 1947.
CHAVANNES, E. (1). *Les Mémoires Historiques de Se-Ma Ts'ien* [*Ssuma Chhien*]. 5 vols. Leroux, Paris, 1895–1905. (Photographically reproduced, in China, without imprint and undated.)
1895 vol. 1 tr. *Shih Chi*, chs. 1 to 4 inclusive.
1897 vol. 2 tr. *Shih Chi*, chs. 5, 6, 7, 8, 9, 10, 11, 12.
1898 vol. 3 (i) tr. *Shih Chi*, chs. 13, 14, 15, 16, 17, 18, 19, 20, 21, 22.
vol. 3 (ii) tr. *Shih Chi*, chs. 23, 24, 25, 26, 27, 28, 29, 30.
1901 vol. 4 tr. *Shih Chi*, chs. 31, 32, 33, 34, 35, 36, 37, 38, 39, 40, 41, 42.
1905 vol. 5 tr. *Shih Chi*, chs. 43, 44, 45, 46, 47.
CHAVANNES, E. (2). 'Le Royaume de Wou [Wu] et de Yue [Yüeh].' *TP*, 1916, **17**, 129.
CHAVANNES, E. (3) (tr.). 'Le Voyage de Song [Sung] Yün dans l'Udyāna et le Gandhāra.' *BEFEO*, 1903, **3**, 379.
CHAVANNES, E. (4) (tr.). *Voyages des Pelèrins Bouddhistes; Les Religieux Eminents qui allèrent chercher la Loi dans les Pays d'Occident; mémoire composé à l'époque de la grande dynastie Thang par I-Tsing* [*I-Ching*]. Leroux, Paris, 1894.
CHAVANNES, E. (5). 'Le T'ai chan [Thai Shan]; Essai de Monographie d'un Culte Chinois.' *BE/AMG*, 1910, no. 21. (With appendix: 'Le Dieu du Sol dans la Chine Antique.')
CHAVANNES, E. (6) (tr.). 'Les Pays d'Occident d'après le *Heou Han Chou*.' *TP*, 1907, **8**, 149. (Ch. 118, on the Western Countries, from *Hou Han Shu*.)
CHAVANNES, E. (12). 'Introduction to the "Documents Chinois découverts par Aurel Stein dans les Sables du Turkestan Oriental",' tr. Mme. Chavannes and H. W. House. *NCR*. 1922, **4**, 341. (Reprinted, with Stein (6, 7) and H. K. Wright (1), in brochure form. Peiping, 1940.)
CHAVANNES, E. (15). 'Les Pays d'Occident d'après le *Wei Lio*' [*Wei Lüeh*]. *TP*, 1905, **6**, 519. (The whole of the passage preserved by Phei Sung-Chih in the commentary following ch. 30 of the *San Kuo Chih*, of which the part on Ta-Chhin was translated by Hirth (1).)
CHAVANNES, E. (16). 'Trois Généraux Chinois de la Dynastie des Han Orientaux.' *TP*, 1906, **7**, 210. (Tr. ch. 77 of the *Hou Han Shu* on Pan Chhao, etc.)
CHÊNG TÊ-KHUN (1). 'A Short History of Szechuan.' *JWCBRS*, 1945, **16**, 1. (West China Union Univ. Museum Offprint Series, no. 8.)
CHÊNG TÊ-KHUN (2) (tr.). 'Travels of the Emperor Mu.' *JRAS/NCB*, 1933, **64**, 124; 1934, **65**, 128.
CHÊNG TÊ-KHUN (4). 'An Introduction to Chinese Civilisation' (mainly prehistory). *ORT*, 1950: Aug. p. 28, 'Early Inhabitants'; Sept. p. 28, 'The Beginnings of Culture'; Oct. p. 29, 'The Building of Culture'.
CHÊNG TÊ-KHUN & LIANG CHHAO-THAO (1). *An Introduction to the South-western Peoples of China.* Chhêngtu, Sze., 1945. (West China Union University Museum Guidebook Series, no. 7.)
CHHEN SHOU-I. (1). 'John Webb, seventeenth century English Sinologist.' *CSPSR*, 1935, **19**, 295.
CHHU TA-KAO (1) (tr.). *Chinese Lyrics*. Cambridge, 1937.
CHHU TA-KAO (2) (tr.). *Tao Tê Ching, a new translation.* Buddhist Lodge, London, 1937.
CHI CHHAO-TING (1). *Key Economic Areas in Chinese History, as revealed in the Development of Public Works for Water-Control.* Allen & Unwin, London, 1936.
CHIANG YI (1). *Chinese Calligraphy*. London, 1938.
CHILDE, V. GORDON (1). *The Bronze Age*. Cambridge, 1930.
CHILDE, V. GORDON (2). *Man Makes Himself*. Watts, London, 1936. (Thinker's Library, no. 87.)
CHILDE, V. GORDON (3). *Progress and Archaeology*. Watts, London, 1944. (Thinker's Library, no. 102.)
CHILDE, V. GORDON (4). *What Happened in History.* Penguin Books, London, 1942; New York, 1946.
CHILDE, V. GORDON (5). *New Light on the Most Ancient East.* Kegan Paul, London, 1934.
CHILDE, V. GORDON (6). *The Dawn of European Civilisation.* 1st ed. 1925. Kegan Paul, London, 1947.
CHILDE, V. GORDON (7). 'India and the West before Darius.' *AQ*, 1939, **13**, 5.
CHILDE, V. GORDON (8). 'The Oriental Background of European Science.' *MQ*, 1938, **1**, 105.
CHILDE, V. GORDON (12). Lecture: 'Discoveries in Siberia and Eastern Russia in Relation to Prehistoric Connections between China and the West.' Cambridge, Oct. 1952.
CHOU PIEN-MING (1). *Internationalising the Chinese Script; Progress in Quokyu Romanisation*, 1937–1945. Amoy University, Amoy, 1945.

CHOU PIEN-MING (2). *Jonghwa Minquo Cienfaa; or the Chinese Constitution in International Costume; and other Specimens in Q.R., with an essay on the Last Lap of Quoyu Romanization.* Amoy University, Amoy, 1948.

CLAPP, F. G. (1). 'Along and Across the Great Wall of China.' *GR*, 1920, **9**, 221.

CLAPP, F. G. (2). 'The Huang Ho.' *GR*, 1922, **12**, 1.

CODRINGTON, K. DE B. (1). 'Geographical Introduction to the History of Central Asia.' *GJ*, 1944, **104**, 27, 73.

COEDÈS, G. (1) (tr.). *Textes d'auteurs grecs et latins relatifs à l'Extrême Orient depuis le 4ème siècle avant J. C. jusqu'au 14ème siècle après J. C.* Leroux, Paris, 1910.

COEDÈS, G. (3). 'Fouilles en Cochin-Chine; le Site de Go-Oc-Eo, Ancient Port du Royaume de Fou-Nan.' *AA*, 1947, **10**, 193.

COMBAZ, G. (1). *L'Inde et l'Orient classique.* 2 vols. Geuthner, Paris, 1937.

COMBAZ, G. (5). 'L'Evolution du Stūpa en Asie': *MCB*, 1932, **2**, 163 (Etude d'Architecture Bouddhique); 1935, **3**, 93 (Contributions Nouvelles et Vue d'Ensemble); 1937, **4**, 1 (Les Symbolismes du Stūpa).

LE COMTE. See Lecomte.

CONRADY, A. (1). 'Indischer Einfluss in China in 4-jahrh. v. Chr.' *ZDMG*, 1906, **60**, 335.

CONRADY, A. (2). 'Alte westöstliche Kulturwörter.' *BVSAW/PH*, 1925, **77**, no. 3.

COOK, T. (1). 'The Independent Lolo of South-West Szechuan.' *JWCBRS*, 1936, **8**, 70.

COOKE, A. H. (1). 'Molluscs' (general introduction) in *CNH*, vol. *Molluscs and Brachiopods*, p. 201.

COOMARASWAMY, A. K. (3). 'Hindu Sculptures at Zayton.' *OAZ*, 1933, **9**, (**19**), 5.

CORDIER, H. (1). *Histoire Générale de la Chine.* 4 vols. Geuthner, Paris, 1920.

CORDIER, H. (2). *Bibliotheca Sinica; Dictionnaire bibliographique des Ouvrages relatifs à l'Empire Chinois.* 3 vols. Ec. des Langues Orientales Vivantes, Paris, 1878–95. 2nd ed. 5 vols. pr. Vienna, 1904–24.

CORDIER, H. (7). 'The Life and Labours of Alexander Wylie.' *JRAS*, 1887, **19**, 351.

CORDIER, H. (8). *Essai d'une Bibliographie des Ouvrages publiés en Chine par les Européens au 17e et au 18e Siècle.* Leroux, Paris, 1883.

CORDIER. See Yule (1).

COSQUIN, E. (1). 'Les Mongols et leur prétendu Rôle dans la Transmission des Contes Indiens vers l'Occident Européen; Etude de Folklore Comparé sur l'introduction du *Siddhi-Kūr* et le Conte du "Magicien et son Apprenti".' *RTP*, 1912. Sep. pub. Clouzot, Niort, 1913.

COUVREUR, F. S. (1) (tr.). *Tch'ouen Ts'iou* [*Chhun Chhiu*] *et Tso Tchouan* [*Tso Chuan*]; *Texte Chinois avec Traduction Française.* 3 vols. Mission Press, Hochienfu, 1914.

COUVREUR, F. S. (2). *Dictionnaire Classique de la Langue Chinoise.* Mission Press, Hsienhsien, 1890; photographically reproduced, Vetch, Peiping, 1947.

COUVREUR, F. S. (3) (tr.). *Li Ki* [*Li Chi*], *ou Mémoires sur les Bienséances et les Cérémonies.* 2 vols. Hochienfu, 1913.

COX, E. H. M. (1). *Plant-Hunting in China, a History of Botanical Exploration in China and the Tibetan Marches.* Collins, London, 1945.

COYAJI, J. C. (1). *Cults and Legends of Ancient Iran and China.* Karani, Bombay, 1936.

CREEL, H. G. (1). *Studies in Early Chinese Culture* (1st series). Waverly, Baltimore, 1937.

CREEL, H. G. (2). *The Birth of China.* Fr. tr. by M. C. Salles, Payot, Paris, 1937. (References are to page numbers of the French ed.)

CREEL, H. G. (3). *Sinism; A Study of the Evolution of the Chinese World-View.* Open Court, Chicago, 1929.

CRESSEY, G. B. (1). *China's Geographic Foundations; A Survey of the Land and its People.* McGraw-Hill, New York, 1934.

CRESSEY, P. F. (1). 'Chinese Traits in European Civilisation: a Study in Diffusion.' *AMSR*, 1945, **10**, 595.

CRESSWELL, K. A. C. (1). 'Dr F. R. Martin's Treatise on Automata.' In *Yearbook of Oriental Art and Culture*, ed. A. Waley, vol. 1, p. 33; vol. 2, Pls. 23 and 24. Benn, London, 1925.

CROWTHER, J. G. (1). *The Social Relations of Science.* Macmillan, London, 1941.

DAREMBERG, C. & SAGLIO, E. (1). *Dictionnaire des Antiquités Grecques et Romains.* Hachette, Paris, 1875–1919.

DARMESTETER, J. (8). 'La Flèche de Nemrod en Perse et en Chine.' *JA*, 1885 (8e sér.), **5**, 229.

DAVIES, R. M. (1). *Yunnan, the Link between India and the Yangtze.* Cambridge, 1909.

DAVIS, J. F. (1). *The Chinese; a general description of China and its Inhabitants.* 3 vols. Knight, London, 1844.

DEFRÉMERY, C. & SANGUINETTI, B. R. (1) (tr.). *Voyages d'ibn Batoutah.* 5 vols. Soc. Asiat., Paris, 1853–9. (Many reprints.)

DELAPORTE, L. (1). *Les Hittites.* Albin Michel, Paris, 1936. (Evol. de l'Hum. Sér.; Préhist. no. 8 bis.)

DELAPORTE, L. (2). *Les Peuples de l'Orient Méditérranéen, I, Le Proche Orient Asiatique.* Presses Univ. de Fr., Paris, 1938. (Clio series no. 1.)

DEYDIER, H. (1). *Contributions à l'Etude de l'Art de Gandhāra.* Maisonneuve, Paris, 1950.

DIELS, H. (2). 'Über die von Prokop beschriebene Kunstuhr von Gaza; mit einem Anhang enthaltende Text und Übersetzung d. ἔκφρασις ὡρολογίου des Prokopios von Gaza.' *APAW/PH*, **1917**, no. 7.

AL-DIMASHQĪ, MUḤAMMAD IBN IBRĀHĪM AL-ANṢĀRĪ AL-ṢŪFĪ. *Nukhbat al-Dahr* (Selection of the Age on the Wonders of Land and Sea). See Mehren.

DIXON, R. B. (1). *The Building of Cultures.* London, 1928.

DOBELL, CLIFFORD (1). *Antony van Leeuwenhoek and his 'Little Animals.'* Bale, London, 1932.

DORÉ, H. (1). *Recherches sur les Superstitions en Chine.* 15 vols. T'u-Se-Wei Press, Shanghai, 1914–29.
Pt. I, vol. 1, pp. 1–146: 'Superstitious' practices, birth, marriage and death customs (*VS*, no. 32).
Pt. I, vol. 2, pp. 147–216: talismans, exorcisms and charms (*VS*, no. 33).
Pt. I, vol. 3, pp. 217–322: divination methods (*VS*, no. 34).
Pt. I, vol. 4, pp. 323–488: seasonal festivals and miscellaneous magic (*VS*, no. 35).
Pt. I, vol. 5, sep. pagination: analysis of Taoist talismans (*VS*, no. 36).
Pt. II, vol. 6, pp. 1–196: Pantheon; Trinities, culture-heroes and Buddhist gods (*VS*, no. 39).
Pt. II, vol. 7, pp. 197–298: Pantheon; Buddhist gods and saints (*VS*, no. 41).
Pt. II, vol. 8, pp. 299–462: Pantheon; Buddhist gods, saints and schools (*VS*, no. 42).
Pt. II, vol. 9, pp. 463–680: Pantheon, Taoist (*VS*, no. 44).
Pt. II, vol. 10, pp. 681–859: Taoist celestial bureaucracy (*VS*, no. 45).
Pt. II, vol. 11, pp. 860–1052: city-gods, field-gods, trade-gods (*VS*, no. 46).
Pt. II, vol. 12, pp. 1053–1286: miscellaneous spirits, stellar deities (*VS*, no. 48).
Pt. III, vol. 13, pp. 1–263: popular Confucianism, sages of the *Wên Miao* (*VS*, no. 49).
Pt. III, vol. 14, pp. 264–606: popular Confucianism, historical figures (*VS*, no. 51).
Pt. III, vol. 15, sep. pagination: popular Buddhism, life of Gautama (*VS*, no. 57).

DRAKE, F. S. (1). (*a*) 'China's North-west Passage. A Chapter in its Opening.' *JRAS/NCB*, 1935, **66**, 42. (*b*) 'The Struggle for the Tarim Basin in the Later Han Dynasty.' *JRAS/NCB*, 1936, **67**, 147.

DRAKE, F. S. (2). 'Mohammedanism in the Thang Dynasty.' *MS*, 1943, **8**, 1.

DUBS, H. H. (1). 'The Reliability of Chinese Histories.' *FEQ*, 1946, **6**, 23.

DUBS, H. H. (2) (tr., with assistance of PHAN LO-CHI and JEN THAI). *History of the Former Han Dynasty, by Pan Ku, a Critical Translation with Annotations.* 2 vols. Waverly, Baltimore, 1938.

DUBS, H. H. (3). 'The Victory of Han Confucianism.' *JAOS*, 1938, **58**, 435. (Reprinted in Dubs (2), vol. 2, pp. 341 ff.)

DUBS, H. H. (4). 'A Chinese Gold Hoard; Wang Mang's Treasury.' *JEH*, 1942, **2**, 36.

DUBS, H. H. (5). 'The Beginnings of Alchemy.' *ISIS*, 1947, **38**, 62.

DUBS, H. H. (6). 'A Military Contact between Chinese and Romans in 36 B.C.' *TP*, 1940, **36**, 64.

DUBS, H. H. (7). *Hsün Tzu; the Moulder of Ancient Confucianism.* Probsthain, London, 1927.

DUBS, H. H. (25). 'Spelled Substitutes for Chinese Writing.' Mimeographed for private circulation, Oxford, 1951.

DUBS, H. H. (26). 'The Date of the Shang Period.' *TP*, 1951, **40**, 322.

DUBS, H. H. (27). 'On the Supposed Monosyllabic Myth' (i.e. Chinese as a monosyllabic language). *JAOS*, 1952, **72**, 82.

DUBS, H. H. (28). 'A Comment on an Illustrated Battle-Account in 35 B.C.' *TP*, 1939, **35**, 211.

DUBS, H. H. (29). 'An Ancient Military Contact between Romans and Chinese.' *AJP*, 1941, **62**, 322.

DUBS, H. H. (30). 'A Roman Influence on Chinese Painting.' *CP*, 1943, **38**, 13.

DUDGEON, J. (2). 'The Beverages of the Chinese' (on tea and wine). *JPOS*, 1895, **3**, 275.

DUYVENDAK, J. J. L. (1). 'Sailing Directions of Chinese Voyages' (a Bodleian Library MS). *TP*, 1938, **34**, 230.

DUYVENDAK, J. J. L. (2). 'Chinese in the Dutch East Indies.' *CSPSR*, 1927, **11**, 1.

DUYVENDAK, J. J. L. (3) (tr.). *The Book of the Lord Shang; a Classic of the Chinese School of Law.* Probsthain, London, 1928.

DUYVENDAK, J. J. L. (8). *China's Discovery of Africa.* Probsthain, London, 1949. (Lectures given at London University, Jan. 1947; rev. P. Paris, *TP*, 1951, **40**, 366.)

DUYVENDAK, J. J. L. (10). 'Ma Huan Re-examined.' *VKAWA/L*, 1933 (n.s.), **32**, no. 3.

DUYVENDAK, J. J. L. (13). 'Early Chinese Studies in Holland.' *TP*, 1936, **32**, 293.

DUYVENDAK, J. J. L. (15). Note on the fragment of the *Shui Hu Chuan* brought back by the Dutch in +1596 and given by Merula to the Bodleian Library, Oxford, in the seventeenth century. *TP*, 1938, **34**, 228.

DUYVENDAK, J. J. L. (16). 'An Illustrated Battle-Account in the History of the Former Han Dynasty. *TP*, 1938, **34**, 249.

EBERHARD, W. (1). 'Kultur und Siedlung der Randvölker Chinas.'. *TP* (Suppl.), 1942, **36**.

EBERHARD, W. (2). 'Lokalkulturen im alten China.' *TP* (Suppl.), 1943, **37**; *MS* Monograph, no. 3, 1942. (crit. H. Wilhelm, *MS*, 1944, **9**, 209.)

EBERHARD, W. (3). 'Early Chinese Cultures and their Development, a Working Hypothesis.' *ARSI*, 1937, 513.

EBERHARD, W. (9). *A History of China from the Earliest Times to the Present Day*. Routledge & Kegan Paul, London, 1950. Tr. from the German ed. (Swiss pub.) of 1948 by E. W. Dickes. Turkish ed. *Čin Tarihi*, Istanbul, 1946. (crit. K. Wittfogel, *AA*, 1950, **13**, 103; J. J. L. Duyvendak, *TP*, 1949, **39**, 369; A. F. Wright, *FEQ*, 1951, **10**, 380.)

ECKE, G. V. (1). 'On a Wei Relief represented in a Rubbing.' *MS*, 1936, **2**, 205.

EDGERTON, W. F. (1). (*a*) 'The Upanishads, what do they seek and why?' *JAOS*, 1929, **49**, 97. (*b*) 'Sources of the Philosophy of the Upanishads.' *JAOS*, 1917, **36**, 197. (*c*) 'Dominant Ideas in the Formation of Indian Culture.' *JAOS*, 1942, **62**, 151.

EDGERTON, W. F. (2). 'Two Notes on the Flying Gallop.' *JAOS*, 1936, **56**, 178.

EDKINS, J. (1). 'A Sketch of the Growth of Science and Arts in China to the Ming Dynasty.' *JPOS*, 1888, **2**, 142.

EDKINS, J. (2). 'What did the ancient Chinese know of the Greeks and Romans?' *JRAS/NCB*, 1883, (n.s.), **18**, 1.

EDKINS, J. (5). 'The Place of Huang Ti in Early Taoism.' *CR*, 1886, **15**, 223.

EDKINS, J. (9). 'Allusions to China in Pliny's *Natural History*.' *JPOS*, 1885, **1**, 1.

EINSTEIN, L. E. (1). 'A Chinese Design in St Mark's at Venice.' *RA*, 1926 (5e sér.), **24**, 28.

EITEL, E. J. (1) (tr.). 'Travels of the Emperor Mu.' *CR*, 1888, **17**, 233, 247.

ELLIS, G. W. (1). *Negro Culture in West Africa*. Neale, New York, 1914.

ERKES, E. (1) (tr.). 'Das Weltbild d. Huai Nan Tzu' (tr. of ch. 4). *OAZ*, 1918, **5**, 27.

ERKES, E. (2). 'Ist die Hsia Dynastie geschichtlich?' *TP*, 1937, **33**, 134.

EZERMAN, J. L. J. F. & VAN WETTUM, B. A. J. (1). 'Alphabetical List of the Emperors of China and of their year-titles or *Nien-hao*, with the date of their reign and duration.' *TP*, 1891, **2**, 357.

FABER, G. A. (1). 'Dress and Dress-Materials in Greece and Rome.' *CIBA/T*, 1938, **1**, (9), 296.

FEBVRE, L. (1) [With L. BATAILLON]. *La Terre et l'Evolution Humaine; Introduction Géographique à l'Histoire*. (L'Evol. de l'Hum. Sér. Sect. 1, Préhist. no. 4.)

FEIFEL, E. (1) (tr.). *Pao Phu Tzu Nei Phien*, chs. 1 to 3. *MS*, 1941, **6**, 113.

FEIFEL, E. (2) (tr.). *Pao Phu Tzu Nei Phien*, ch. 4. *MS*, 1944, **9**, 1.

FÊNG TA-JAN & KILBORN, L. G. (1). 'Nosu and Miao Arrow Poisons.' *JWCBRS*, 1937, **9**, 130.

FÊNG YU-LAN (1). *A History of Chinese Philosophy*: Vol. 1, *The Period of the Philosophers (from the beginnings to c. B.C. 100)* tr. D. Bodde; Vetch, Peiping, 1937; Allen & Unwin, London, 1937. Vol. 2, *The Period of Classical Learning (from the 2nd century B.C. to the 20th century A.D.)* tr. D. Bodde; Princeton Univ. Press, Princeton, N.J., 1953. Translations by Bodde of parts of vol. 2 had appeared earlier in *HJAS*. (See Fêng Yu-Lan, *1*.)

FÊNG YU-LAN (2). *The Spirit of Chinese Philosophy*, tr. E. R. Hughes. Kegan Paul, London, 1947.

FÊNG YU-LAN (5) (tr.). *Chuang Tzu; a new selected translation with an exposition of the philosophy of Kuo Hsiang*. Commercial Press, Shanghai, 1933.

FERGUSON, J. C. (1). 'China's Earliest Culture. *CSPSR*, 1938, **22**, 1.

FERGUSON, J. C. (4). 'The Southern Migration of the Sung Dynasty.' *JRAS/NCB*, 1924, **55**, 14.

FERGUSON, J. C. (5). 'Painters among Catholic Missionaries and their Helpers in Peking.' *JRAS/NCB*, 1934, **65**, 21.

FERRAND, G. (1). *Relations de Voyages et Textes Géographiques Arabes, Persans et Turcs relatifs à l'Extrême Orient, du 8e au 18e siècles, traduits, revus et annotés etc.* 2 vols. Leroux, Paris, 1913.

FERRAND, G. (2) (tr.). *Voyage du marchand Sulaymān en Inde et en Chine redigé en +851; suivi de remarques par Abū Zayd Ḥasan (vers +916)*. Bossard, Paris, 1922.

FESTUGIÈRE, A. G. (1). *La Révélation d'Hermès Trismégiste, I. L'Astrologie et les Sciences Occultes*. Gabalda, Paris, 1944. (See Filliozat, 5.)

FESTUGIÈRE, A. G. (2). 'Trois Rencontres entre la Grèce et l'Inde.' *RHR/AMG*, 1943, **125**, 32.

FÉVRIER, J. (1). 'L'Alphabet.' In *Encyclopédie Française*, 1937, vol. 1, pt. 2, ch. 4, 1.44–1.

FILLIOZAT, J. (1). *La Doctrine Classique de la Médecine Indienne*. Imp. Nat., CNRS and Geuthner, Paris, 1949.

FILLIOZAT, J. (2). 'Les Origines d'une Technique Mystique Indienne.' *RP*, 1946, **136**, 208.

FILLIOZAT, J. (3). 'Taoisme et Yoga.' *DVN*, 1949, **3**, 1.
FILLIOZAT, J. (4). 'Les Echanges de l'Inde et de l'Empire Romain aux premiers siècles de l'ère Chrétienne.' *RH*, 1949, **201**, 1.
FILLIOZAT, J. (5). Review of Festugière (1) (*q.v.*). *JA*, 1944, **234**, 349.
FILLIOZAT, J. (6). 'La Doctrine des Brahmanes d'après St Hippolyte.' *JA*, 1945, **234**, 451; *RHR/AMG*, 1945, **128**, 59.
FISHER, B. (1). 'The *Qanāts* of Persia.' *GR*, 1928, **18**, 302.
FITZGERALD, C. P. (1). *China; a Short Cultural History*. Cresset Press, London, 1935.
FITZGERALD, C. P. (2). 'The Yunnan-Burma Road.' *GJ*, 1940, **95**, 161.
FITZGERALD, C. P. (3). 'The Tiger's Leap.' *GJ*, 1941, **98**, 147.
FITZGERALD, C. P. (4). 'The Tali District of W. Yunnan.' *GJ*, 1942, **99**, 50.
FITZGERALD, C. P. (5). 'The Northern Marches of Yunnan.' *GJ*, 1943, **102**, 49.
FOCHLER-HAUKE, G. (1). *Die Mandschurei; eine geographische-geopolitische Landeskunde*. Vowinckel, Heidelberg, 1941.
FOORD, E. (1). 'China and the Destruction of the Roman Empire.' *CTR*, 1908, **94**, 207.
FOREMAN, G. (1). *Sequoyah, a Biography*. Univ. Oklahoma Press, 1938.
FORGUE, E. (1). Art. 'Histoire de la Chirurgie', in Laignel-Lavastine, *Histoire Générale de la Médecine* (*q.v.*), vol. 2, p. 350.
FORKE, A. (1). 'Ta Ts'in [Ta-Chhin] das Römische Reich.' *OAZ*, 1927, **14**, 48.
FORKE, A. (3) (tr.). *Me Ti [Mo Ti] des Sozialethikers und seiner Schüler philosophische Werke*. Berlin, 1922. (*MSOS*, Beiband, **23** to **25**.)
FORKE, A. (4) (tr.). *Lun Hêng, Philosophical Essays of Wang Chhung*: Vol. 1, 1907. Kelly & Walsh, Shanghai; Luzac, London; Harrassowitz, Leipzig. Vol. 2, 1911 (with the addition of Reimer, Berlin). (*MSOS*, Beibände, **10** and **14**.) (crit. P. Pelliot, *JA*, 1912 (10e sér.), **20**, 156.)
FORKE, A. (9). *Geschichte d. neueren chinesischen Philosophie* (i.e. from beg. of Sung to modern times). de Gruyter, Hamburg, 1938. (Hansische Univ. Abhdl. a. d. Geb. d. Auslandskunde, no. 46 (Ser. B, no. 25).)
FORKE, A. (12). *Geschichte d. mittelälterlichen chinesischen Philosophie* (i.e. from beg. of Former Han to end of Wu Tai). de Gruyter, Hamburg, 1934. (Hamburg. Univ. Abhdl. a. d. Geb. d. Auslandskunde, no. 41 (Ser. B, no. 21).)
FORKE, A. (13). *Geschichte d. alten chinesischen Philosophie* (i.e. from antiquity to beg. of Former Han). de Gruyter, Hamburg, 1927. (Hamburg. Univ. Abhdl. a. d. Geb. d. Auslandskunde, no. 25 (Ser. B, no. 14).)
FORREST, R. A. D. (1). *The Chinese Language*. Faber & Faber, London, 1948.
FOUCHER, A. (1). *La Vieille Route de l'Inde de Bactres à Taxila* (Mém. de la Déleg. Archéol. Franç. en Afghanistan). 2 vols. Paris, 1940–7.
FOUCHER, A. (2). *L'Art Gréco-Bouddhique de Gandhāra*. Paris, 1905–22.
FOX, H. M. (1). 'Lunar Periodicity in Reproduction.' *PRSB*, 1924, **95**, 523.
FOX, R. (1). *Genghis Khan*. Lane, London, 1936.
FRANKE, H. (4). 'Gustav Haloun (1898–1951): In Memoriam.' *ZDMG*, 1952, **102**, 1.
FRANKE, O. (1). *Geschichte d. chinesischen Reiches*. 5 vols. de Gruyter, Berlin, 1930–53.
FRANKE, O. (2). 'Der Ursprung der chinesischen Geschichtsschreibung.' *SPAW/PH*, 1925, 276.
FRANKE, O. (3). *Studien z. Geschichte des Konfuzianischen Dogmas und d. chinesischen Staatsreligion, das Problem des Tsch'un-ts'iu [Chhun Chhiu] und Tung Tschung-schu's Tsch'un-ts'iu-fan-lu*. Hamburg, 1920. (Hamburg. Univ. Abhdl. a. d. Geb. d. Auslandskunde, no. 1.)
FRANKE, O. (4). *Li Tschi [Li Chih], ein Beitrag z. Geschichte d. chinesisches Geisteskämpfe in 16-Jahrh. APAW/PH*, 1938, no. 10.
FRANKE, O. (5). 'Zur Frage der Einführung des Buddhismus in China.' *MSOS*, 1910, **13**, 295.
FRANKE, O. (8). *Aus Kultur und Geschichte Chinas, Vorträge und Abhandlungen aus den Jahren 1902–1942*. Deutschland Institut, Peiping, 1945.
FRANKE, O. (9). 'Zwei wichtige literarische Erwerbungen des Seminars für Sprache und Kultur Chinas zu Hamburg' (on the *Yung-Lo Ta Tien* and the *Thu Shu Chi Chhêng*). *Jahrb. d. Hamburgischen wissenschaftl. Anstalten*, 1914, **32**. (Reprinted in Franke (8), p. 91.)
FRANKE, O. (10). 'Der Sinn der chinesischen Geschichtsschreibung.' *FF*, 1937, **13**, no. 13. (Reprinted in Franke (8), p. 374.)
FRANKE, W. (1). 'Juan Yuan.' *MS*, 1944, **9**, 53.
FREEDMAN, M. (1). 'Nan Yang; Chinese South-east Asia.' *AH*, 1948, **1**, 13.
FUCHS, W. (4). 'Huei-Ch'ao's [Hui Chhao] Pilgerreise durch Nordwest-Indien und Zentral-Asien um +726.' *SPAW/PH*, 1938, **30**, 426.
FULLER, M. L. & CLAPP, F. G. (1). 'Loess and Rock Dwellings of Shensi.' *GR*, 1924, **14**, 215.
FURON, R. (1). *Manuel de Préhistoire Générale*. Payot, Paris, 1943.

VON GABAIN, A. (1) (tr.). 'Ein Fürstenspiegel; Das *Sin-Yü* [*Hsin Yü*] des Lu Kia. [Lu Chia].' Inaug. Diss. Berlin, 1930; and *MSOS*, 1930, **33**, 1.
VON DER GABELENTZ, G. (1). *Chinesische Grammatik*. Leipzig, 1881.
GALE, J. S. (1). 'The Korean Alphabet.' *JRAS/KB*, 1912, **4**, 13.
LE GALL, S. (1). *Le Philosophe Tchou Hi* [*Chu Hsi*], *sa Doctrine, son Influence*. T'u-Se-Wei, Shanghai, 1894 (*VS*, no. 6). (Incl. tr. of part of ch. 49 of *Chu Tzu Chhüan Shu*.)
GALT, H. S. (1). *A History of Chinese Educational Institutions*. Vol. 1. *To the End of the Five Dynasties* (+960). Probsthain, London, 1951.
GARDNER, C. S. (1). *A Modern System for the Romanisation of Chinese*. Harvard, 1930.
GARDNER, C. S. (2). 'The Western Transcription of Chinese.' *JRAS/NCB*, 1931, **62**, 137.
GARDNER, C. S. (3). *Chinese Traditional Historiography*. Harvard Univ. Press, Cambridge, Mass. 1938. (Harvard Histor. Monographs, no. 11.)
GAUBIL, A. (2). *Histoire Abrégée de l'Astronomie Chinoise* (with Appendices 1, Des Cycles des Chinois; 2, Dissertation sur l'Eclipse Solaire rapporté dans le *Chou-King* [*Shu Ching*]; 3, Dissertation sur l'Eclipse du Soleil rapportée dans le *Chi-King* [*Shih Ching*]; 4, Dissertation sur la première Eclipse du Soleil rapportée dans le *Tchun-Tsieou* [*Chhun Chhiu*]; 5, Dissertation sur l'Eclipse du Soleil, observée en Chine l'an trente-et-unième de Jésus-Christ; 6, Pour l'Intelligence de la Table du *Yue-Ling* [*Yüeh Ling*]; 7, Sur les Koua; 8, Sur le *Lo-Chou* (recognition of *Lo Shu* as magic square)). In *Observations Mathématiques, Astronomiques, Géographiques, Chronologiques et Physiques, tirées des anciens Livres Chinois ou faites nouvellement aux Indes, à la Chine, et ailleurs, par les Pères de la Compagnie de Jésus*, ed. E. Souciet. Rollin, Paris, 1732, vol. 2.
GAUBIL, A. (3). *Traité de l'Astronomie Chinoise*. In *Observations Mathématiques, etc.*, ed. E. Souciet. Rollin, Paris, 1732, vol. 3.
GEIL, W. E. (1). *The Sacred Five of China* [*mountains*]. Murray, London, 1926.
GEIL, W. E. (2). *The Eighteen Capitals of China*. Constable, London, 1911.
GEIL, W. E. (3). *The Great Wall of China*. Shanghai and London, 1909.
GERINI, G. E. (1). *Researches on Ptolemy's Geography of Eastern Asia* (*Further India and Indo-Malay Peninsula*). Royal Asiatic Society and Royal Geographic Society, London, 1909. (Asiatic Society Monographs, no. 1.)
GIBB, H. A. R. (1). 'The Embassy of Hārūn al-Rashīd to Chhang-An.' *BLSOAS*, 1922, **2**, 619.
GIBBON, EDWARD (1). *The History of the Decline and Fall of the Roman Empire*. 12 vols. Strahan, London, 1790 (1st ed. 1776–88).
GILES, H. A. (1). *A Chinese Biographical Dictionary*. 2 vols. Kelly & Walsh, Shanghai, 1898; Quaritch, London, 1898.
GILES, H. A. (2). *Chinese-English Dictionary*. 2 vols. Quaritch, London, 1892; 2nd. ed. 1912. Shanghai, 1909–12.
GILES, H. A. (3) (tr.). *The Travels of Fa-Hsien*. Cambridge, 1923.
GILES, L. (1). 'A Note on the *Yung-Lo Ta Tien*.' *NCR*, 1920, **2**, 137.
GILES, L. (2). *An Alphabetical Index to the Chinese Encyclopaedia* (*Chhin Ting Ku Chin Thu Shu Chi Chhêng*). British Museum, London, 1911.
GILES, L. (3). 'Two Parallel Anecdotes from Greek and Chinese Sources.' *BLSOAS*, 1922, **2**, 609.
GILES, L. (4) (tr.). *Taoist Teachings from the Book of Lieh Tzu*. Murray, London, 1912; 2nd ed. 1947.
GILES, L. (5). *Six Centuries at Tunhuang*. China Society, London, 1944.
GOODRICH, L. CARRINGTON (1). *Short History of the Chinese People*. Harper, New York, 1943.
GOODRICH, L. CARRINGTON (2). *The Literary Inquisition of Chhien-Lung*. Waverly, Baltimore, 1935.
GOODRICH, L. CARRINGTON (3). 'Cotton in China.' *ISIS*, 1943, **34**, 408.
GOODRICH, L. CARRINGTON & CHHÜ THUNG-TSU (1). 'Foreign Music at the Court of Sui Wên Ti.' *JAOS*, 1949, **69**, 148.
GOOSENS, R. (1). 'Un Texte Grec relatif à l'Aśvamedha' (in the *Life of Apollonios of Tyana* by Philostratos). *JA*, 1930, **217**, 280.
GOTCH, F. (1). 'On Some Aspects of the Scientific Method.' Essay in *Lectures on the Method of Science*, ed. T. B. Strong. Oxford, 1906.
GOUROU, P. (1). (*a*) *La Terre et l'Homme en Extrême-Orient*. Colin, Paris, 1947. (*b*) 'Notes on China's Unused Uplands.' *PA*, 1948, **21**, 227.
GRAHAM, D. C. (1). 'Historic Notes on the Pho Jen (Beh Ren); the last group of the Pre-Chinese Thai people to remain in Szechuan.' *JWCBRS*, 1936, **8**, 82.
GRAHAM, D. C. (2). 'Customs of the Chhuan Miao.' *JWCBRS*, 1937, **9**, 13.
GRAHAM, D. C. (3). 'Note about Couvade' and correspondence. *JWCBRS*, 1936, **8**, 180; 1937, **9**, 231.

GRANET, M. (1). *Danses et Légendes de la Chine Ancienne.* 2 vols. Alcan, Paris, 1926.
GRANET, M. (2). *Fêtes et Chansons Anciennes de la Chine.* Alcan, Paris, 1926; 2nd ed. Leroux, Paris, 1929.
GRANET, M. (3). *La Civilisation Chinoise.* Renaissance du Livre, Paris, 1929; 2nd ed. Albin Michel, Paris, 1948. (Evol. de l'Hum. sér. no. 25.) (crit. Ting Wên-Chiang, *MSOS*, 1931, **34**, 161.)
GREGORY, J. W. (1). 'Is the Earth drying up?' *GJ*, 1914, **43**, 148, 293.
GREIFENHAGIUS. See Müller, Andreas (Greifenhagius).
GRENARD, F. (1). *Haute Asie.* Vol. 8, pt. 2 of *Géographie Universelle.* Colin, Paris, 1929.
GRENARD, F. (2). 'Sur quelques Routes Anciennes et Modernes.' In J. L. Dutreuil de Rhins, *Mission Scientifique dans la Haute Asie.* Paris, 1898, pp. 212 ff.
DE GROOT, J. J. M. (1). *Chinesische Urkunde z. Geschichte Asiens.* Vol. 1, *Die Hunnen d. vorchristlichen Zeit*; Vol. 2, *Die Westlände Chinas in d. vorchristl. Zeit*, ed. O. Franke. de Gruyter, Berlin, 1921 & 1926. Crit. E. von Zach, *AM*, 1924, **1**, 125.
GROSIER, J. B. G. A. (1). *De la Chine; ou Description Générale de cet Empire*, etc. 7 vols. Pillet, Paris, 1818–20.
GROUSSET, R. (1). *Histoire de l'Extrême-Orient.* 2 vols. Geuthner, Paris, 1929. (Also appeared in *BE/AMG*, nos. 39 & 40.)
GROUSSET, R. (2). *L'Empire Mongol (1e Phase).* Boccard, Paris, 1941. (*Histoire du Monde*, ed. E. Cavaignac, vol. 8 (2).) See Bouvat.
GROUSSET, R. (3). *Sur les Traces du Bouddha.* Plon, Paris, 1929.
GROUSSET, R. (4). *Bilan de l'Histoire.* Plon, Paris, 1946.
GROUSSET, R. (7). *De la Grèce à la Chine.* Documents d'Art, Monaco, 1948.
GRUBE, W. (1) (tr.). *Die Metamorphosen der Götter* (*Fêng Shen Yen I*, chs. 1–46 with summary of chs. 47–100). 2 vols. Brill, Leiden, 1912.
GRUBE, W. (2). *Die sprachgeschichtliche Stellung des Chinesischen.* Weigel, Leipzig, 1881.
GRUMM-GRJMAILO (1). 'History of the Introduction of the Grape Vine to China.' *Archives d'Hist. de la Technol. USSR*, 1935, **5**, no. 15.
DE GUIGNES, C. L. J. (1). *Idée Générale du Commerce et des Liaisons que les Chinois ont eus avec les Nations Occidentales.* Mémoires de Littérature tirés des Registres de l'Académie des Inscriptions et Belles-Lettres, 1784, **46**, 555.
DE GUIGNES, JOSEPH (1). *Mémoire dans lequel on prouve, que les Chinois sont une Colonie Egyptienne.* Desaint & Saillant, Paris, 1760. (A paper to the Académie des Inscriptions, 14 Nov. 1758; *MRAI*, 1764, **29**, 1.)
DE GUIGNES, JOSEPH (2). 'Mémoire dans lequel on entreprend de fixer la Situation de quelques Peuples Scythes dont il est parlé dans Hérodote, et de rechercher si du temps de cet Historien on connaissait la Chine.' *MRAI*, 1770, **35**, 539.

'H., A. R.' (1). 'The Romanisation of Chinese Place-Names.' *GJ*, 1943, **102**, 67.
HADI HASAN (1). *A History of Persian Navigation.* Methuen, London, 1928.
HAGERTY, M. J. (1) (tr.). 'Han Yen-Chih's *Chü Lu* (Monograph on the Oranges of Wênchow, Chekiang)', with introduction by P. Pelliot. *TP*, 1923, **22**, 63.
DU HALDE, J. B. (1). *Description Géographique, Historique, Chronologique, Politique et Physique de l'Empire de la Chine et de la Tartarie Chinoise.* 4 vols. Paris, 1735. The Hague, 1736; Eng. tr. R. Brookes, London, 1736, 1741.
HALOUN, G. (1). 'Chinese Script.' *WR*, Sept. 1942, p. 27.
HALOUN, G. (4). 'Zur Üe-Tsï [Yüeh-chih] Frage.' *ZDMG*, 1937, **91**, 243.
HALOUN, G. (6). 'Die Rekonstruktion der chinesischen Urgeschichte durch die Chinesen.' *JDZWT*, 1925, **3**, 243.
HALOUN, G. (7). 'Seit wann kannten die Chinesen die Tocharer oder Indogermanen überhaupt?' *AM*, 1924, **1**, 1.
HAMBIS, L. (1). 'Le chapitre 107 du *Yuan Che* [*Yuan Shih*]; les généalogies impériales Mongoles dans l'Histoire Chinoise Officielle de la Dynastie Mongole.' *TP* (Suppl.), 1945, **38**, 1.
HANBURY, DANIEL (1). *Science Papers, chiefly Pharmacological and Botanical.* Macmillan, London, 1876.
DE HARLEZ, C. (1). *Le Yih-King* [*I Ching*], *Texte Primitif Rétabli, Traduit et Commenté.* Hayez, Bruxelles, 1889.
HARRISON, H. S. (2). 'Opportunism and the Factors of Invention.' *AAN*, 1930, **32**, 106.
HAVRET, H. (1). 'Conversion des Dates Cycliques (Années et Jours) en Dates Juliennes.' *TP*, 1898, **9**, 142.
VAN HÉE, L. (10). 'The *Chhou Jen Chuan* of Juan Yuan (+1764 to +1849).' *ISIS*, 1926, **8**, 103.

HERRMANN, A. (1). *Historical and Commercial Atlas of China.* Harvard-Yenching Institute, Cambridge, Mass. 1935. (Harvard-Yenching Institute, Monograph series, no. 1.)

HERRMANN, A. (2). *Die alten Seidenstrassen zw. China u. Syrien. Beitr. z. alten Geographie Asiens, I* (with excellent map). Berlin, 1910. (Quellen u. Forschungen z. alten Gesch. u. Geographie, no. 21); photographically reproduced Tientsin, 1941.

HERRMANN, A. (3). 'Die alten Verkehrswege zw. Indien u. Süd-China nach Ptolemäus.' *ZGEB*, 1913, 771.

HERRMANN, A. (4). 'Ein alter Seeverkehr zw. Abessinien u. Süd-China zum Beginn unserer Zeitrechnung.' *ZGEB*, 1913, 553.

HERRMANN, A. (5). 'Die Seidenstrassen vom alten China nach dem Römischen Reich.' *MGGW*, 1915, **58**, 472.

HERRMANN, A. (6). *Die Verkehrswege zw. China, Indien und Rom um etwa 100 nach Chr.* Leipzig, 1922. (Veröffentl. d. Forschungsinstitut f. vergleich. Religionsgeschichte Univ. Leipzig, no. 7.)

HERRMANN, A. (7). 'Die Lage des Landes Ta Ts'in [Ta-Chhin].' *OAZ*, 1927, **14**, 196.

HERRMANN, A. (12). 'Das geographische Bild Chinas im Altertum.' *SA* (Forke-Festschrift Sonderausgabe), 1937, 72.

HIRTH, F. (1). *China and the Roman Orient.* Kelly & Walsh, Shanghai; G. Hirth, Leipzig and Munich, 1885; photographically reproduced in China with no imprint, 1939.

HIRTH, F. (2) (tr.). 'The Story of Chang Chhien, China's Pioneer in West Asia.' *JAOS*, 1917, **37**, 89. (Translation of ch. 123 of the *Shih Chi*, containing Chang Chhien's Report; from §18–52 inclusive and 101 to 103. §98 runs on to §104, 99 and 100 being a separate interpolation. Also tr. of ch. 111 containing the biogr. of Chang Chhien.)

HIRTH, F. (3). *Ancient History of China; to the end of the Chou Dynasty.* New York, 1908; 2nd ed. 1923.

HIRTH, F. (4). 'Words introduced from the Chinese into European languages.' *CR*, 1873, **2**, 95.

HIRTH, F. (9). *Über fremde Einflüsse in der chinesischen Kunst.* G. Hirth, München u. Leipzig, 1896.

HIRTH, F. (10). 'Biographisches nach eigenen Aufzeichnungen.' *AM* (Hirth Presentation Volume), 1922, **1**, 1. (With complete bibliography of the writings of F. Hirth appended.)

HIRTH, F. (11). 'Die Länder des Islam nach chinesischen Quellen.' *TP*, 1894, **5** (Suppl.). (Translation of, and notes on, the relevant parts of the *Chu Fan Chih* of Chao Ju-Kua; subsequently incorporated in Hirth & Rockhill.)

HIRTH, F. & ROCKHILL, W. W. (tr.). *Chao Ju-Kua; His work on the Chinese and Arab Trade in the twelfth and thirteenth centuries, entitled 'Chu Fan Chih.'* Imp. Acad. Sci., St Petersburg, 1911. (crit. G. Vacca, *RSO*, 1913, **6**, 209; P. Pelliot, *TP*, 1912, **13**, 446; E. Schaer, *AGNT*, 1913, **6**, 329; O. Franke, *OAZ*, 1913, **2**, 98; A. Vissière, *JA*, 1914, (11e sér.), **3**, 196.)

HITTI, P. K. (1). *History of the Arabs.* 4th ed. Macmillan, London, 1949.

HOANG, P. (1). *De Calendario Sinici Variae Notiones; Calendarii Sinicae et Europaei Concordantia*, Zikkawei (Hsü-Chia-Wei), Shanghai, 1885; *A Notice of the Chinese Calendar and a Concordance with the European Calendar*, Zikkawei (Hsü-Chia-Wei), Shanghai, 1904.

HOANG, P. (4). *Concordance des Chronologies Néomeniques Chinoise et Européenne.* Shanghai, 1910 (*VS*, no. 29).

HODOUS, L. (1). *Folkways in China.* Probsthain, London, 1929.

HOPKINS, L. C. (1). (*a*) 'Shang Rulers.' *JRAS*, 1917, 69. (*b*) 'The Royal Genealogies on the Honan Relics and the Record of the Shang Dynasty.' *AM* (Hirth Presentation Volume), 1922, **1**, 194.

HOPKINS, L. C. (2). (*a*) 'The Shaman or *Wu*.' *NCR*, 1920, **2**, 423. (*b*) 'The Shaman or Chinese *Wu*; his inspired dancing and versatile character.' *JRAS*, 1945, 3.

HOPKINS, L. C. (3). *The Development of Chinese Writing.* China Society, London, n.d.

HOPKINS, L. C. (4). 'L'Ecriture dans l'Ancienne Chine.' *SCI*, 1920, **27**, 19.

HOPKINS, L. C. (5). 'Pictographic Reconnaissances, I.' *JRAS*, 1917, 773.

HOPKINS, L. C. (6). 'Pictographic Reconnaissances, II.' *JRAS*, 1918, 387.

HOPKINS, L. C. (7). 'Pictographic Reconnaissances, III.' *JRAS*, 1919, 369.

HOPKINS, L. C. (8). 'Pictographic Reconnaissances, IV.' *JRAS*, 1922, 49.

HOPKINS, L. C. (9). 'Pictographic Reconnaissances, V.' *JRAS*, 1923, 383.

HOPKINS, L. C. (10). 'Pictographic Reconnaissances, VI.' *JRAS*, 1924, 407.

HOPKINS, L. C. (11). 'Pictographic Reconnaissances, VII.' *JRAS*, 1926, 461.

HOPKINS, L. C. (12). 'Pictographic Reconnaissances, VIII.' *JRAS*, 1927, 769.

HOPKINS, L. C. (13). 'Pictographic Reconnaissances, IX' (and Index to Chinese characters examined in this series and other papers in the same Journal 1916–28). *JRAS*, 1928, 327.

HOPKINS, L. C. (14). 'Archaic Chinese Characters, I.' *JRAS*, 1937, 27.

HOPKINS, L. C. (15). 'Archaic Chinese Characters, II.' *JRAS*, 1937, 209.

HOPKINS, L. C. (16). 'Archaic Chinese Characters, III.' *JRAS*, 1937, 409.

HOPKINS, L. C. (17). 'The Dragon Terrestrial and the Dragon Celestial, I, A Study of the *Lung* (terrestrial).' *JRAS*, 1931, 791.
HOPKINS, L. C. (18). 'The Dragon Terrestrial and the Dragon Celestial, II, A Study of the *Chhen* (celestial).' *JRAS*, 1932, 91.
HOPKINS, L. C. (19). 'The Human Figure in Archaic Chinese Writing; a Study in Attitudes.' *JRAS*, 1929, 557.
HOPKINS, L. C. (20). 'The Human Figure in Archaic Chinese Writing.' *JRAS*, 1930, 95.
HOPKINS, L. C. (21). 'The Ancestral Message.' *JRAS*, 1938, 413.
HOPKINS, L. C. (22). 'The Wind, the Phoenix, and a String of Shells.' *JRAS*, 1917, 377.
HOPKINS, L. C. (23). 'The Honan Relics; a new investigator and some Results.' *JRAS*, 1921, 29.
HOPKINS, L. C. (24). 'On a newly-discovered Early Chou Inscribed Bronze.' *JRAS*, 1924, 1.
HOPKINS, L. C. (25). 'Metamorphic Stylisation and the Sabotage of Significance; a Study in Ancient and Modern Chinese Writing.' *JRAS*, 1925, 451.
HOPKINS, L. C. (26). 'Where the Rainbow Ends.' *JRAS*, 1931, 603.
HOPKINS, L. C. (27). 'Archaic Sons and Grandsons; a Study of a Chinese Complication Complex.' *JRAS*, 1934, 57.
HOPKINS, L. C. (28). 'Symbols of Parentage in Archaic Chinese, I.' *JRAS*, 1940, 351.
HOPKINS, L. C. (29). 'Symbols of Parentage in Archaic Chinese, II.' *JRAS*, 1941, 204.
HOPKINS, L. C. (30). 'Sunlight and Moonshine.' *JRAS*, 1942, 102.
HOPKINS, L. C. (31). 'The *Cas-Chrom* v. the *Lei-Ssu*; A Study of the Primitive Forms of Plough in Scotland and Ancient China.' I, *JRAS*, 1935, 707. II, *JRAS*, 1936, 45.
HOPKINS, L. C. (32). 'The Rescue of the Chinese Rhinoceros.' *JRAS*, 1939, 253.
HOPKINS, L. C. (33). 'The Bearskin, another Pictographic Reconnaissance from Primitive Prophylactic to Present-Day Panache; a Chinese Epigraphic Puzzle.' *JRAS*, 1943, 110.
HOPKINS, L. C. (34). 'A Cryptic Message and a New Solution.' *JRAS*, 1947, 191.
HOPKINS, L. C. (35). 'The Chinese Numerals and their Notational Systems.' *JRAS*, 1916, 315, 737.
HORA, S. L. (1). (*a*) 'The History of Science and Technology in India and South-east Asia.' *N*, 1951, **168**, 1047. (*b*) (ed.) 'Symposium on History of Sciences in South Asia.' *PNISI*, 1952, **18**, 323.
D'HORMON, A. (1) (ed.). *Lectures Chinoises.* Ecole Franco-Chinoise, Peiping, 1945–.
HORNBLOWER, G. D. (1). 'Round the International Exhibition of Persian Art.' *IC*, 1931, **5**, 588 (599); 1932, **6**, 67 (75, 76).
HORWITZ, H. T. (1). 'Ein Beitrag zu den Beziehungen zwischen ostasiatischer und europaïscher Technik.' *ZOIAV*, 1913, **65**, 390.
HOURANI, G. F. (1). *Arab Seafaring in the Indian Ocean in Ancient and Early Medieval Times.* Princeton Univ. Press, Princeton, 1951. (Princeton Oriental Studies, no. 13.)
HOURANI, G. F. (2). 'Direct Sailing between the Persian Gulf and China in pre-Islamic Times.' *JRAS*, 1947, 157.
HSIAO YÜ (1). 'Recherches sur Mong Kiang Niu.' (History of the *Mêng Chiang Nü* ballad.) *S*, 1948, **1**, 189.
HU HUAN-YUNG (1). 'A Geographical Sketch of Chiangsu Province.' *GR*, 1947, **37**, 609.
HU SHIH (1). *The Chinese Renaissance.* Univ. of Chicago Press, Chicago, 1934.
HUART, C. & DELAPORTE, L. (1). *L'Iran Antique, Elam et Perse, et la Civilisation Iranienne.* Albin Michel, Paris, 1943. (Evol. de l'Hum. Sér.; Préhist. no. 24.)
HUBER, E. (1). 'Itineraire du Pèlerin Ki Ye [Chi-Yeh] dans l'Inde.' *BEFEO*, 1902, **2**, 256.
HUBERT, H. (1). *Les Celtes et l'Expansion Celtique jusqu'à l'Epoque de la Tène.* Albin Michel, Paris, 1932. (Evol. de l'Hum. Sér.; Préhist. no. 21.)
HUBERT, H. (2). *Les Celtes depuis l'Epoque de la Tène, et la Civilisation Celtique.* Albin Michel, Paris, 1932. (Evol. de l'Hum. Sér.; Préhist. no. 21 bis.)
HUDSON, G. F. (1). *Europe and China; A Survey of their Relations from the Earliest Times to 1800.* Arnold, London, 1931.
HUGHES, E. R. (1). *Chinese Philosophy in Classical Times.* Dent, London, 1942. (Everyman Library, no. 973.)
HUGHES, E. R. (2) (tr.). *The Great Learning and the Mean-in-Action.* Dent, London, 1942.
HUGHES, E. R. (3). *The Invasion of China by the Western World.* Black, London, 1937.
HUGHES, E. R. (4). 'Concerning the Importance and Reliability of the *I Wên Chih*' (bibliography in the *Chhien Han Shu*). *MCB*, 1939, **6**, 173.
HÜLLE, H. (1). 'Geschichte u. Inhalt d. Wissenschaft von Ostasien.' *CB*, 1925, **42**, 545.
HUMMEL, A. W. (1). 'Phonetics and the Scientific Method.' *ARLC/DO*, 1940, 169.
HUMMEL, A. W. (2) (ed.). *Eminent Chinese of the Chhing Period.* 2 vols. Library of Congress, Washington, 1944.

HUMMEL, A. W. (4). 'Notes on Li Chih (Ming), unorthodox scholar and martyr.' *ARLC/DO*, 1931–2, 190; 1938, 232; 1939, 248.

HUNTINGTON, ELLSWORTH (1). *The Pulse of Asia.* Boston, 1907, 1910.

HUNTINGTON, ELLSWORTH (2). *Civilisation and Climate.* 3rd ed. New Haven, 1924.

HUZZAYIN, S. A. (1). *Arabia and the Far East; their commercial and cultural relations in Graeco-Roman and Irano-Arabian times.* Soc. Royale de Geogr. Cairo, 1942.

IBN BAṬṬŪṬAH, ABŪ 'ABDALLĀH AL-LUWĀṬI AL-ṬANGHĪ (1). *Riḥla ibn Baṭṭūṭah* (Journey of ibn Baṭṭūṭah). See Defrémery & Sanguinetti (1).

IMPEY, L. (1). 'Shangtu (Xanadu), the summer Capital of Kublai Khan.' *GR*, 1925, **15**, 584.

'ABD AL-JALĪL, J. M. (1). *Brève Histoire de la Littérature Arabe.* Maisonneuve, Paris, 1943; 2nd ed. 1947.

JANSE, O. (1). 'Notes sur quelques Epées Anciennes trouvées en Chine.' *BMFEA*, 1930, **2**, 67.

JANSE, O. (2). 'Tubes et Boutons cruciformes trouvés en Eurasie.' *BMFEA*, 1932, **4**, 187.

JANSE, O. (3). 'L'Empire des Steppes et les Relations entre l'Europe et l'Extrême-Orient dans l'Antiquité.' *RAA/AMG*, 1935, **9**, 9.

JANSE, O. (4). 'Quelques Antiquités Chinoises d'un caractère Hallstattien.' *BMFEA*, 1930, **2**, 177.

JANSE, O. (5). *Archaeological Research in Indo-China* (Harvard-Yenching Monograph Series, no. 7). Harvard Univ. Press, Cambridge, Mass. 1947. (Also in *RAA/AMG*, 1935, **9**, 144, 209; 1936, **10**, 42.)

JENSEN, H. (1). 'Indogermanisch u. Chinesisch.' Art. in *Germanen u. Indogermanen* (H. Hirt Festschrift), ed. H. Arntz. Winter, Heidelberg, 1936, p. 139. (Indogermanische Bibliothek, no. 15 (2).)

JOURDAIN, A. (1). 'Mémoire sur l'Observatoire de Méragha et les Instruments employés pour y observer.' *Magasin Encyclopédique*, 1809 (6), **84**, 43; and sep. Paris, 1810.

JULIEN, STANISLAS (1) (tr.). *Voyages des Pèlerins Bouddhistes.* 3 vols. Impr. Imp., Paris, 1853–8. (Vol. 1 contains Hui-Li's life of Hsüan-Chuang; vols. 2 and 3 contain Hsüan-Chuang's *Hsi Yü Chi.*)

JULIEN, STANISLAS (2). 'Mélanges de Géographie Asiatique et de Philologie Sinico-Indienne.' Paris, 1864, pp. 166ff. (See note in *JA*, 1864 (6e sér.), **3**, 372). A collection of offprints from *JA*, including 'Notices sur les Pays et les Peuples Etrangers, tirées des Géographies et des Annales Chinoises, V. Thien-Tchou, l'Inde.' *JA*, 1847 (4e sér.), **10**, 81 (109) (essentially a translation of part of *Wên Hsien Thung Khao*, ch. 338, p. 14*a*).

KAHLE, P. (1). 'China as described by Turkish Geographers from Iranian Sources.' *PIS*, 1940, **2**, 4.

KAHLE, P. (2). 'Eine Islamische Quelle über China um +1500 (Das *Khiṭai-nāmah* des 'Alī Ekber).' *AO*, 1933, **12**, 91.

KALLENBERG, C. (1). 'Der Handel mit Perlen und Perlmutterschalen in Rothen Meere.' *OMO*, 1884 (March), p. 86.

KANDA, SIGERU (1). 'Ancient Records of Sunspots and Aurorae in the Far East, and the Variation of the Period of Solar Activity.' *PIAJ*, 1933, **9**, 293.

KANTER, H. (1). 'Der Löss in China.' *MGGH*, 1922, **34**, 99.

KARLGREN, B. (1). *Grammata Serica; Script and Phonetics in Chinese and Sino-Japanese. BMFEA*, 1940, **12**, 1. (Photographically reproduced as separate volume, Peiping, 1941.)

KARLGREN, B. (2). 'Legends and Cults in Ancient China.' *BMFEA*, 1948, **18**, 199.

KARLGREN, B. (3). *The Romanisation of Chinese.* China Society, London, 1928.

KARLGREN, B. (4). *Sound and Symbol in Chinese.* Oxford , 1923; repr. 1946. (Eng. tr. of *Ordet och Pennan i Mittens Rike.* Stockholm, 1918.)

KARLGREN, B. (5). *Philology and Ancient China.* Aschehong (Nygaard), Oslo, 1926. (Institutet for Sammenlignende Kulturforskning; A, Forelesninger, no. 8.)

KARLGREN, B. (6). 'Le Proto-Chinois, Langue Flexionelle.' *JA*, 1920 (11e sér.), **15**, 205.

KARLGREN, B. (7). *Etudes sur la Phonologie Chinoise.* 3 vols. Stockholm, 1915–19.

KARLGREN, B. (8). 'On the Authenticity and Nature of the *Tso Chuan.*' *GHA*, 1926, **32**, no. 3 (crit. H. Maspero, *JA*, 1928, **212**, 159).

KARLGREN, B. (10). 'The Authenticity of Ancient Chinese Texts.' *BMFEA*, 1929' **1**, 165.

KARLGREN, B. (12) (tr.). 'The Book of Documents' (*Shu Ching*). *BMFEA*, 1950, **22**, 1.

KARLGREN, B. (14) (tr.). *The Book of Odes*; *Chinese Text, Transcription and Translation.* Museum of Far Eastern Antiquities, Stockholm, 1950. (A reprint of the translation only from his papers in *BMFEA*, **16** and **17**.)

KARPINSKI, L. C. (1) (tr.). *Robert of Chester's Latin Translation of the Algebra of al-Khwārizmī, with an Introduction, Critical Notes, and an English Version.* Univ. of Michigan Studies, New York, 1915.

KAYE, G. R. (3). *Indian Mathematics.* Thacker & Spink, Calcutta, 1915.

KENNEDY, J. (1). 'Buddhist Gnosticism; the System of Basilides.' *JRAS*, 1902, 377.

KENNEDY, J. (2). 'The Gospels of the Infancy, the *Lalita Vistara*, and the *Vishṇu Purāṇa*: or the Transmission of Religious Ideas between India and the West.' *JRAS*, 1917, 209, 469.

KING, F. H. (1). 'The Wonderful Canals of China.' *NGM*, 1912, **23**, 931.

KIRCHER, ATHANASIUS (1). *China Monumentis qua Sacris qua Profanis Illustrata.* Amsterdam, 1667. (French tr. Amsterdam, 1670.)

KLEINGUNTHER, A. (1). ΠΡΩΤΟΣ ΕΥΡΕΤΗΣ (Greek lists of inventors and technic deities). *PL*, 1933, Suppl. **26**, 26.

KLIENE, C. (1). *Anglo-Chinese Calendar, 1751–2000* A.D. Shanghai and London, 1895.

KLUNZINGER, C. B. (1). 'Die Umgegend von Qoseir am Rothen Meere.' (On coral.) *ZGEB*, 1879, **14**, 401.

KOESTER, H. (1). 'Four thousand hours over China.' *NGM*, 1938, **73**, 571.

KONDAKOV, N. P. (1). 'Les Costumes Orientaux à la Cour Byzantine.' *B*, 1924, **1**, 7.

KOSAMBI, D. D. (1). 'Chronological Order of the Punch-Marked Coins, I. A Re-Examination of the Older Taxila Hoard.' *JRAS/BOM*, 1949 (n.s.), **24/25**, 33.

KROEBER, A. L. (2). 'Stimulus Diffusion.' *AAN*, 1940, **42**, 1.

KROEBER, A. L. (3). 'The Ancient Oikoumene as an Historic Culture Aggregate' (Huxley Memorial Lecture). *JRAI*, 1945.

KROEBER, A. L. (4). *Configurations of Culture Growth.* Univ. Calif., Berkeley and Los Angeles, 1944.

KROPOTKIN, PRINCE PETER (1). 'The Desiccation of Eurasia.' *GJ*, 1904, **23**, 722.

KU CHIEH-KANG (1). *Autobiography of a Chinese Historian* (preface to *Ku Shih Pien* (*q.v.*)), tr. A. W. Hummel. Leiden, 1931. (Sinica Leidensia series, no. 1.)

KU HUNG-MING (1) (tr.). *The Discourses and Sayings of Confucius.* Kelly & Walsh, Shanghai, 1898.

KUWABARA, JITSUZO (1). 'On Phu Shou-Kêng, a man of the Western Regions, who was the Superintendent of the Trading Ships Office in Chhüan-chou towards the end of the Sung Dynasty, together with a general sketch of the Trade of the Arabs in China during the Thang and Sung eras.' *MRDTB*, 1928, **2**, 1; 1935, **7**, 1 (rev. P. Pelliot, *TP*, 1929, **26**, 364; S. E[lisséev], *HJAS*, 1936, **1**, 265).

LACH, D. F. (2). 'The Chinese Studies of Andreas Müller.' *JAOS*, 1940, **60**, 568.

LACH, D. F. (3). 'China and the Era of the Enlightenment.' *JMH*, 1942, **14**, 209.

LACH, D. F. (4). *Contributions of China to German Civilisation.* Inaug. Diss., Chicago, 1944 (photo-printed excerpt).

DE LACOUPERIE, TERRIEN (1). *The Western Origin of Chinese Civilisation.* London, 1894.

LAIGNEL-LAVASTINE, M. (1) (ed.). *Histoire Générale de la Médecine, de la Pharmacie, de l'Art Dentaire et de l'Art Vétérinaire.* 3 vols. Albin Michel, Paris, 1938.

LARSON, F. A. (1). *Duke of Mongolia.* Boston, 1930.

LATTIMORE, O. (1). *Inner Asian Frontiers of China.* Oxford Univ. Press, London and New York, 1940. (Amer. Geogr. Soc. Research Monograph Series, no. 21.)

LATTIMORE, O. (2). 'Origins of the Great Wall of China; A Frontier Concept in Theory and Practice.' *GR*, 1937, **27**, 529.

LATTIMORE, O. (3). 'An Inner Asian Approach to the Historical Geography of China.' *GJ*, 1947, **110**, 180.

LATTIMORE, O. (4). 'Caravan Routes of Inner Asia.' *GJ*, 1928, **72**, 523.

LATTIMORE, O. (5). 'The Desert Road to Turkestan.' *NGM*, 1929, **55**, 661.

LAUFER, B. (1). *Sino-Iranica; Chinese Contributions to the History of Civilisation in Ancient Iran. FMNHP/AS*, 1919, **15**, no. 31 (Pub. no. 201). (rev. and crit. Chang Hung-Chao, *MGSC*, 1925 (ser. B), no. 5.)

LAUFER, B. (2). 'The Hsi-Hsia Language.' *TP*, 1916, **17**, 1.

LAUFER, B. (3). *Chinese Pottery of the Han Dynasty* (Pub. of the East Asiatic Cttee. of the Amer. Mus. Nat. Hist.). Brill, Leiden, 1909. (Photographically reproduced Tientsin, 1940.)

LAUFER, B. (6). 'The Story of the *Pinna*.' *JAFL*, 1915, **28**, 103.

LAUFER, B. (25). 'The Bird-Chariot in China and Europe.' In *Anthropological Papers written in honour of Franz Boas and presented to him on the 25th anniversary of his doctorate*, ed. B. Laufer. Stechert, New York, 1926, pp. 410 ff.

LAUFER, B. (27). 'Malabathron.' *JA*, 1918 (11e sér.), **12**, 5.

LAUFER, B. (28). 'Christian Art in China.' *MSOS*, 1910, **13**, 100.

LECOMTE, LOUIS (1). *Nouveaux Mémoires sur l'Etat présent de la Chine*. Anisson, Paris, 1696. (Eng. tr. *Memoirs and Observations, Topographical, Physical, Mathematical, Mechanical, Natural, Civil and Ecclesiastical, made in a late journey through the Empire of China, and published in several letters, particularly upon the Chinese Pottery and Varnishing, the Silk and other Manufactures, the Pearl Fishing, the History of Plants and Animals, etc. translated from the Paris edition etc.* 2nd ed. London, 1698. Germ. tr. Frankfurt, 1699–1700.)

LEE, H. (1). *The Vegetable Lamb of Tartary; a curious Fable of the Cotton Plant.* London, 1887.

LEE, J. S. See Li Ssu-Kuang.

LEGGE, J. (1) (tr.). *The Texts of Confucianism, translated. Pt. 1, The Shu King, the Religious portions of the She King, the Hsiao King* [*Shu Ching, Shih Ching*, and *Hsiao Ching*]. Oxford, 1879. (*SBE*, no. 3); reprinted in various eds.; Com. Press, Shanghai.

LEGGE, J. (2) (tr.). *The Chinese Classics, etc.*: Vol. 1. *Confucian Analects, The Great Learning, and the Doctrine of the Mean.* Legge, Hongkong, 1861; Trübner, London, 1861.

LEGGE, J. (3) (tr.). *The Chinese Classics, etc.*: Vol. 2. *The Works of Mencius.* Legge, Hongkong, 1861; Trübner, London, 1861.

LEGGE, J. (4) (tr.). *A Record of Buddhistic Kingdoms; an account by the Chinese monk Fa-Hsien of his travels in India and Ceylon (+399 to +414) in search of the Buddhist books of discipline.* Oxford, 1886.

LEGGE, J. (5) (tr.). *The Texts of Taoism.* (Contains (*a*) *Tao Tê Ching*, (*b*) *Chuang Tzu*, (*c*) *Thai Shang Kan Ying Phien*, (*d*) *Chhing Ching Ching*, (*e*) *Yin Fu Ching*, (*f*) *Jih Yung Ching*.) 2 vols. Oxford, 1891; photolitho reprint, 1927. (*SBE*, nos. 39 and 40.)

LEGGE, J. (7) (tr.). *The Texts of Confucianism, Pt. 3. The Li Ki* [*Li Chi*]. 2 vols. Oxford, 1885; reprint, 1926 (*SBE*, nos. 27 and 28).

LEGGE, J. (8) (tr.). *The Chinese Classics, etc.* Vol. 4, Pts. 1 and 2. *She-King* [*Shih Ching*]; *The Book of Poetry.* Lane Crawford, Hongkong; and Trübner, London, 1871. Reprint, without notes, Com. Press, Shanghai, n.d.

LEGGE, J. (9) (tr.). *The Texts of Confucianism, Pt. 2. The Yi King* [*I Ching*]. Oxford, 1882, 1899. (*SBE*, no. 16.)

LEGGE, J. (10) (tr.). *The Chinese Classics, etc.* Vol. 3, Pts. 1 and 2. *The Shoo King* (*Shu Ching*). Hongkong, and Trübner, London. 1865.

LEGGE, J. (11) (tr.). *The Chinese Classics, etc.* Vol. 5, Pts. 1 and 2. *The Ch'un Ts'eu with the Tso Chuen* (*Chhun Chhiu and Tso Chuan*). Hongkong, and Trübner, London, 1872.

LEI HAI-TSUNG (1). *The Warring States (−473 to −221); the Modern Period in Ancient China.* War Area Service Corps Information Pamphlets, series A, no. 1, Kunming, 1943.

LEI HAI-TSUNG (2). 'Periodisation in Chinese and World History.' *CSPSR*, 1937, **20**, 461.

LEROI-GOURHAN, ANDRÉ (1). *Evolution et Techniques.* Vol. 1. *L'Homme et la Matière*, 1943; Vol. 2. *Milieu et Techniques*, 1945. Albin Michel, Paris.

LÉVI, S. (1). 'Les Missions de Wang Hiuen-Ts'e [Wang Hsüan-Tshê] dans l'Inde.' *JA*, 1900 (9e sér.), **15**, 297, 401.

LÉVI, S. (2). 'Ceylan et la Chine.' *JA*, 1900 (9e sér.), **15**, 411.

LÉVI, S. (5). 'Les Marchands de Mer et leur Rôle dans le Bouddhisme Primitif.' *BAFAO*, 1929, 19.

LÉVI, S. & CHAVANNES, E. (1). 'Voyages des Pèlerins Bouddhistes; l'Itinéraire d'Ou-K'ong' (+751 to +790; Wu-Khung), *JA*, 1895 (9e sér.), **6**, 341, with additional note (p. 371). 'Le Ki-Pin' (the location of Chi-Pin, Kashmir, etc.).

LI CHI (1). 'Chinese Archaeology.' Art. in *Symposium on Chinese Culture*, ed. Sophia Zen. IPR, Shanghai, 1931, pp. 184 ff.

LI CHI (2). *The Formation of the Chinese People; an Anthropological Enquiry.* Harvard Univ. Press, Cambridge, Mass. 1928.

LI SSU-KUANG (J. S. LEE) (1). *The Geology of China.* Murby, London, 1939.

LIANG CHHI-CHHAO (1). *History of Political Thought during the Early Tsin* [*Chhin*] *Period.* Kegan Paul, London, 1930.

LIN MOU-SHÊNG (1). 'The Revolution in the History of Chinese History.' *CIB*, 1938, **3**, 12.

LIN YU-THANG (1) (tr.). *The Wisdom of Laotse* [*and Chuang Tzu*] *translated, edited and with an introduction and notes.* Random House, New York, 1948.

LINDBLOM, G. (1). 'Vessels with Star-shaped Lids.' *ETH*, 1942, **1**, 55.

LIPS, J. E. (1). 'Foreigners in Chinese Plastic Art.' *ASIA*, 1941, **41**, 377.

LITTRÉ, E. (1) (ed. and tr.). *Hippocrates, Opera Omnia.* Baillière, Paris, 1863.

LIU, S. C. H. (1). 'Introduction of Modern Science into China.' *CJ*, 1941, **34**, 120, 210.

LIVI, R. (1). *La Schiavitù Domestica nei Tempi di Mezzo.* Padua, 1928.

DE LA LOUBÈRE, S. (1). *A New Historical Relation of the Kingdom of Siam, by Monsieur de la Loubère, Envoy-Extraordinary from the French King to the King of Siam in the Years 1687 and 1688, wherein a full and curious Account is given of the Chinese Way of Arithmetick and Mathematick Learning*, tr. A. P., Gen[t ?] R.S.S. [i.e. F.R.S.]. Horne Saunders & Bennet, London, 1693 (from the French edition, Paris, 1691).

LOVEJOY, A. O. & BOAS, G. (1). *A Documentary History of Primitivism and Related Ideas*. Vol. 1. *Primitivism and Related Ideas in Antiquity*. Johns Hopkins Univ. Press, Baltimore, 1935.

LU CHIH-WEI (1). *On Chinese Poetry*. Peiping, 1935.

DE LUBAC, H. (1). *Textes Alexandrins et Bouddhiques*. Paris, 1938.

MCCRINDLE, J. W. (1). *Ancient India as described in Classical Literature; being a Collection of Greek and Latin texts relating to India, extracted from Herodotus, Strabo, Diodorus Siculus, Pliny, Aelian (Kosmas Indicopleustes), Philostratus, Dion Chrysostom, Porphyry, Stobaeus, the Itinerary of Alexander the Great, the Periegesis of Dionysius, the Dionysiaka of Nonnus, the Romance History of Alexander, and other works*. Constable, Westminster, 1901.

MCCRINDLE, J. W. (2). *Ancient India as described by Ktesias the Knidian; being a translation of the abridgement of his Indica by Photios, and of the fragments of that work preserved in other Writers, with notes, etc.* Thacker & Spink, Calcutta, 1882.

MCCRINDLE, J. W. (3). *Ancient India as described by Megasthenes and Arrian, being a translation of the fragments of the Indica of Megasthenes collected by Dr Schwanbeck, and of the first part of the Indica of Arrian.* Thacker, Bombay, 1877. (Reprinted with additions from *IAQ*, 1876–7.)

MCCRINDLE, J. W. (4). *The Commerce and Navigation of the Erythraean Sea; being a translation of the 'Periplus Maris Erythraeus', of an anonymous writer, and of Arrian's account of the Voyage of Nearchos from the mouth of the Indus to the Head of the Persian Gulf, with notes, etc.* Thacker & Spink, Calcutta, 1879.

MCCRINDLE, J. W. (5). *The Invasion of India by Alexander the Great, as described by Arrian, Quintus Curtius, Diodorus, Plutarch, and Justin.* Constable, Westminster, 1896.

MCCRINDLE, J. W. (6). *Ancient India as described by Ptolemy, being a translation of the chapters which describe India and Central and East Asia, in the Treatise on Geography written by Klaudios Ptolemaios the celebrated astronomer, with notes, etc.* Thacker & Spink, Calcutta, 1885.

MCCRINDLE, J. W. (7). *The Christian Topography of Cosmas (Indicopleustes), an Egyptian monk.* London, 1897. (Hakluyt Society Publications, series 1, no. 98.)

MCDOUGALL, W. (1). *The Group Mind.* Cambridge, 1920.

MCGILLIVRAY, D. (1). *A Mandarin-Romanised Dictionary of Chinese.* Presbyt. Miss. Press, Shanghai, 1907.

MCGOVERN, W. M. (1). *Early Empires of Central Asia.* Univ. of North Carolina Press, Chapel Hill, 1939.

MCKAY, E. (1). *Early Indus Civilisations*, ed. D. McKay. Luzac, London, 2nd ed. 1948.

DE MAILLA, J. A. M. DE MOYRIAC (1) (tr.). *Histoire Générale de la Chine, ou Annales de cet Empire, traduites du Tong Kien Kang Mou [Thung Chien Kang Mu].* Ed. Grosier & le Roux des Hautesrayes; Pierres & Clousier, Paris, 1777–85. 13 vols. (This translation was made from the edition of +1708; Hummel (2), p. 689.)

MARGOULIÈS, G. (1) (tr.). *Le Kou Wên [Ku Wên] Chinois; Recueil de Textes avec Introduction et Notes.* Geuthner, Paris, 1925 (Inaug. Diss., Paris). (rev. H. Maspero, *JA*, 1928, **212**, 174.)

MARSHALL, SIR JOHN (1). *Taxila. An Illustrated Account of Archaeological Excavations carried out at Taxila under the orders of the Government of India between the years 1913 and 1934.* 3 vols. Cambridge, 1951.

MARTIN, W. A. P. (1). *The Chinese, their Education, Philosophy and Letters.* Harper, New York, 1881.

MARTIN, W. A. P. (2). *The Lore of Cathay; or, the Intellect of China.* Oliphant, Edinburgh and London, 1901.

MARTIN, W. A. P. (3). *Hanlin Papers.* 2 vols. Vol. 1. Trübner, London, 1880, Harper, New York, 1880; Vol. 2. Kelly & Walsh, Shanghai, 1894.

MARTIN, W. A. P. (4). 'Plato and Confucius; a curious Coincidence.' *JPOS*, 1887, **1**, 199; *JAOS*, 1890, **14**, xxxi. (repr. in Martin (3), vol. 2, p. 199.)

MASON, I. (1). 'The Mohammedans of China; when and how they first came.' *JRAS/NCB*, 1929, **60**, 42.

MASON, I. (2) (tr.). *The Arabian Prophet, a life of Mohammed from Chinese and Arabic Sources* (tr. of Liu Chih's *Chih Shêng Shih Lu* of +1724). Com. Press, Shanghai, 1921.

MASPERO, H. (1). 'La Composition et la date du *Tso Chuan*.' *MCB*, 1931, **1**, 137.

MASPERO, H. (2). *La Chine Antique.* Boccard, Paris, 1927. (*Histoire du Monde*, ed. E. Cavaignac, vol 4.) (rev. B. Laufer, *AHR*, 1928, **33**, 903.) Not to be confused with the book of his brother G. Maspero, *La Chine* (2 vols. Delagrave, Paris, 1918 and 1925), which deals with modern history.

MASPERO, H. (3). 'L'Astronomie Chinoise avant les Han.' *TP*, 1929, **26**, 267. (Abstract by Vacca (5).)

MASPERO, H. (4). 'Les Instruments Astronomiques des Chinois au temps des Han.' *MCB*, 1939, **6**, 183.

MASPERO, H. (5). 'Le Songe et l'Ambassade de l'Empereur Ming.' *BEFEO*, 1910, **10**, 95, 629.

MASPERO, H. (6). 'Les Langues d'Extrême-Orient.' In *Encyclopédie Française*, 1937, vol. 1, pt. 2, ch. 3, 1.40–1 ff.

MASPERO, H. (12). *Les Religions Chinoises; Mélanges Posthumes sur les Religions et l'Histoire de la Chine*, vol. 1, ed. P. Demiéville. Civilisations du Sud, Paris, 1950. (Publ. du Mus. Guimet, Biblioth. de Diffusion, no. 57.)

MASPERO, H. (19). 'Communautés et Moines Bouddhistes Chinois au 2e et 3e siècles.' *BEFEO*, 1910, **10**, 222.

MASSON-OURSEL, P., DE WILLMAN-GRABOWSKA, H. & STERN, P. (1). *L'Inde Antique et la Civilisation Indienne*. Albin Michel, Paris, 1933. (Evol. de l'Hum. Sér.; Préhist. no. 26.)

AL-MAS'ŪDĪ. See de Meynard & de Courteille.

MATHEWS, R. H. (1). *Chinese-English Dictionary*. China Inland Mission, Shanghai, 1931; Harvard-Yenching Inst., Harvard, 1947.

MAYERS, W. F. (1). *Chinese Reader's Manual*. Presbyterian Press, Shanghai, 1874; reprinted, 1924.

MAYERS, W. F. (2). 'Bibliography of the Chinese Imperial Collections of Literature' (i.e. *Yung-Lo Ta Tien*; *Thu Shu Chi Chhêng*; *Yuan Chien Lei Han*; *Phei Wên Yuan Fu*; *Phien Tzu Lei Pien*; *Ssu Ku Chhüan Shu*). *CR*, 1878, **6**, 213, 285.

MAYERS, W. F. (3). 'Chinese Explorations of the Indian Ocean during the fifteenth century.' *CR*, 1875, **3**, 219, 331; 1875, **4**, 61.

MEHREN, A. F. M. (1) (tr.). *Manuel de la Cosmographie du Moyen Age*. Copenhagen, 1874.

MEI YI-PAO (1) (tr.). *The Ethical and Political Works of Motse* [*Mo Tzu*]. Probsthain, London, 1929.

MEILE, P. (1). 'Apollonius de Tyane et les Rites Védiques.' *JA*, 1945, **234**, 451.

MEILE, P. (2). 'Les Yavanas dans l'Inde Tamoule.' *JA*, 1940, **232**, 85.

MELLOR, J. W. (2). *Comprehensive Treatise on Inorganic and Theoretical Chemistry*. 15 vols. Longmans Green, London, 1923.

DE MÉLY, F. (1). *Les Lapidaires Chinois*. Vol. 1 of *Les Lapidaires de l'Antiquité et du Moyen Age*. Leroux, Paris, 1896. (Contains facsimile reproduction of the mineralogical section of (*Ho Han*) *San|Tshai Thu Hui*, chs. 59 and 60, from a Japanese edition.) (rev. M. Berthelot, *JS*, 1896, 573.)

DE MÉLY, F. (5). 'Les Pierres de Foudre chez les Chinois et les Japonais.' *RA*, 1895 (3e sér.), **27**, 326.

DE MÉLY, F. (7). *De Périgueux au Fleuve Jaune*. Geuthner, Paris, 1927.

MENZIES, J. M. (1). *Oracle Records from the Waste of Yin*. Shanghai, 1917.

MÉTRAUX, A. (1). 'The King's Alphabet; a Comment on "Race Superiority".' *UNN*, 1951, **6**, 18.

DE MEYNARD, C. BARBIER & DE COURTEILLE, P. (1) (tr.). *Les Prairies d'Or* (the *Murūj al-Dhabab* of al-Mas'ūdī, +947). 9 vols. Paris, 1861–77.

MICHEL, H. (9). 'Un Service de l'Heure Millennaire.' *CET*, 1952, **68**, 1.

MIELI, ALDO (1). *La Science Arabe, et son Rôle dans l'Evolution Scientifique Mondiale*. Brill, Leiden, 1938.

MIELI, ALDO (2). *Panorama General de Historia de la Ciencia*. Vol. 1, *El Mundo Antiguo; griegos y romanos*; Vol. 2, *El Mundo Islámico e el Occidente Medieval Cristiano*. Espasa-Calpe, Buenos Aires, 1946. (Nos. 1 and 5 respectively of Colección Historia y Filosofia de la Ciencia, ed. J. Rey Pastor.)

MIKAMI, Y. (1). *The Development of Mathematics in China and Japan*. Teubner, Leipzig, 1913. (Abhdl. z. Gesch. d. math. Wissenschaften mit Einschluss ihrer Anwendungen, no. 30.) (rev. H. Bosmans, *RQS*, 1913, **74**, 641.)

MILLS, J. V. (3). 'Notes on Early Chinese Voyages.' *JRAS*, 1951, 3.

MINNS, E. H. (1). *Scythians and Greeks; a Survey of Ancient History and Archaeology on the North Coast of the Euxine from the Danube to the Caucasus*. Cambridge, 1913.

MINORSKY, V. (1). 'Maragha.' *EI*, 1930, vol. 3, p. 261.

MINORSKY, V. (2). 'The Middle East in Western Politics in the thirteenth, fourteenth and fifteenth centuries.' *JRCAS*, 1940, **27**, 427.

VON MÖLLENDORFF, P. G. & VON MÖLLENDORFF, O. F. (1). *Manual of Chinese Bibliography*. Kelly & Walsh, Shanghai, 1876.

MORGAN, E. (1) (tr.). *Tao the Great Luminant; Essays from Huai Nan Tzu, with introductory articles, notes and analyses*. Kelly & Walsh, Shanghai, n.d. (1933?).

MORSE, W. R. & YEN, Y. (1). 'Ancient Historical Aboriginal Ethnic Groups of Szechuan Province.' *JWCBRS*, 1936, **8**, 106.

MOULE, A. C. (1). *Christians in China before the year 1550*. SPCK, London, 1930.
MOULE, A. C. (2). 'Thai Shan.' *JRAS/NCB*, 1912, **43**, 3.
MOULE, A. C. & PELLIOT, P. (1) (tr. and annot.). *Marco Polo (+1254 to +1325); The Description of the World*. Routledge, London, 1938.
MÜLLER, A. 'Andreas Müller Greifenhagius' (+1630 to +1694). *ZDMG*, 1881, **35**, iii.
MÜLLER, ANDREAS (GREIFENHAGIUS) (1). *Historia Sinensis Abdallae cogn. Abu Said Beidawi*. Jena, 1677 (with title *Hist. Sin. Abdallae Beidavaei*); 1679; 1689. (In Persian and Latin.)
MÜLLER, ANDREAS (GREIFENHAGIUS) (2). *Disquisitio Geographica et Historica; de Chataja, In Qua (1) Praecipue Geographorum nobilis illa Controversia: Quaenam Chataja sit, et an sit idem ille terrarum tractus, quem Sinas, et vulgo Chinam vocant, aut pars ejus aliqua? latissime tractatur, (2) Eadem vero opera pleraque rerum, quae unquam de Chataja, deque Sinis memorabilia fuerunt, atque etiam nunc sunt, compendiose narrantur*. Berlin, 1670.
MULLIKIN, M. A. (1). 'Thai Shan, Sacred Mountain of the East.' *NGM*, 1945, **87**, 699.
MUMFORD, LEWIS (1). *Technics and Civilisation*. Routledge, London, 1934.
MÜNSTERBERG, O. (1). 'Influences Occidentales dans l'Art de l'Extrême-Orient.' *REES*, 1909, **1**.
MÜNSTERBERG, O. (2). 'Leonardo da Vinci und der chinesische Landschaftsmalerei.' *OAV*, 1910, **1**, 92. (crit. E. Chavannes, *TP*, 1911, **12**, 102.)

NACKEN, J. (1). 'A Chinese Webster; *Lu-Shu-Ku* [*Liu Shu Ku*]; or the Six Classes of Characters and their substantiation, by Tae-Tung [Tai Tung] of the 13th century.' *CR*, 1873, **2**, 175, 215, 354.
NAGASAWA, K. (1). *Geschichte der Chinesischen Literatur, und ihrer gedanklichen Grundlage*, tr. from the Japanese by E. Feifel. Fu-jen Univ. Press, Peiping, 1945.
NAGEL, P. (1). 'Umrechnung d. zyklischen Daten des chinesischen Kalenders in europäische Daten.' *MSOS*, 1931, **34**, 153.
NANJIO, B. (1). *A Catalogue of the Chinese Translations of the Buddhist Tripiṭaka*. Oxford, 1883.
NAU, F. (1). 'Notes d'Astronomie Syrienne' (sect. 3: La Plus Ancienne Mention Orientale des Chiffres Indiens) (Severus Sebokht, +662). *JA*, 1910 (10e sér.), **16**, 209 (225).
NEEDHAM, JOHN TURBERVILLE (1). *De Inscriptione quadam Ægyptiaca Taurini inventa et Characteribus Ægyptiis olim et Sinis communibus exarata idolo cuidam antiquo in regia universitate servato ad utrasque Academias Londinensem et Parisensem rerum antiquarum investigationi et studio praepositas data Epistola*. Palearini, Rome, 1761.
NEEDHAM, JOSEPH (1). *Chemical Embryology*. 3 vols. Cambridge, 1931.
NEEDHAM, JOSEPH (2). *A History of Embryology*. Cambridge, 1934.
NEEDHAM, JOSEPH (6). *History is on our side: a contribution to political religion and scientific faith*. Allen & Unwin, London, 1946.
NEEDHAM, JOSEPH (26). 'The History of Science and Technology in India and South-east Asia' (review of symposium and note on dating). *N*, 1951, **168**, 64, 1048. (See Hora, 1.)
NEEDHAM, JOSEPH & LIAO HUNG-YING (1) (tr.). 'The Ballad of Mêng Chiang Nü weeping at the Great Wall.' *S*, 1948, **1**, 194.
NEEDHAM, JOSEPH & NEEDHAM, DOROTHY M. (1) (ed.). *Science Outpost*. Pilot Press, London, 1948.
NEHRU, JAWAHARLAL (1). *The Discovery of India*. Meridian, London, 1946.
NEUGEBAUER, O. (6). 'Über eine Methode zur Distanzbestimmung Alexandria-Rom bei Heron.' *KDVS/HFM*, 1939, **26**, no. 2 (p. 21) and no. 7.

OBRUCHEV, V. A. (1). 'Das Lössland des Nordwestens Chinas.' *GZ*, 1895, **1**, 263.
OGBURN, W. F. (1). *Social Change*. London, 1923.
OGBURN, W. F. (2) (ed.). *Technological Trends and National Policy, including the Social Implications of New Inventions* (Report of the Sub-Committee on Technology of the (U.S.) National Resources Committee). Govt. Printing Office, Washington D.C., 1937.
D'OHSSON, MOURADJA (1). *Histoire des Mongols dépuis Tchinguiz* [*Chingiz*] *Khan jusqu'à Timour Bey ou Tamerlan*. The Hague and Amsterdam, 1834–52.
D'OLLONE, H., VISSIÈRE, A., BLOCHET, E. *et al.* (1). *Recherches sur les Mussulmans Chinois*. Leroux, Paris, 1911. (Mission d'Ollone, 1906–1909; cf. d'Ollone, H., *In Forbidden China*, tr. B. Miall, London, 1912.)
OLSCHKI, L. (1). *Marco Polo's Precursors*. Baltimore, 1943.
OLSCHKI, L. (4). *Guillaume Boucher; a French Artist at the Court of the Khans*. Johns Hopkins Univ. Press, Baltimore, 1946. (rev. H. Franke, *OR*, 1950, **3**, 135.)
OLSCHKI, L. (6). 'Asiatic Exoticism in Italian Art of the Early Renaissance.' *AB*, 1944, **26**, 95.

PARRY, T. W. (1). 'The Art of Trephining among Prehistoric and Primitive Peoples; their motives for its practice and their methods of procedure.' *JBAA*, 1916 (n.s.), **22**, 33.

PARTINGTON, J. R. (1). *Origins and Development of Applied Chemistry*. Longmans Green, London, 1935.

PAULY-WISSOWA (PAULY, A. ed. G. WISSOWA) (1). *Real-Encyclopädie d. classischen Altertumswiss.* Metzler, Stuttgart, 1893–.

PAUTHIER, J. P. G. (1). 'Examen Méthodique des Faits concernant le Thien-Tchu ou l'Inde. *JA*, 1839 (3e sér.), **8**, 401.

DE PAUW, C. (1). *Recherches Philosophiques sur les Egyptiens et les Chinois*, vols. 4 and 5 of *Œuvres Philosophiques*, 1774. 2nd ed., Bastien, Paris, Rep. An III (1795).

PEAKE, C. H. (1). 'Some Aspects of the Introduction of Modern Science into China. *ISIS*, 1934, **22**, 173.

PEAKE, H. & FLEURE, H. J. (1). *The Steppe and the Sown*. London, 1928.

PELLIOT, P. (1). Critical Note on the Earliest Reference to Tea. *TP*, 1922, **21**, 436.

PELLIOT, P. (2). 'Les Grands Voyages Maritimes Chinois au début du 15e Siècle.' *TP*, 1933, **30**, 237; 1935, **31**, 274.

PELLIOT, P. (3). On the Bodhidharma Legend, in 'Artistes des Six Dynasties et des Thang.' *TP*, 1923, **22**, 253.

PELLIOT, P. (4). Critique of Watters (1). *BEFEO*, 1905, **5**, 423.

PELLIOT, P. (5). 'Les Plus Anciens Monuments de l'Ecriture Arabe en Chine.' *JA*, 1913 (11e sér.), **2**, 177.

PELLIOT, P. (6). 'Note sur les Anciens Itinéraires Chinois dans l'Orient Romain.' *JA*, 1921 (11e sér.), **17**, 139.

PELLIOT, P. (7). 'Les Influences Iraniennes en Asie Centrale.' *RI*, 1912 (n.s.), **18**, 1.

PELLIOT, P. (8). 'Autour d'une Traduction Sanskrite du *Tao-tö-king* [*Tao Tê Ching*].' *TP*, 1912, **13**, 350.

PELLIOT, P. (10). 'Les Mongols et le Papauté. *ROC*, 1922 (3e sér.), **3**, 3; **4**, 225.

PELLIOT, P. (25). *Les Grottes de Touen-Hoang* [*Tunhuang*]; *Peintures et Sculptures Bouddhiques des Epoques des Wei, des Thang et des Song.* (Mission Pelliot en Asie Centrale, six portfolios of plates.) Paris, 1920–24.

PELLIOT, P. (27). *Les Influences Européennes sur l'Art Chinois au 17e et au 18e siècle*. Imp. Nat., Paris, 1948. (Conférence faite au Musée Guimet, Feb. 1927.)

PELLIOT, P. (28). 'La Peinture et la Gravure Européennes en Chine au temps de Matthieu Ricci.' *TP*, 1921, **20**, 1.

PELLIOT, P. (31). 'Li-Kien, autre nom du Ta-Ts'in (Orient Méditerranéen)' (Ta-Chhin). *TP*, 1915, **16**, 690.

PELLIOT, P. (32). 'Des Artisans Chinois à la Capitale Abbaside en +751/+762.' *TP*, 1928, **26**, 110.

PFIZMAIER, A. (1) (tr.). 'Notiz u.d. Geschichtswerk *Tso Tschuen*.' *DWAW/PH*, 1850, **1**, 37. (Tr. chs. 1–5 *Tso Chuan*, selections.)

PFIZMAIER, A. (2) (tr.). 'Die Zeiten des Fürsten Yin von Lu.' *SWAW/PH*, 1854, **13**, 292. (Partial tr. ch. 1 *Tso Chuan*.)

PFIZMAIER, A. (3) (tr.). 'Die Zeiten des Fürsten Hoan [Huan], Tschuang [Chuang] und Min, von Lu.' *SWAW/PH*, 1854, **13**, 430, 449, 469. (Partial tr. chs. 2, 3, 4 *Tso Chuan*.)

PFIZMAIER, A. (4) (tr.). 'Die Zeiten des Fürsten Hi [Hsi] von Lu.' *SWAW/PH*, 1854, **14**, 425. (Partial tr. ch. 5 *Tso Chuan*.)

PFIZMAIER, A. (5) (tr.). 'Die Zeiten des Fürsten Wên von Lu.' *SWAW/PH*, 1855, **15**, 424. (Partial tr. ch. 6 *Tso Chuan*.)

PFIZMAIER, A. (6) (tr.). 'Die Zeiten des Fürsten Siuen [Hsüan] von Lu.' *SWAW/PH*, 1855, **17**, 12. (Partial tr. ch. 7 *Tso Chuan*.)

PFIZMAIER, A. (7) (tr.). 'Notizen a.d. Gesch. d. chin. Reich von j. 590 bis 572 v. Chr.' *SWAW/PH*, 1855, **17**, 255. (Partial tr. ch. 8 *Tso Chuan*.)

PFIZMAIER, A. (8) (tr.). 'Notizen a.d. Gesch. d. chin. Reich von j. 572 bis 546 v. Chr.' *SWAW/PH*, 1856, **18**, 115. (Partial tr. ch. 9 *Tso Chuan*.)

PFIZMAIER, A. (9) (tr.). 'Notizen a.d. Gesch. d. chin. Reich von j. 545 bis 538 v. Chr.' *SWAW/PH*, 1856, **20**, 487. (Partial tr. chs. 9, cont'd, 10 *Tso Chuan*.)

PFIZMAIER, A. (10) (tr.). 'Notizen a.d. Gesch. d. chin. Reich von j. 537 bis 529 v. Chr.' *SWAW/PH*, 1857, **21**, 156. (Partial tr. ch. 10 *Tso Chuan* contd.)

PFIZMAIER, A. (11) (tr.). 'Notizen a.d. Gesch. d. chin. Reich von j. 528 bis 510 v. Chr.' *SWAW/PH*, 1858, **25**, 61. (Partial tr. ch. 10 *Tso Chuan*, contd.)

PFIZMAIER, A. (12) (tr.). 'Notizen a.d. Gesch. d. chin. Reich von j. 509 bis 468 v. Chr.' *SWAW/PH*, 1858, **27**, 113. (Partial tr. chs. 11, 12 *Tso Chuan*.)

PFIZMAIER, A. (13) (tr.). 'Die Geschichte des Reiches U' (Wu). *DWAW/PH*, 1857, **8**, 123. (Tr. ch. 31 *Shih Chi*; cf. Chavannes (1), vol. 4.)

PFIZMAIER, A. (14) (tr.). 'Die Geschichte des Hauses Thai Kung' (of Chhi). *SWAW/PH*, 1862, **40**, 645. (Tr. ch. 32 *Shih Chi*; cf. Chavannes (1), vol. 4.)

PFIZMAIER, A. (15) (tr.). 'Die Geschichte des Hauses Tscheu Kung' (Chou Kung). *SWAW/PH*, 1863, **41**, 90. (Tr. ch. 33, *Shih Chi*; cf. Chavannes (1), vol. 4.)

PFIZMAIER, A. (16) (tr.). 'Die Geschichte des Hauses Schao-Kung u. Khang-Scho' (of Yen and Wei). *SWAW/PH*, 1863, **41**, 435, 454. (Tr. chs. 34, 37 *Shih Chi*; cf. Chavannes (1), vol. 4.)

PFIZMAIER, A. (17) (tr.). 'Die Geschichte des Fürstenlandes Tsin' (Chin). *SWAW/PH*, 1863, **43**, 74. (Tr. ch. 39 *Shih Chi*; cf. Chavannes (1), vol. 4.)

PFIZMAIER, A. (18) (tr.). 'Die Geschichte des Fürstenlandes Tsu' (Chhu). *SWAW/PH*, 1863, **44**, 68. (Tr. ch. 40 *Shih Chi*; cf. Chavannes (1), vol. 4.)

PFIZMAIER, A. (19) (tr.). 'Keu-Tsien, Konig von Yue, und dessen Haus' (Kou Chien of Yüeh and Fan Li). *SWAW/PH*, 1863, **44**, 197. (Tr. ch. 41 *Shih Chi*; cf. Chavannes (1), vol. 4.)

PFIZMAIER, A. (20) (tr.). 'Geschichte d. Hauses Tschao' (Chao). *DWAW/PH*, 1859, **9**, 45. (Tr. ch. 43 *Shih Chi*; cf. Chavannes (1), vol. 5.)

PFIZMAIER, A. (21) (tr.). 'Das Leben des Feldherrn U-Khi' (Wu Chhi). *SWAW/PH*, 1859, **30**, 267. (Tr. ch. 65 *Shih Chi*; not in Chavannes (1).)

PFIZMAIER, A. (22) (tr.). 'Der Landesherr von Schang' (Shang Yang). *SWAW/PH*, 1858, **29**, 98. (Tr. ch. 68 *Shih Chi*; not in Chavannes (1). Cf. Duyvendak (3).)

PFIZMAIER, A. (23) (tr.). 'Das Rednergeschlecht Su' (Su Chhin). *SWAW/PH*, 1860, **32**, 642. (Tr. ch. 69 *Shih Chi*; not in Chavannes (1).)

PFIZMAIER, A. (24) (tr.). 'Der Redner Tschang I und einige seiner Zeitgenossen' (Chang I and Chhu Li Tzu). *SWAW/PH*, 1860, **33**, 525, 566. (Tr. chs. 70, 71 *Shih Chi*; not in Chavannes (1).)

PFIZMAIER, A. (25) (tr.). 'Wei Jen, Fürst von Jang.' *SWAW/PH*, 1859, **30**, 155. (Tr. ch. 72 *Shih Chi*; not in Chavannes (1).)

PFIZMAIER, A. (26) (tr.). 'Zur Geschichte von Entsatzes von Han Tan' *SWAW/PH*, 1859, **31**, 65, 87, 104, 120. (Tr. chs. 75, 76, 78, 83 *Shih Chi*; includes life of the Prince of Phing Yuan; not in Chavannes (1).)

PFIZMAIER, A. (27) (tr.). 'Das Leben des Prinzen Wu Ki [Wu Chi] von Wei.' *SWAW/PH*, 1858, **28**, 171. (Tr. ch. 77 *Shih Chi*; not in Chavannes (1).)

PFIZMAIER, A. (28) (tr.). 'Das Leben des Redners Fan Hoei' (Fan Hui). *SWAW/PH*, 1859, **30**, 227. (Tr. ch. 80 (in part), *Shih Chi*; not in Chavannes (1).)

PFIZMAIER, A. (29) (tr.). 'Die Feldherren des Reiches Tschao' (Chao). *SWAW/PH*, 1858, **28**, 55, 65, 69 (Tr. chs. 80 (in part), 81, 82 *Shih Chi*; not in Chavannes (1).)

PFIZMAIER, A. (30) (tr.). 'Li Sse, der Minister des ersten Kaisers' (Li Ssu). *SWAW/PH*, 1859, **31**, 120, 311. (Tr. chs. 83, 87 *Shih Chi*; not in Chavannes (1). Cf. Bodde (1).)

PFIZMAIER, A. (31) (tr.). 'Das Ende Mung Tien's' (Mêng Thien). *SWAW/PH*, 1860, **32**, 134. (Tr. ch. 88 *Shih Chi*; not in Chavannes (1). Cf. Bodde (1).)

PFIZMAIER, A. (32) (tr.). 'Die Genossen des Königs Tschin Sching' (Chang Erh and Chhen Yü). *SWAW/PH*, 1860, **32**, 333. (Tr. ch. 89 *Shih Chi*, ch. 32 *Chhien Han Shu*; not in Chavannes (1).)

PFIZMAIER, A. (33) (tr.). 'Die Nachkommen der Könige von Wei, Tsi [Chhi] und Han.' *SWAW/PH*, 1860, **32**, 529, 533, 542, 551, 562, 567. (Tr. chs. 90, 93, 94, 97 *Shih Chi*, ch. 33 *Chhien Han Shu*; not in Chavannes (1).)

PFIZMAIER, A. (34) (tr.). 'Die Feldherren Han Sin, Pêng Yue, und King Pu' (Han Hsin, Phêng Yüeh and Ching Pu). *SWAW/PH*, 1860, **34**, 371, 411, 418. (Tr. chs. 90 (in part), 91, 92 *Shih Chi*, ch. 34 *Chhien Han Shu*; not in Chavannes (1).)

PFIZMAIER, A. (35) (tr.). 'Der Abfall des Königs Pi von U' (Wu). *SWAW/PH*, 1861, **36**, 17. (Tr. ch. 106 *Shih Chi*; not in Chavannes (1).)

PFIZMAIER, A. (36) (tr.). 'Sse-ma Ki-Tschü, der Wahrsager von Tschang-ngan' (Ssuma Chi-Chu, in the chapter on diviners, *Jih Chê Lieh Chuan*). *SWAW/PH*, 1861, **37**, 408. (Tr. ch. 127, *Shih Chi*; not in Chavannes (1).)

PFIZMAIER, A. (37) (tr.). 'Die Gewaltherrschaft Hiang Yü's' (Hsiang Yü). *SWAW/PH*, 1860, **32**, 7. (Tr. ch. 31 *Chhien Han Shu*.)

PFIZMAIER, A. (38) (tr.). 'Die Anfänge des Aufstandes gegen das Herrscherhaus Thsin' (Chhin). *SWAW/PH*, 1860, **32**, 273. (Tr. ch. 33 *Chhien Han Shu*.)

PFIZMAIER, A. (39) (tr.). 'Die Könige von Hoai Nan aus dem Hause Han' (Huai Nan Tzu). *SWAW/PH*, 1862, **39**, 575. (Tr. ch. 44 *Chhien Han Shu*.)

PFIZMAIER, A. (40) (tr.). 'Das Erreigniss des Wurmfrasses der Beschwörer.' *SWAW/PH*, 1862, **39**, 50, 55, 58, 65, 76, 89. (Tr. chs. 45, 63, 66, 74 *Chhien Han Shu*.)

PFIZMAIER, A. (41) (tr.). 'Worte des Tadels in dem Reiche der Han.' *SWAW/PH*, 1861, **35**, 206. (Tr. ch. 51 *Chhien Han Shu.*)

PFIZMAIER, A. (42) (tr.). 'Die Heerführer Li Kuang und Li Ling.' *SWAW/PH*, 1863, **44**, 511. (Tr. ch. 54 (first part) *Chhien Han Shu.*)

PFIZMAIER, A. (43) (tr.). 'Die Geschichte einer Gesandtschaft bei den Hiung-Nu's' (Su Wu). *SWAW/PH*, 1863, **44**, 581. (Tr. ch. 54 (second part), *Chhien Han Shu.*)

PFIZMAIER, A. (44) (tr.). 'Die Heerführer Wei Tsing und Ho Khiu-Ping' (Wei Chhing and Ho Chhü-Ping). *SWAW/PH*, 1864, 45, 139. (Tr. ch. 55 *Chhien Han Shu.*)

PFIZMAIER, A. (45) (tr.). 'Die Antworten Tung Tschung-Schü's [Tung Chung-Shu] auf die Umfragen des Himmelssohnes.' *SWAW/PH*, 1862, **39**, 345. (Tr. ch. 56 *Chhien Han Shu.*)

PFIZMAIER, A. (46) (tr.). 'Zwei Statthalter der Landschaft Kuei Ki.' *SWAW/PH*, 1861, **37**, 304. (Tr. ch. 64A (first part) *Chhien Han Shu.*)

PFIZMAIER, A. (47) (tr.). 'Die Bevorzugten des Allhalters Hiao Wu' (emperor Hsiao Wu Ti). *SWAW/PH*, 1861, **38**, 213, 234. (Tr. ch. 64A (second part), 64B *Chhien Han Shu.*)

PFIZMAIER, A. (48) (tr.). 'Die Würdenträger Tsiuen Pu-I, Su Kuang, Yü Ting-Kue, und deren Gesinnungsgenossen' (Chien Pu-I, Su Kuang, Yü Ting-Kuo). *SWAW/PH*, 1862, **40**, 131. (Tr. ch. 71 *Chhien Han Shu.*)

PFIZMAIER, A. (49) (tr.). 'Tschin Thang, Fürst-Zertrümmerer von Hu' (Chhen Thang). *SWAW/PH*, 1862, **40**, 396. (Tr. ch. 70 (in part) *Chhien Han Shu.*)

PFIZMAIER, A. (50) (tr.). 'Die Menschenabtheilung der wandernden Schirmgewaltigen' (*yu hsia*; knights errant or soldiers of fortune). *SWAW/PH*, 1861, **37**, 103. (Tr. ch. 92 *Chhien Han Shu.*)

PFIZMAIER, A. (51) (tr.). 'Die Eroberung der beiden Yue [Yüeh] und des Landes Tschao Sien [Chao-Hsien, Korea] durch Han.' *SWAW/PH*, 1864, **46**, 481. (Tr. ch. 95 *Chhien Han Shu.*)

PFIZMAIER, A. (52) (tr.). 'Zur Geschichte d. Zwischenreiches von Han.' *SWAW/PH*, 1869, **61**, 275, 309. (Tr. chs. 41, 42 *Hou Han Shu.*)

PFIZMAIER, A. (53) (tr.). 'Die Aufstände Wei-Ngao's und Kungsun Scho's' (Wei Ao and Kungsun Shu). *SWAW/PH*, 1869, **62**, 159. (Tr. ch. 43 *Hou Han Shu.*)

PFIZMAIER, A. (54) (tr.). 'Aus der Geschichte d. Zeitraumes Yuen-Khang von Tsin' (+292 to +299). *SWAW/PH*, 1876, **82**, 179, 205, 212, 223, 230, 232. (Tr. chs. 31 (in part), 40, 58, 60, 100 *Chin Shu.*)

PFIZMAIER, A. (55) (tr.). 'Aus der Geschichte des Hofes von Tsin.' *SWAW/PH*, 1876, **81**, 545, 561, 568. (Tr. chs. 31 (in part), 53, 59 *Chin Shu.*)

PFIZMAIER, A. (56) (tr.). 'Über einige Wundermänner Chinas' (magicians and technicians such as Chhen Hsün, Tai Yang, Wang Chia, Shunyu Chih, etc.). *SWAW/PH*, 1877, **85**, 37. (Tr. ch. 95 *Chin Shu.*)

PFIZMAIER, A. (57) (tr.). 'Die Machthaber Hoan Wen und Hoan Hiuen' (Huan Wên and Huan Hsüan). *SWAW/PH*, 1877, **85**, 603, 632. (Tr. chs. 98, 99 *Chin Shu.*)

PFIZMAIER, A. (58) (tr.). 'Ungewöhnliche Erscheinungen und Zufälle in China um die Zeiten der Südlichen Sung.' *SWAW/PH*, 1875, **79**, 362. (Tr. chs. 30–4 (*Wu Hsing Chih*) of (*Liu*) *Sung Shu.*)

PFIZMAIER, A. (59) (tr.). 'Die letzten Zeiten d. Reiches der Tsch'in' (Chhen). *SWAW/PH*, 1881, **98**, 701, 714, 718, 746, 751, 759, 775; 1882, **101**, 264. (Tr. chs. 6, 7, 11, 28, 31, 35, 36 *Chhen Shu.*)

PFIZMAIER, A. (60) (tr.). 'Nachrichten aus der Geschichte d. nördl. Thsi.' *DWAW/PH*, 1884, **34**, 1, 12, 24, 35, 43, 48, 64, 77, 84, 90, 92, 95. (Tr. chs. 8, 9, 10, 11, 12, 13, 17, 28, 39, 49, 50 *Pei Chhi Shu.*)

PFIZMAIER, A. (61) (tr.). 'Darlegungen a.d. Gesch. d. Hauses Sui.' *SWAW/PH*, 1881, **97**, 627, 649, 653, 658, 686, 702. (Tr. chs. 36, 40, 45, 48, 79 *Sui Shu.*)

PFIZMAIER, A. (62) (tr.). 'Lebensbeschreibungen von Heerführern und Würdenträgern des Hauses Sui.' *DWAW/PH*, 1882, **32**, 281, 301, 320, 351, 369. (Tr. chs. 37, 38, 39, 40 (in part), 41 (in part) *Sui Shu.*)

PFIZMAIER, A. (63) (tr.). 'Fortsetzungen a.d. Gesch. d. Hauses Sui.' *SWAW/PH*, 1882, **101**, 187, 201, 207, 230, 249. (Tr. chs. 41 (in part), 70, 73 (in part), 74, 85 *Sui Shu.*)

PFIZMAIER, A. (64) (tr.). 'Die fremdländischen Reiche zu den Zeiten d. Sui.' *SWAW/PH*, 1881, **97**, 411, 418, 422, 429, 444, 477, 483. (Tr. chs. 64, 81, 82, 83, 84 *Sui Shu.*)

PFIZMAIER, A. (65) (tr.). 'Die Classe der Wahrhaftigen in China.' *SWAW/PH*, 1881, **98**, 983, 1001, 1036. (Tr. chs. 71, 73, 77 *Sui Shu.*)

PFIZMAIER, A. (66) (tr.). 'Zur Geschichte d. Aufstände gegen das Haus Sui.' *SWAW/PH*, 1878, **88**, 729, 743, 766, 799. (Tr. chs. 1, 84, 85, 86 *Hsin Thang Shu.*)

PFIZMAIER, A. (67) (tr.). 'Seltsamkeiten aus den Zeiten d. Thang' I and II. I, *SWAW/PH*, 1879, **94**, 7, 11, 19. II, *SWAW/PH*, 1881, **96**, 293. (Tr. chs. 34–36 (*Wu Hsing Chih*), 88, 89 *Hsin Thang Shu.*)

PFIZMAIER, A. (68) (tr.). 'Darlegung der chinesischen Ämter.' *DWAW/PH*, 1879, **29**, 141, 170, 213; 1880, **301**, 305, 341. (Tr. chs. 46, 47, 48, 49A *Hsin Thang Shu*; cf. des Rotours (1).)

PFIZMAIER, A. (69) (tr.). 'Die Sammelhäuser der Lehenkönige Chinas.' *SWAW/PH*, 1880, **95**, 919. (Tr. ch. 49B *Hsin Thang Shu*; cf. des Rotours (1).)

PFIZMAIER, A. (70) (tr.). 'Über einige chinesische Schriftwerke des siebenten und achten Jahrhunderts n. Chr.' *SWAW/PH*, 1879, **93**, 127, 159. (Tr. chs. 57, 59 (in part: *I Wên Chih* including agriculture, astronomy, mathematics; war, five-element theory) *Hsin Thang Shu.*)

PFIZMAIER, A. (71) (tr.). 'Die philosophischen Werke Chinas in dem Zeitalter der Thang.' *SWAW/PH*, 1878, **89**, 237. (Tr. ch. 59 (in part: *I Wên Chih*, philosophical section, including Buddhism) *Hsin Thang Shu.*)

PFIZMAIER, A. (72) (tr.). 'Der Stand der chinesische Geschichtsschreibung in dem Zeitalter der Thang' (original has Sung as misprint). *DWAW/PH*, 1877, **27**, 309, 383. (Tr. chs. 57 (in part), 58 (*I Wên Chih*, history and classics section) *Hsin Thang Shu.*)

PFIZMAIER, A. (73) (tr.). 'Zur Geschichte d. Gründung d. Hauses Thang.' *SWAW/PH*, 1878, **91**, 21, 46, 71. (Tr. chs. 86 (in part), 87, 88 (in part) *Hsin Thang Shu.*)

PFIZMAIER, A. (74) (tr.). 'Nachrichten von Gelehrten Chinas.' (Scholars such as Khung Ying-Ta, Ouyang Hsün, etc.) *SWAW/PH*, 1878, **91**, 694, 734, 758. (Tr. chs. 198, 199, 200 *Hsin Thang Shu.*)

PHEI WÊN-CHUNG (1). 'Preliminary Report on the late Palaeolithic Cave of Chou-khou-tien.' *BGSC*, 1934, **13**, 3.

PHELPS, D. L. (1). 'The Place of Music in the Platonic and Confucian Systems of Moral Education.' *JRAS/NCB*, 1928, **59**, 128.

PINOT, V. (1). *La Chine et la Formation de l'Esprit Philosophique en France (1640–1740)*. Geuthner, Paris, 1932.

POUJADE, J. (1). *La Route des Indes et ses Navires.* Payot, Paris, 1946.

POUZYNA, I. V. (1). *La Chine, l'Italie, et les Débuts de la Renaissance (13ᵉ–14ᵉ siècles).* Ed. d'Art et d'Hist. Paris, 1935. (rev. J. B[uhot], *RAA/AMG*, 1935, **9**, 237.)

[POWELL, THOMAS.] *Humane Industry; or, a History of most Manual Arts, deducing the Original, Progress, and Improvement of them; furnished with variety of Instance and Examples, shewing forth the excellency of Humane Wit.* Herringman, London, 1661.

PRASAD, I. (1). *L'Inde du 7ᵉ au 16ᵉ siècles*, tr. H. de Saugy. Boccard, Paris, 1930. (*Histoire du Monde*, ed. E. Cavaignac, vol. 8 (1).)

PRICE, M. T. (1). 'The Assumed Isolation of China and Autochthony of her Culture.' *AMSR*, 1945, **10**, 38.

PRZYŁUSKI, J. (1). 'Les Unipédes.' *MCB*, 1933, **2**, 307.

PRZYŁUSKI, J. (5). *La Légende de l'Empereur Açoka (Açoka-Avadana) dans les Textes Indiens et Chinois.* Geuthner, Paris, 1923. (Inaug. Diss., Paris.)

PUINI C. (1). 'I Muraglione della Cina.' *RGI*, 1915, **22**, 481.

PUMPELLY, R. (1). 'Geological Researches in China, Mongolia and Japan, during the years 1862 1865.' *SCK*, 1866, **202**, 77.

PURCELL, VICTOR (1). *The Chinese in South-east Asia.* Oxford, 1951.

AL-QAZWĪNĪ, ḤAMDALLAH AL-MUSTAUFĪ (1). *Nuzhat al-qulūb* (Delight of the Hearts). See le Strange (1) and Stephenson, J. (1).

QUATREMÈRE, E. M. (1) (tr.). *Histoire des Mongols de la Perse; écrite en Persan par Raschid-el-din* (part of the *Jāmi' al-Tawārīkh*). Imp. Roy., Paris, 1836.

RADHAKRISHNAN, S. (1). *India and China.* Hind Kitab, Bombay, 1947.

RAGLAN, LORD (1). *How came Civilisation?* Methuen, London, 1939.

RAWLINSON, H. G. (1). *Intercourse between India and the Western World from the Earliest Times to the Fall of Rome.* Cambridge, 1916; 2nd ed. 1926.

RAWLINSON, H. G. (2). *Bactria; The History of a Forgotten Empire.* Probsthain, London, 1912.

RAY, P. C. (1). *A History of Hindu Chemistry, from the Earliest Times to the middle of the sixteenth century A.D., with Sanskrit Texts, Variants, Translation and Illustrations.* 2 vols. Chuckerverty & Chatterjee, Calcutta, 1904, 1925.

READ, BERNARD E. (1) [with LIU JU-CHHIANG]. *Chinese Medicinal Plants from the 'Pên Tshao Kang Mu' A.D. 1596...a Botanical Chemical and Pharmacological Reference List* (Publication of the Peking Nat. Hist. Bull.). French Bookstore, Peiping, 1936. (Chs. 12 to 37 of *Pên Tshao Kang Mu.*) (rev. W. T. Swingle, *ARLC/DO*, 1937, 191.)

READ, BERNARD E. (2) [with LI YÜ-THIEN]. *Chinese Materia Medica; Animal Drugs.*

		Serial nos.	Corresp. with chaps. of *Pên Tshao Kang Mu*
Pt. I	Domestic Animals	322–349	50
II	Wild Animals	350–387	51*A* and *B*
III	Rodentia	388–399	51*B*
IV	Monkeys and Supernatural Beings	400–407	51*B*
V	Man as a Medicine	408–444	52

PNHB, 1931, **5** (no. 4), 37–80; **6** (no. 1), 1–102. (Sep. issued, French Bookstore, Peiping, 1931.)

READ, BERNARD E. (3) [with LI YÜ-THIEN]. *Chinese Materia Medica; Avian Drugs.*

Pt. VI	Birds	245–321	47, 48, 49

PNHB, 1932, **6** (no. 4), 1–101. (Sep. issued, French Bookstore, Peiping, 1932.)

READ, BERNARD E. (4) [with LI YÜ-THIEN]. *Chinese Materia Medica; Dragon and Snake Drugs.*

Pt. VII	Reptiles	102–127	43

PNHB, 1934, **8** (no. 4), 297–357. (Sep. issued, French Bookstore, Peiping, 1934.)

READ, BERNARD E. (5) [with YU CHING-MEI]. *Chinese Materia Medica; Turtle and Shellfish Drugs.*

Pt. VIII	Reptiles and Invertebrates	199–244	45, 46

PNHB (Suppl.), 1939, 1–136. (Sep. issued, French Bookstore, Peiping, 1937.)

READ, BERNARD E. (6) [with YU CHING-MEI]. *Chinese Materia Medica; Fish Drugs.*

Pt. IX	Fishes (incl. some amphibia, octopoda and crustacea)	128–198	44

PNHB (Suppl.), 1939. (Sep. issued, French Bookstore, Peiping, n.d. prob. 1939.)

READ, BERNARD E. (7) [with YU CHING-MEI]. *Chinese Materia Medica; Insect Drugs.*

Pt. X	Insects (incl. arachnida etc.)	1–101	39, 40, 41, 42

PNHB (Suppl.), 1941. (Sep. issued, Lynn, Peiping, 1941.)

READ, BERNARD E. (8). *Famine Foods listed in the 'Chiu Huang Pên Tshao'.* Lester Institute, Shanghai, 1946.

READ, BERNARD E. & PAK, C. (PHU CHU-PING) (1). *A Compendium of Minerals and Stones used in Chinese Medicine, from the 'Pên Tshao Kang Mu'. PNHB*, 1928, **3** (no. 2), i–vii, 1–120. (Revised and enlarged, issued separately, French Bookstore, Peiping, 1936 (2nd ed.).) Serial nos. 1–135, corresp. with chs. of *Pên Tshao Kang Mu*, 8, 9, 10, 11.

REICHWEIN, A. (1). *China and Europe; Intellectual and Artistic Contacts in the Eighteenth Century.* Kegan Paul, London, 1925. Tr. from the German edition, Berlin, 1923.

REINACH, S. (1). (*a*) *La Représentation du Galop dans l'Art Ancienne et Moderne.* Leroux, Paris, 1901, 2nd ed. 1925. (*b*) *RA*, 1900 (3ᵉ sér.), **36**, 216, 441; **37**, 244; 1901, **38**, 27, 224; **39**, 1. (See also *Amalthée; Mélanges d'Archéologie et Histoire.* 3 vols. Leroux, Paris, 1930, vol. 1, p. 19.)

REINAUD, J. T. (1) (tr.). *Relation des Voyages faits par les Arabes et les Persans dans l'Inde et la Chine le 9e siècle de l'ère Chrétienne.* 2 vols. Paris, 1845. (Re-translation of, and commentary on, the MSS translated more than a century earlier by E. Renaudot (*q.v.*).)

REINAUD, J. T. (2). *Relations Politiques et Commerciales de l'Empire Romain avec l'Asie Orientale (l'Hyrcanie, l'Inde, la Bactriane et la Chine) pendant les cinq premiers siècles de l'ère chrétienne, d'après les Témoignages Latins, Grecs, Arabes, Persans, Indiens et Chinois.* Impr. Imp., Paris, 1863. (Reprinted from *JA*, 1863 (6ᵉ sér.), **1**, 93, 297.)

REINAUD, J. T. & GUYARD, S. (1) (tr.). *'Taqwīm al-Buldān' of Abū'l-Fidā.* Paris, vol. 1, 1848 (Reinaud); vol. 2, 1883 (Guyard).

RÉMUSAT, J. P. A. (1) (tr.). *Fa Hian, 'Foe Koue Ki', traduit par Rémusat, etc.* Paris, 1836. Eng. tr. *The Pilgrimage of Fa Hian; from the French edition of the 'Foe Koue Ki' of Rémusat, Klaproth and Landresse with additional notes and illustrations.* Calcutta, 1848. (Fa-Hsien's *Fo Kuo Chi.*)

RÉMUSAT, J. P. A. (2). *Programme du Cours de Langue et de Littérature Chinoises et de Tartare-Mandchou, précédé du Discours prononcé à la première Séance de ce Cours, dans l'une des salles du Collège Royal de France, le 16 janvier 1815.* Charles, Paris, 1815.

RÉMUSAT, J. P. A. (9). 'Vie de Yelou Tchou-Tsai [Yehlü Chhu-Tshai].' In *Nouveaux Mélanges Asiatiques*, vol. 2, p. 64. Schubart and Heideloff, Paris, 1829.

[RENAUDOT, EUSEBIUS] (1) (tr.). *Anciennes Relations des Indes et de la Chine de deux Voyageurs Mahometans, qui y allèrent dans le Neuvième Siècle, traduites d'Arabe, avec des Remarques sur les principaux Endroits de ces Relations.* (With four Appendices, as follows: (i) Eclaircissement touchant la Prédication de la Religion Chrestienne à la Chine; (ii) Eclaircissement touchant l'Entrée des Mahometans dans la Chine; (iii) Eclaircissement touchant les Juifs qui ont esté trouvez à la Chine; (iv) Eclaircissement sur les Sciences des Chinois.) Coignard, Paris, 1718. Eng. tr. London, 1733. (The title of Renaudot's book, which was presented partly to counter the claims of the pro-Chinese party in religious and learned circles (the Jesuits, Golius, Vossius, etc., see Pinot (1), pp. 109, 160, 229, 237), was misleading. The two documents translated were: (*a*) The account of Sulaimān al-Tājir (Sulaiman the Merchant), written by an anonymous author in +851. (*b*) The completion *Silsilat al-Tawārīkh* of +920 by Abū Zayd al-Ḥasan al-Shīrāfī, based on the account of Ibn Wahb al-Baṣrī, who was in China in +876 (see Mieli (1), pp. 13, 79, 81, 115, 302; al-Jalīl (1), p. 138; Hitti (1), pp. 343, 383; Yule (2), vol. 1, pp. 125–33).) Cf. Reinaud (1); Sauvaget (2).

RENOU, L. (1). *Anthologie Sanscrite.* Payot, Paris, 1947.

REY, ABEL (1). *La Science dans l'Antiquité.* Vol. 1: *La Science Orientale avant les Grecs*, 1930, 2nd ed. 1942; Vol. 2: *La Jeunesse de la Science Grecque*, 1933; Vol. 3: *La Maturité de la Pensée Scientifique en Grèce*, 1939; Vol. 4: *L'Apogée de la Science Technique Grecque* (*Les Sciences de la Nature et de l'Homme, les Mathématiques, d'Hippocrate à Platon*), 1946. Albin Michel, Paris. (Evol. de l'Hum. Sér. complémentaire.)

RICHARD, L. (1). *Comprehensive Geography of the Chinese Empire.* Eng. tr. M. Kennelly. Thu-Se-Wei Press, Shanghai, 1908.

RICHARDSON, H. L. (1). 'Szechuan during the War' (World War II). *GJ*, 1945, **106**, 1.

VON RICHTHOFEN, F. (1). 'On the Mode of Origin of the Loess.' *GM*, 1882 (n.s.), **9**, 293.

VON RICHTHOFEN, F. (2). *China: Ergebnisse eigener Reisen und darauf gegründeter Studien.* 5 vols. 2 atlases. Berlin, 1877–1912.

VON RICHTHOFEN, F. (3). 'Über den Seeverkehr nach und von China in Altherthum und Mittelalter.' *VGEB*, 1876, **3**, 86.

VON RICHTHOFEN, F. (4). 'Über die centralasiatischen Seidenstrassen bis zum 200 n. Chr.' *VGEB*, 1877, **4**, 96.

RIDDELL, W. H. (1). '*Hoshi-no-Tama*; Buddhist sacred Jewel etc.' *AQ*, 1946, **20**, 113.

ROBERTS, F. M. (1). *Western Travellers to China.* Kelly & Walsh, Shanghai, 1932.

ROBERTSON, W. (1). *An Historical Disquisition concerning the Knowledge which the Ancients had of India; and the Progress of Trade with that Country prior to the Discovery of the Passage by the Cape of Good Hope.* London, 1791.

DE ROSNY, L. (1) (tr.). *Chan-Hai-King* [*Shan Hai Ching*]*; Antique Géographie Chinoise.* Maisonneuve, Paris, 1891.

DE ROSNY, L. (2) (tr.). *Hiao-King* [*Hsiao Ching*]*; Livre Sacré de la Piété Filiale.* Maisonneuve, Paris, 1889.

ROSS, E. D. (1). 'The Invasions and Immigrations of the Tatars.' *JRCAS*, 1928, **15**, 133.

ROSS, E. D. (2). 'Nomadic Movements in Asia.' *JRSA*, 1929, **77**, 1077.

ROSS, J. (1). *The Origins of the Chinese People.* Oliphant, Edinburgh, 1916.

VON ROSTHORN, A. (1). *Die Ausbreitung d. chinesischen Macht in südwestliche Richtung bis zum 4 jahrhundert n. Chr.; eine histor.-geogr. Studie.* Inaug. Diss., Leipzig, 1895.

ROSTOVTZEV, M. I. (1). *Iranians and Greeks in South Russia.* Oxford, 1922.

ROSTOVTZEV, M. I. (2). 'Parthian Art and the Motive of the Flying Gallop.' In *Independence, Convergence and Borrowing, in Institutions, Thought and Art.* Harvard Tercentenary Publication, Harvard Univ. Press, Cambridge, Mass., 1937, p. 44.

ROSTOVTZEV, M. I. (4). 'La Porte-Epée des Iraniens et des Chinois.' In *L'Art Byzantin chez les Slaves, les Balkans*, commemoration volume for T. Uspensky (vol. 4 of *Orient et Byzance*, ed. G. Millet). Geuthner, Paris, 1930.

ROSTOVTZEV, M. I. (5). *Social and Economic History of the Roman Empire.* Oxford, 1926.

ROSTOVTZEV, M. I. (6). 'L'Art Gréco-Sarmate et l'Art Chinois de l'Epoque des Han.' *ARTHA*, 1924 (no. 3), 1.

ROSTOVTZEV, M. I. (7). *The Animal Style in South Russia and China.* Princeton Univ. Press, Princeton, 1929 (no index). (Princeton Monographs in Art and Archaeol. no. 14.)

ROSTOVTZEV, M. I. (8). *Caravan Cities.* Tr. D. and T. Talbot Rice, Oxford, 1932.

DES ROTOURS, R. (1). *Traité des Fonctionnaires et Traité de l'Armée, traduits de la Nouvelle Histoire des Thang* (chs. 46–50). 2 vols. Brill, Leiden, 1948. (Bibl. de l'Inst. des Hautes Etudes Chinoises, no. 6.) (rev. P. Demiéville, *JA*, 1950, **238**, 395.)

ROXBY, P. M. (1). 'The Terrain of Early Chinese Civilisation.' *G*, 1938, **23**, 225.

ROXBY, P. M. (2). 'The Major Regions of China'. *G*, 1938, **23**, 9.

SACHAU, E. (1) (tr.). *Alberuni's India*. 2 vols. London, 1888; reprinted, 1910.

SAEKI, P. Y. (1). *The Nestorian Monument in China*. SPCK, London, 1916.

SALLES, G., FARAL, E., FILLIOZAT, L., VAILLANT, L., DEMIÉVILLE, P., DENY, J. & HAMBIS, L. (1). *Paul Pelliot*. Soc. Asiatique, Paris, 1947.

SALMONY, A. (3). *Europa-Ostasien; Religiöse Skulpturen*. Kiepenheuer, Potsdam, 1922.

SANSOM, SIR GEORGE (1). *Japan: a Short Cultural History*. Cresset Press, London, 1931; 2nd ed. 1946.

SARGENT, C. B. (1). 'Subsidised History; Pan Ku and the Historical Records of the Former Han Dynasty. *FEQ*, 1944, **3**, 119.

SARKAR, B. K. (1). *Hindu Achievements in the Exact Sciences*. Longmans Green, New York and Calcutta, 1918.

SARTON, GEORGE (1). *Introduction to the History of Science*. Vol. 1, 1927; vol. 2, 1931 (2 parts); vol. 3, 1947 (2 parts). Williams & Wilkins, Baltimore (Carnegie Institution Publ. no. 376.).

SARTON, GEORGE. (6). 'Arabic Scientific Literature.' In *Ignace Goldzieher Memorial Volume*. Budapest, 1948, pt. 1, p. 55.

DE SAUSSURE, L. (1). *Les Origines de l'Astronomie Chinoise*. Maisonneuve, Paris, 1930. (Commentaries by E. Zinner, *VAG*, 1931, **66**, 21; A. Pogo, *ISIS*, 1932, **17**, 267. This book (posthumously issued) contains eleven of the most important original papers of de Saussure on Chinese astronomy (3, 6, 7, 8, 9, 10, 11, 12, 13, 14). It omits, however, the important addendum to (3), 3*a*, as well as the valuable series (16). Unfortunately the editing was slovenly. Although the reprinted papers were re-paged, the cross-references in the footnotes were unaltered; Pogo, however, (*loc. cit.*) has provided a table of corrections by the use of which de Saussure's cross-references can be readily utilised.)

SAUVAGET, J. (1). *Introduction à l'Etude de l'Orient Mussulman; Elements de Bibliographie*. Maisonneuve, Paris, 1946. (Inst. d'Et. Islamiques de l'Univ. de Paris; sér. Initiation à l'Islam, no. 1.)

SAUVAGET, J. (2)(tr.). *Relation de la Chine et de l'Inde, redigée en +851*. (*Akhbār as-Ṣīn wa'l-Hind*). Belles Lettres, Paris, 1948. (Budé Association; Arab Series.)

SCHAEFER, G. (1). 'The Loom of Ancient Egypt.' *CIBA/T*, 1938, **2** (16), 546.

SCHAFER, E. H. (2). 'Iranian Merchants in Thang Dynasty Tales.' *SOS*, 1951, **11**, 403.

SCHEFER, C. (1). 'Trois Chapitres du Khitay Namah.' In *Mélanges Orientaux* (Internat. Congr. Orientalists, Leiden, 1883). Pub. Ec. Lang. Or. Viv. (2e sér.), no. 9.

SCHINDLER, B. (4). 'Preliminary Account of the Work of Henri Maspero concerning the Chinese Documents on Wood and Paper discovered by Sir Aurel Stein on his third expedition in Central Asia.' *AM*, 1950 (n.s.), **1**, 216. (Contains an account of the earliest, +6th cent., block printing.)

SCHLEGEL, G. (1). 'Alphabetical List of the Mikados and Shoguns of Japan, and...their year-titles or *nen-go*.' *TP*, 1891, **2**, 381.

SCHLEGEL, G. (2). 'Hennins or conical Ladies' Hats in Asia, China, and Europe.' *TP*, 1891, **2**, 422.

SCHLEGEL, G. (3). 'Chinesche Mouches.' *TITLV*, 1864, **14**, 569.

SCHLEGEL, G. (4). *Chinesische Bräuche u. Spiele in Europa*. Breslau, 1869.

SCHLEGEL, G. (6). 'The *Shui-Yang* or Water-Sheep.' *Actes du 8e Congrès Internat. des Orientalistes*. Stockholm, 1899, p. 22.

SCHMIDT, L. (1). 'Die Ursachen der Völkerwanderung.' *NJKA*, 1903, **11**, 344.

SCHMITTHENNER, H. (1). 'Die chinesische Lösslandschaft.' *GZ*, 1919, **25**, 308.

SCHNEIDER, O. (1). 'Zur Bernsteinfrage.' In *Naturwissenschaftliche Beiträge z. Geogr. u. Kulturgesch.* Dresden, 1883, pp. 177 ff.

SCHOFF, W. H. (1). *Parthian Stations by Isidore of Charax; an account of the overland Trade Routes between the Levant and India in the 1st century B.C.* Philadelphia, 1914.

SCHOFF, W. H. (2). *Early Communication between China and the Mediterranean*. Philadelphia, 1921.

SCHOFF, W. H. (3). *'The Periplus of the Erythraean Sea'; Travel and Trade in the Indian Ocean by a Merchant of the First Century, translated from the Greek and annotated, etc.* Longmans Green, New York, 1912.

SCHOFF, W. H. (4). 'Navigation to the Far East under the Roman Empire.' *JAOS*, 1917, **37**, 240.

SCHOFF, W. H. (5). 'Some Aspects of the Overland Oriental Trade at the (Beginning of the) Christian Era.' *JAOS*, 1915, **35**, 31.

SCHOMBERG, R. C. F. (1). 'Alleged Changes in the Climate of Southern [Chinese] Turkestan'. *GJ*, 1932, **80**, 133.
SCHRAMEIER, D. (1). 'On Martin Martini' (and his *Novus Atlas Sinensis* of 1655). *JPOS*, 1888, **2**, 99.
SCHUSTER, C. (1). 'A Comparative Study of Motives in Western Chinese Folk Embroideries.' *MS*, 1936, **2**, 21, with addendum 'The Triumphant Equestrian', 437.
SCHWANBECK, E. A. (1). *De Megasthene Rerum Indicarum Scriptore.* Bonn, 1845.
SCHWARZ, E. H. L. (1). 'Chinese Connections with Africa.' *JRAS/B*, 1938 (3rd ser.), **4**, 175.
SEEMANN, H. J. (1). 'Die Instrumente der Sternwarte zu Maragha nach den Mitteilungen von al-Urdi.' *SPMSE*, 1928, **60**, 15.
SELIGMAN, C. G. (1). 'Bird-Chariots and Socketed Celts in Europe and China.' *JRAI*, 1920, **50**, 153.
SELIGMAN, C. G. (2). 'The Roman Orient and the Far East.' *AQ*, 1937, **11**, 5. *ARSI*, 1938, 547.
SELIGMAN, C. G. (3). 'Chinese Socketed Celts.' *AQ*, 1938, **12**, 86.
SEMUSHKIN, TIKHON (1). *Children of the Soviet Arctic.* Hutchinson, London, n.d. (1947).
SHIAH, N. (1). 'Chinese Socketed Celts.' *AQ*, 1939, **13**, 96.
SHOR, F. & SHOR, J. (1). 'The Caves of the Thousand Buddhas.' *NGM*, 1951, **99**, 383.
SHRYOCK, J. K. (1). *Origin and Development of the State Cult of Confucius.* Appleton-Century, New York, 1932.
SIGERIST, H. (1). *A History of Medicine.* 8 vols. Oxford University Press, New York, 1951–.
SIGGEL, A. (1). 'Die Indischen Bücher aus dem *Paradies d. Weisheit über d. Medizin* des 'Alī Ibn Sahl Rabban al-Ṭabarī.' Steiner, Wiesbaden, 1950. (Akad. d. Wiss. u. d. Lit. in Mainz; Abhdl. d. geistes- und sozial-wissenschaftlichen Klasse, no. 14.)
SINGER, C. (1). *A Short History of Biology.* Oxford, 1931.
SINGER, C. (2). *A Short History of Science, to the Nineteenth Century.* Oxford, 1941.
SINHA, JADUNATH (1). *Indian Realism.* Kegan Paul, London, 1938.
SINOR, D. (1). 'Autour d'une Migration de Peuples au 5e siécle.' *JA*, 1947, **235**, 1.
SION, J. (1). *Asie des Moussons.* Vol. 9 of *Géographie Universelle.* Colin, Paris, 1928.
SMITH, ANTHONY (1). *Blind White Fish in Persia.* Allen & Unwin, London, 1953.
SMITH, D. E. (1). *History of Mathematics.* Vol. 1. *General Survey of the History of Elementary Mathematics*, 1923. Vol. 2. *Special Topics of Elementary Mathematics*, 1925. Ginn, New York.
SMITH, V. A. (1). *Oxford History of India, from the earliest times to 1911.* 2nd ed. ed. S. M. Edwardes. Oxford, 1923.
SOOTHILL, W. E. (1). *The Student's Four Thousand Character and General Pocket Dictionary.* Presbyterian Mission Press, Shanghai, 1899; often reprinted.
SOOTHILL, W. E. (2). *China and the West; a Sketch of their Intercourse.* Oxford, 1925.
SORRE, M. (1). *Les Fondements de la Géographie Humaine.* Vol. 1. *Les Fondements Biologiques*, 1947; Vol. 2. *Les Fondements Techniques*, 1948. Colin, Paris.
SOULIER, G. (1). *Les Influences Orientales dans la Peinture Toscane.* Paris, 1924.
SOWERBY, A. DE C. (1). *Nature in Chinese Art* (with two appendices on the Shang pictographs by H. E. Gibson). Day, New York, 1940.
STANGE, H. O. H. (1) (tr.). 'Die Monographie über Wang Mang' (ch. 99 of the *Chhien Han Shu*). *AKML*, 1938, **22**, no. 3.
STEIN, SIR AUREL (1). *Ruins of Desert Cathay.* 2 vols. Macmillan, London, 1912.
STEIN, SIR AUREL (2). *Innermost Asia.* 2 vols. text; 1 vol. plates; 1 box maps. Oxford, 1928.
STEIN, SIR AUREL (3). *On Ancient Central Asian Tracks.* Macmillan, London, 1933.
STEIN, SIR AUREL (4). *Serindia.* 3 vols. text; 1 vol. plates; 1 box maps. Oxford, 1921.
STEIN, SIR AUREL (5). 'Innermost Asia; its Geography as a Factor in History.' *GJ*, 1925, **65**, 377, 473.
STEIN, SIR AUREL (6). 'Notes on Ancient Chinese Documents, discovered along the Han Frontier Wall in the Desert of Tunhuang.' *NCR*, 1921, **3**, 243. (Reprinted, with Stein (7), Chavannes (12), and Wright, H. K. (1), in brochure form, Peiping, 1940.)
STEIN, SIR AUREL (7). 'A Chinese Expedition across the Pamirs and Hindukush, +747.' *NCR*, 1922, **4**, 161. (Reprinted, with Stein (6), Chavannes (12), and Wright, H. K. (1), in brochure form, Peiping, 1940.)
STEIN, SIR AUREL (8). 'The Indo-Iranian Borderlands; their Prehistory in the Light of Geography and of Recent Explorations.' *JRAI*, 1934, **64**, 188 (196).
STEIN, SIR AUREL & BINYON, L. (1). *The Thousand Buddhas; Ancient Buddhist Paintings from the Cave-Temples of Tunhuang on the Western Frontier of China, recovered and described.* 1 vol. text; 2 vols. plates. Quaritch, London, 1921.
STEPHENSON, C. (1). 'In Praise of Medieval Tinkers.' *JEH*, 1948, **8**, 26.
STEPHENSON, J. (1) (tr.). *The Zoological parts of al-Qazwīnī's 'Nuzhat'.* London, 1928.
LE STRANGE, G. (1) (tr.). *The Geographical part of al-Qazwīnī's 'Nuzhat'.* London, 1919.
STUART, G. A. (1). *Chinese Materia Medica; Vegetable Kingdom: extensively revised from Dr F. Porter Smith's work.* Presbyterian Mission Press, Shanghai, 1911.

SÜHEYL ÜNVER, A. (1). *Tanksuknamei Ilhan der Fünunu Ulumu Hatai Mukaddimesi* (Turkish tr.). Istanbul, 1939. (T. C. Istanbul Universitesi Tib Tarihi Enstitusu, Adet: 14.)

SÜHEYL ÜNVER, A. (2). *Wang Shu-ho eseri hakkinda.* Tib Fak. Mecmuasi Yil 7, Sayr 2, Umumi no. 28 (Turkish with Eng. summary). Istanbul, 1944.

SÜHEYL ÜNVER, A. (3). 'Umuni Tib Tarihi: Bazi Resimler ve Vesikalar (Allgemeine Geschichte der Medizin; einige Bilder und Dokumente [from Arabic and Persian sources])' (Turkish and German). Istanbul, 1943. (T. C. Istanbul Universitesi Tib Tarihi Enstitusu, Adet: 24.)

SWANN, NANCY L. (1). *Food and Money in Ancient China; the Earliest Economic History of China to +25* (with tr. of [*Chhien*] *Han Shu*, ch. 24 and related texts [*Chhien*] *Han Shu*, ch. 91 and *Shih Chi*, ch. 129). Princeton Univ. Press, Princeton, 1950.

SWINGLE, W. T. (1). 'Noteworthy Chinese Works on Wild and Cultivated Food Plants.' *ARLC/DO*, 1935, 193.

AL-ṬABARĪ, ABŪ'L-ḤASAN 'ALĪ (1). *Firdaus al-Ḥikmah* (The Paradise of Wisdom), ed. Muḥammad Zubayr al-Siddiqī. Berlin, 1928. (See Siggel, 1.)

DE TAKACS, Z. (1). 'Congruences between the Arts of the Eurasiatic Migration Periods.' *AA*, 1945, **8**, 269.

TAKAKUSU, J. (1) (tr.). *A Record of the Buddhist Religion as practised in India and the Malay Archipelago (+671 to +695), by I-Tsing (I-Ching).* Oxford, 1896.

TAKAKUSU, J. (3). 'Le Voyage de Kanshin Chien-Chen] en Orient.' *BEFEO*, 1928, **28**, 1, 441; 1929, **29**, 47.

TARN, W. W. (1). *The Greeks in Bactria and India.* Cambridge, 1951.

TARN, W. W. (2). *Alexander the Great.* 2 vols. (1) Narrative; (2) Sources and Studies. Cambridge, 1948.

TAYLOR, GRIFFITH (1). (*a*) 'Ecological Basis of Anthropology.' *E*, 1934, **15**, 223. (*b*) 'Racial Migration Zones and their Significance.' *HB*, 1930, **2**, 34. (*c*) *The Block Diagram and its Ecological Uses.* Romerow Festschrift, Towarzystwo Geograficzne, Lwów, 1934.

TCHANG. See Chang.

TEGGART, F. J. (1). *Rome and China; A Study of Correlations in Historical Events.* Univ. of California Press, Berkeley, Calif., 1939.

TEICHMAN, SIR ERIC (1). 'The Motor Route from Peiping to Kashgar.' *GJ*, 1937, **89**, 297.

TÊNG SSU-YÜ & BIGGERSTAFF, K. (1). *An Annotated Bibliography of Selected Chinese Reference Works.* Harvard-Yenching Instit. Peiping, 1936. (Yenching Journ. Chin. Studies, Monograph no. **12**.)

THOMAS, F. W. (3). 'Aśoka, the Imperial Patron of Buddhism.' In ch. 20, *CHI*, vol. 1, p. 499.

THORNDIKE, L. (1). *A History of Magic and Experimental Science.* 6 vols. Columbia Univ. Press, New York: vols. 1 and 2, 1923; 3 and 4, 1934; 5 and 6, 1941.

THORNDIKE, L. (8). 'Some Unfamiliar Aspects of Medieval Science.' In *Science and Civilisation*, ed. R. C. Stauffer. Centennial Celebration Volume of the Univ. of Wisconsin, Univ. of Wisconsin Press, Madison, 1949, p. 33.

TING WÊN-CHIANG (V. K. TING) (1). 'How China acquired her civilisation.' Art. in *Symposium on Chinese Culture*, ed. Sophia Zen. IPR, Shanghai, 1931, pp. 9 ff.

DE TIZAC, H. D'ARDENNE (1). *Les Hautes Epoques de l'Art Chinois d'après les Collections du Musée Cernuschi.* Nilsson, Paris, n.d.

TJAN TJOE-SOM. See Tsêng Chu-Sên.

TOMASCHEK, W. (1). 'Kritik d. ältesten Nachrichten ü.d. skythischen Norden. I. Ü.d. Arimaspische Gedicht des Aristeas.' *SWAW/PH*, 1888, **116**, 715. 'II. Die Nachrichten Herodots ü.d. skythischen Karawanenweg nach Innerasien.' *SWAW/PH*, 1889, **117**, 1. 'Die alten Thraker.' *SWAW/PH*, 1893, **128**, 1.

TOYNBEE, A. J. (1). *A Study of History.* 6 vols. RIIA, London, 1935–9.

TSÊNG CHU-SÊN (TJAN TJOE-SOM) (1). *Po Hu T'ung* [*Pai Hu Thung*]; *The Comprehensive Discussions in the White Tiger Hall.* Brill, Leiden, 1949. (Sinica Leidensia ser. no. 6.)

TSÊNG CHU-SÊN (TJAN TJOE-SOM) (2). 'The Date of Kao Tsu's First Court Ceremonial.' In *India Antiqua, a Volume of Studies presented by his friends and pupils to J. P. Vogel...*, p. 304. 1947.

VACCA, G. (1). 'Some Points on the History of Science in China.' *JRAS/NCB*, 1930, **61**, 10.

VACCA, G. (5). *Due Astronomi Cinesi del 4 Sec. AC e i loro cataloghi stellari.* Zanichelli, Bologna, 1934. (Offprint from *Calendario del r. Osservatorio Astronomico di Roma*, n.s., 1934, **10**.)

DE LA VALLÉE POUSSIN, L. (1). *L'Inde au Temps des Mauryas.* Boccard, Paris, 1930. (*Histoire du Monde*, ed. E. Cavaignac, vol. 6.)

DE LA VALLÉE POUSSIN, L. (2). *Dynasties et Histoire de l'Inde depuis Kanishka jusqu'aux Invasions Mussulmanes.* Boccard, Paris, 1935. (*Histoire du Monde*, ed. E. Cavaignac, vol. 6*b*.)

VAVILOV, N. I. (1). 'The Problem of the Origin of the World's Agriculture in the Light of the Latest Investigations.' In *Science at the Cross-Roads.* Papers read to the 2nd International Congress of the History of Science and Technology. Kniga, London, 1931.

VENTURI, P. T. (1) (ed.). *Opera Storiche del P. Matteo Ricci.* 2 vols. Giorgetti, Macerata, 1911.

VERGIL, POLYDORE (1). *De Rerum Inventoribus.* 1512, and many editions subsequently. Engl. tr. from 1546 onwards, e.g. *An Abridgement of the Works of the most Learned Polidore Virgil, being an History of the Inventors, and Original Beginning of all Antiquities, Arts, Mysteries, Sciences, Ordinances, Orders, Rites and Ceremonies, both Civil and Religious—also, of all Sects and Schisms. A work very useful for Divines, Historians, and all manner of Artificers. Compendiously gathered, by T[homas] Langley.* Streater, London, 1659.

VERHAEREN, H. (1). *L'Ancienne Bibliothèque du Pé-T'ang (Pei Thang).* Lazaristes Press, Peking, 1940.

VLADIMIRTSOV, B. (1). *Chingis Khan,* tr. D. Mirsky. London, 1930.

VLADIMIRTSOV, B. (2). *The Social Structure of the Mongols; Mongol Nomad Feudalism.* Academy of Sciences, Leningrad, 1934.

WACE, A. J. B. (1). 'Weaving or Embroidery?' *AJA,* 1948, **52**, 51.

WADE, SIR THOMAS F. (1). *A Progressive Course designed to assist the Student of Colloquial Chinese as spoken in the Capital and the Metropolitan Department.* 2 vols. 2nd ed. 1886; 3rd (abridged) ed., with W. C. Hillier; Kelly & Walsh, Shanghai, 1903.

WALEY, A. (1) (tr.). *The Book of Songs.* Allen & Unwin, London, 1937.

WALEY, A. (2) (tr.). *One Hundred and Seventy Chinese Poems.* Constable, London, 1918; often reprinted.

WALEY, A. (3) (tr.). *More Translations from the Chinese.* Allen & Unwin, London, 1919; often reprinted.

WALEY, A. (4) (tr.). *The Way and its Power; a study of the Tao Tê Ching and its Place in Chinese Thought.* Allen & Unwin, London, 1934.

WALEY, A. (5) (tr.). *The Analects of Confucius.* Allen & Unwin, London, 1938.

WALEY, A. (12). *The Life and Times of Po Chü-I (+772 to +846).* Allen & Unwin, London, 1949.

WALEY, A. (16). *The Real Tripitaka* (life of Hsüan-Chuang, and other essays). Allen & Unwin, London, 1952.

WALEY, A. (17) (tr.). *Monkey, by Wu Chhêng-Ên.* Allen & Unwin, London, 1942.

WALKER, R. L. (1). 'August Pfizmaier's Translations from the Chinese.' *JAOS,* 1949, **69**, 215.

WANG, THEODORE & LYMAN, E. R. (1). *Chronological Tables of Chinese Dynasties.* Shanghai Printing Co., Shanghai, 1902.

WANG YÜ-CHHÜAN (1). *Early Chinese Coinage.* Amer. Numismatic Soc., New York, 1951. (Numismatic Notes and Monographs, no. 122.)

WARD, F. KINGDON (1). 'Tibet as a Grazing Land.' *GJ,* 1947, **110**, 60.

WARMINGTON, E. H. (1). *The Commerce between the Roman Empire and India.* Cambridge, 1928.

WARNER, LANGDON (1). *The Long Old Road in China.* Arrowsmith, London, 1927.

WARWICK, A. (1). '1000 miles along the Great Wall.' *NGM,* 1923, **43**, 113.

WATTERS, T. (1). *On Yuan Chwang's [Hsüan-Chuang] Travels in India (+629 to +645).* 2 vols. ed. T. W. R. Davis & S. W. Bushell. RAS, London, 1904–5. (Oriental Transl. Fund, vols. 14 and 15.) (See Pelliot (4).)

WEBB, JOHN (1). *Historical Essay endeavouring a Probability that the Language of the Empire of China is the Primitive Language.* London, 1669.

WEIDENREICH, G. (1). (*a*) '*Sinanthropus* population of Chou-khou-tien, with a Preliminary Report on New Discoveries.' *BGSC,* 1935, **14**, 4. (*b*) 'Some Problems dealing with Ancient Man.' *AAN,* 1940, **42**, 375.

WERNER, E. T. C. (1). *Myths and Legends of China.* Harrap, London, 1922.

WESTON, S. (1). *A Chinese Chronicle by Abdalla of Beyza translated from the Persian.* London, 1820. (See Müller, Andreas (1).)

WHEELER, R. E. M. (1) [with GHOSH, A. & DEVA, K.]. 'Arikamedu; an Indo-Roman Trading Station on the East Coast of India.' *ANI,* 1946 (no. 2), 17.

WHEELER, R. E. M. (2). 'Archaeology and the Transmission of Ideas.' *AQ,* 1952, **26**, 180.

WHEELER, R. E. M. (3). 'Romano-Buddhist Art; an Old Problem Restated.' *AQ,* 1949, **23**, 4.

WHEWELL, WILLIAM (1). *History of the Inductive Sciences.* Parker, London, 1847. 3 vols. (crit. G. Sarton, *A/AIHS,* 1950, **3**, 11.)

WHITE, LYNN (1). 'Technology and Invention in the Middle Ages.' *SP,* 1940, **15**, 141.

WIEGER, L. (1). *Textes Historiques.* 2 vols. (Ch. and Fr.) Mission Press, Hsienhsien, 1929.

WIEGER, L. (2). *Textes Philosophiques.* (Ch. and Fr.) Mission Press, Hsienhsien, 1930.

WIEGER, L. (3). *La Chine à travers les Ages; Précis, Index Biographique et Index Bibliographique.* Mission Press, Hsienhsien, 1924. (Eng. tr. E. T. C. Werner.)

WIEGER, L. (4). *Histoire des Croyances Religieuses et des Opinions Philosophiques en Chine depuis l'origine jusqu'à nos jours.* Mission Press, Hsienhsien, 1917.

WIEGER, L. (5). *Chinois Ecrit; Précis, Grammaire, Phraséologie.* Tientsin, 1929.

WIEGER, L. (6). *Taoisme.* Vol. 1. *Bibliographie Générale*: (1) Le Canon (Patrologie); (2) Les Index Officiels et Privés. Mission Press, Hsienhsien, 1911. (crit. P. Pelliot, *JA*, 1912 (10e sér.), **20**, 141.)

WIEGER, L. (7). *Taoisme.* Vol. 2. *Les Pères du Système Taoiste* (tr. selections of Lao Tzu, Chuang Tzu, Lieh Tzu). Mission Press, Hsienhsien, 1913.

WILBUR, C. M. (1). *Slavery in China during the Former Han Dynasty (−206 to +25). FMNHP/AS*, 1943, **34**, 1. (Pub. no. 525.) Also *JEH*, 1943, **3**, 56.

WILHELM, HELLMUT (1). *Chinas Geschichte; Zehn einführende Vorträge.* Vetch, Peiping, 1942.

WILHELM, HELLMUT (8). 'The Background of Tsêng Kuo-Fan's Ideology.' *ASEA*, 1950, **3**, 90.

WILHELM, RICHARD (1). *Short History of Chinese Civilisation*, tr. J. Joshua. Harrap, London, 1929.

WILHELM, RICHARD (2) (tr.). *I Ging [I Ching]: Das Buch der Wandlungen.* 2 vols. (3 books, pagination of 1 and 2 continuous in first volume). Diederichs, Jena, 1924. Eng. tr. C. F. Baynes (2 vols.) 1950. Bollingen, Pantheon, New York.

WILHELM, RICHARD (3) (tr.). *Frühling u. Herbst d. Lü Bu-We* (the *Lü Shih Chhun Chhiu*). Diederichs, Jena, 1928.

WILHELM, RICHARD (4) (tr.). *Liä Dsi [Lieh Tzu]; Das Wahre Buch vom Quellenden Urgrund; Tschung Hü Dschen Ging [Chhung Hsü Chen Ching]; Die Lehren der Philosophen Liä Yü Kou und Yang Dschu.* Diederichs, Jena, 1921.

WILHELM, RICHARD (6) (tr.). *Li Gi [Li Chi]*. Diederichs, Jena, 1930.

WILKINS, W. J. (1). *Hindu Mythology, Vedic and Puranic.* Thacker & Spink, Calcutta, 1913.

WILLIAMS, S. WELLS (1). *The Middle Kingdom; A Survey of the Geography, Government, Education, Social Life, Arts, Religion, etc. of the Chinese Empire and its Inhabitants.* 2 vols. Wiley, New York, 1848; later eds. 1861, 1900, London, 1883.

WINTERSTEIN, H. (1) (ed.). *Handbuch d. vergleichende Physiologie.* Fischer, Jena, 1924.

WITTFOGEL, K. A. (1). 'Chinese Society and the Dynasties of Conquest.' In *China*, ed. H. F. McNair. Univ. of Calif. Press, Berkeley, 1946, p. 112.

WITTFOGEL, K. A., FÊNG CHIA-SHÊNG *et al.* (1). 'History of Chinese Society (Liao), +907 to +1125.' *TAPS*, 1948, **36**, 1–650. (rev. P. Demiéville, *TP*, 1950, **39**, 347; E. Balazs, *PA*, 1950, **23**, 318.)

WOODROFFE, SIR J. (pseudonym: A. Avalon) (1). *Śakti and Śakta.* Ganesh, Madras, 1929; Luzac, London, 1929.

WOOLLEY, L. (1). *Ur of the Chaldees.* London, 1929; Penguin Books, London, 1938.

WORCESTER, G. R. G. (1). *Junks and Sampans of the Upper Yangtze.* Inspectorate-General of Customs, Shanghai, 1940. (China Maritime Customs Publ., Ser. III, Miscellaneous, no. 51.)

WORCESTER, G. R. G. (3). *The Junks and Sampans of the Yangtze; a study in Chinese Nautical Research.* Vol. 1. *Introduction, and Craft of the Estuary and Shanghai Area.* Vol. 2. *The Craft of the Lower and Middle Yangtze and Tributaries.* Inspectorate-General of Customs, Shanghai, 1947, 1948. (China Maritime Customs Publ., Ser. III, Miscellaneous, nos. 53, 54.) (rev. D. W. Waters, *MMI*, 1948, **34**, 134.)

WRIGHT, H. K. (1). 'The Thousand Buddhas of the Tunhuang Caves.' *NCR*, 1922, **4**, 401. (Reprinted, with Stein (6, 7) and Chavannes (12), in brochure form, Peiping, 1940.)

WU CHIN-TING (1). *Prehistoric Pottery in China.* Courtauld Institute & Kegan Paul, London, 1938.

WU CHING-CHHAO (1). 'Chinese Immigration in the Pacific Area.' *CSPSR*, 1928, **12**, 543; 1929, **13**, 50, 161.

Wu. G. D. See Wu Chin-Ting.

WU KHANG (1). *Les Trois Politiques du Tchouen Tsieou [Chhun Chhiu] interprétées par Tong Tchong-Chou [Tung Chung-Shu] d'après les principes de l'école de Kong-Yang [Kungyang].* Leroux, Paris, 1932. (Includes tr. of ch. 121 of *Shih Chi*, the biography of Tung Chung-Shu.)

WU KHANG (2). *Histoire de la Bibliographie Chinoise.* Leroux, Paris, 1938. (Bibliothèque de l'Institut des Hautes Etudes Chinoises, no. 5.)

WU LU-CHHIANG & DAVIS, T. L. (2) (tr.). 'An Ancient Chinese Alchemical Classic; Ko Hung on the Gold Medicine, and on the Yellow and the White; being the fourth and sixteenth chapters of *Pao Phu Tzu*,' *PAAAS*, 1935, **70**, 221.

WÜST, W. (1). 'Buddhismus und Christentum auf vorderasiatisch-antiken Boden.' *FF*, 1932, **8**, 19.

WYLIE, A. (1). *Notes on Chinese Literature.* First ed. Shanghai, 1867. Ed. here used Vetch, Peiping, 1939 (photographed from the Shanghai 1922 ed.).

WYLIE, A. (2). 'History of the Hsiung-Nu' (transl. of the chapter on the Huns in the *Chhien Han Shu*, ch. 94). *JRAI*, 1874, **3**, 401; 1875, **5**, 41.

WYLIE, A. (3). 'The History of the South-western Barbarians and Chao Sëen [Chao-Hsien, Korea]' (tr. of ch. 95 of the *Chhien Han Shu*). *JRAI*, 1880, **9**, 53.
WYLIE, A. (10) (tr.). 'Notes on the Western Regions, translated from the *Ts'een Han Shoo* [*Chhien Han Shu*] Bk. 96.' *JRAI*, 1881, **10**, 20; 1882, **11**, 83. (Chs. 96A and B, as also the biography of Chang Chhien in ch. 61, pp. 1–6, and the biography of Chhen Thang, in ch. 70.)
YANG CHIA-LO (ed.) (1). *Encyclopaedia Quatuor Bibliothecarum.* World Book Co., Shanghai, 1946. Cf. Bibliography B.
YANG CHING-CHI (1). *Among the Aboriginal Tribes of Kuangsi.* Chinese Folklore Society, Sun Yat-Sen University, Pingshek, Kuangtung, n.d., prob. 1944.
YANG LIEN-SHÊNG (3). *Money and Credit in China; a Short History.* Harvard Univ. Press, Cambridge, Mass., 1952. (Harvard-Yenching Institute, Monograph Series, no. 12.)
YATES, J. (1). *Textrinum Antiquorum; an account of Weaving among the Ancients.* London, 1843.
YETTS, W. P. (1). 'The Shang-Yin Dynasty and the Anyang Finds.' *JRAS*, 1933, 657.
YETTS, W. P. (2). (*a*) 'Taoist Tales.' *NCR*, 1919, **1**, 169; 1920, **2**, 37. (*b*) 'Taoist Tales; a Rejoinder.' *NCR*, 1921, **3**, 65.
YETTS, W. P. (3). 'Links between Ancient China and the West.' *GR*, 1926, **16**, 614.
YETTS, W. P. (13). 'The Horse; a Factor in Early Chinese History.' *ESA*, 1934, **9**, 231.
YETTS, W. P. & HOPKINS, L. C. (1). 'A Chinese Bronze Ritual Vessel.' *JRAS*, 1933, 107.
YULE, SIR HENRY (1) (ed.). *The Book of Ser Marco Polo the Venetian, concerning the Kingdoms and Marvels of the East, tr. and ed., with Notes, by H.Y....*, ed. H. Cordier, Murray, London, 1903 (reprinted, 1921), 2 vols., with a third volume *Notes and Addenda to Sir Henry Yule's Edition of Marco Polo*, by H. Cordier, Murray, London, 1920.
YULE, SIR HENRY (2). *Cathay and the Way Thither; being a Collection of Medieval Notices of China.* Hakluyt Society Pubs. (2nd ser.) 1913–15. (1st ed. 1866.) Revised by H. Cordier. 4 vols. Vol. 1 (no. 38), *Introduction; Preliminary Essay on the Intercourse between China and the Western Nations previous to the Discovery of the Cape Route.* Vol. 2 (no. 33), *Odoric of Pordenone.* Vol. 3 (no. 37), *John of Monte Corvino and others.* Vol. 4 (no. 41), *Ibn Battuta and Benedict of Goes.* (Photographically reproduced, Peiping, 1942.)

ZANELLI, A. (1). *Le Schiave Orientali in Firenze nei Sec. 14 e 15.* Florence, 1885.

ADDENDA TO BIBLIOGRAPHY C

BROWN, C. H. (1). 'Some Structural Proteins in *Mytilus edulis.*' *QJMS*, 1952, **93**, 487.
BRÜCKNER, E. (1). 'Klimaschwankungen und Völkerwanderungen in der alten Welt.' *MGGW*, 1915, **58**, 210.
BRÜCKNER, E. (2). 'Klimaschwankungen seit 1700 nebst Bemerkungen ü.d. Klimaschwankungen der Diluvial-Zeit.' *Abhdl. herausgeg. v. A. Penck.* Vienna, 1890, **4**, no. 2.

COHEN, J. (1). 'On the Project of a Universal Character.' *M*, 1954, **63**, 49.
CORDIER, H. (9). 'Notes pour servir à l'Histoire des Etudes Chinoises en Europe, jusqu'à l'époque de Fourmont l'ainé.' In *Nouveaux Mélanges Orientaux; Mémoires, Textes et Traductions publiés par les professeurs de l'Ecole Spéciale des Langues Orientales Vivantes à l'occasion du 7ème Congrès International des Orientalistes réuni à Vienne.* Paris, 1886, p. 399. Reprinted in *Mélanges d'Histoire et de Géographie Orientales.* Paris, 1914, vol. 1, p. 75.
CORDIER, H. (10). 'Fragments d'une Histoire des Etudes Chinoises au 18e siècle.' In *Centenaire de l'Ecole des Langues Orientales Vivantes.* Imp. Nat. Paris, 1895. (On Fourmont, de Prémare, Vandermonde, J. T. Needham, d'Incarville, etc.)
CURRY, J. C. (1). 'Climate and Migrations.' *AQ*, 1928, **2**, 292.

FERRAND, G. (5). 'Les Relations de la Chine avec le Golfe Persique avant l'Hégire.' Art. in *Mélanges offerts à Gaudefroy-Demombynes par ses amis et anciens élèves, 1935–1945.* Maisonneuve, Paris, 1945.

HOPKINS, L. C. (36) (tr.). *The Six Scripts or the Principles of Chinese Writing*, by Tai Thung. Amoy, 1881. Reprinted by photolitho, with a memoir of the translator by W. Perceval Yetts, Cambridge, 1954.

MASPERO, H. (14). *Etudes Historiques; Mélanges Posthumes sur les Religions et l'Histoire de la Chine*, vol. 3, ed. P. Demiéville. Civilisations du Sud, Paris, 1950. (Publ. du Mus. Guimet, Biblioth. de Diffusion, no. 59.)

MASPERO, H. (20). 'Les Origines de la Communauté Bouddhiste de Loyang.' *JA*, 1934, **225**, 87.

MASPERO, H. (22). 'Un Texte Taoiste sur l'Orient Romain.' *MIFC*, 1937, **17**, 377 (*Mélanges G. Maspero*, vol. 2). Reprinted in Maspero (14), pp. 95ff.

MEDHURST, W. H. (1) (tr.). *The Shoo King* [*Shu Ching*] *or Historical Classic, being the most ancient authentic record of the Annals of the Chinese Empire, illustrated by later Commentators.* Shanghai, 1846. (Word by word translation with inserted Chinese characters.)

PANIKKAR, K. M. (1). 'Asia and Western Dominance.' Allen & Unwin, London, 1953.

PULLEYBLANK, E. G. (1). *The Background of the Rebellion of An Lu-Shan.* Oxford, 1954.

ROWLAND, B. (2). *The Art and Architecture of India—Buddhist, Hindu, Jain.* Penguin Books, London, 1953. (Pelican History of Art series.)

SAINSON, C. (1) (tr.). *Histoire particulière du Nan Tchao; Nan Tchao Ye Che* [*Nan Chao Yeh Shih*]. Imp. Nat. Paris, 1904. (Pub. Ec. Lang. Or. Viv. (5e sér.) no. 4.)

DE SAUSSURE, R. (1). 'Léopold de Saussure (1866–1925).' *ISIS*, 1937, **27**, 286.

SELIGMAN, C. G. (4). 'Further Notes on Bird-Chariots in Europe and China.' *JRAI*, 1928, **58**, 247.

SOOTHILL, W. E. (5). *The Hall of Light; a Study of Early Chinese Kingship.* Lutterworth, London, 1951. (Includes a translation of the *Hsia Hsiao Chêng*.)

LE STRANGE, G. (2). *Palestine under the Moslems.* London, 1890.

WIENS, H. J. (1). 'The "Shu Tao", or Road to Szechuan.' *GR*, 1949, **39**, 584.

WINFIELD, G. (1). *China; the Land and the People.* Sloane, New York, 1948. (crit. J. Needham, *PA*, 1949, **22**, 282.)

YONGE, C. M. (1). 'Form and Habit in *Pinna carnea* Gmelin.' *PTRSB*, 1953, **237**, 335.

YÜ PING-YAO & GILLIS, I. V. (1). *Title Index to the Ssu Khu Chhüan Shu.* French Bookstore, Peiping, 1934.

GENERAL INDEX

by MURIEL MOYLE

NOTES

(1) Articles (such as 'the', 'al-', etc.) occurring at the beginning of an entry, and prefixes (such as 'de', 'van', etc.) are ignored in the alphabetical sequence.
(2) The various parts of hyphenated words are treated as separate words in the alphabetical sequence. It should be remembered that, in accordance with the conventions adopted, some Chinese proper names are written as separate syllables while others may be written as one word, e.g. An-Hsi and Anhsi below.
(3) In the arrangement of Chinese words, Chh- and Hs- follow normal alphabetical sequence, and *ü* is treated as equivalent to *u*.
(4) References to footnotes are not given except for certain special subjects with which the text does not deal. They are indicated by brackets containing the superscript letter of the footnote.
(5) Explanatory words in brackets indicating fields of work are added for Chinese scientific and technological persons (and occasionally for some of other cultures), but not for political or military figures (except kings and princes).